THE *SLOW* MOON *CLIMBS*

•

THE
SLOW
MOON
CLIMBS

•

The SCIENCE, HISTORY,
and MEANING of
MENOPAUSE

SUSAN P.
MATTERN

PRINCETON UNIVERSITY PRESS
PRINCETON & OXFORD

Copyright © 2019 by Princeton University Press

Published by Princeton University Press
41 William Street, Princeton, New Jersey 08540
6 Oxford Street, Woodstock, Oxfordshire OX20 1TR

press.princeton.edu

LCCN 2019935935
ISBN 9780691171630

British Library Cataloging-in-Publication Data is available

Editorial: Rob Tempio and Matt Rohal
Production Editorial: Natalie Baan
Text Design: Leslie Flis
Jacket: Amanda Weiss
Production: Merli Guerra
Publicity: Sarah Henning-Stout and Katie Lewis
Copyeditor: Sarah Vogelsong

This book has been composed in Arno Pro and Baskerville

Printed on acid-free paper. ∞

Printed in the United States of America

10 9 8 7 6 5 4 3 2 1

To my mother,

Nancy Garland Mattern,

and to the memory of my grandmother,

Josephine Chatfield Garland.

CONTENTS

•

Chapter 12
A Cultural Syndrome?

Epilogue
Good-Bye to All That

ACKNOWLEDGMENTS

●

I THANK Rob McQuilkin and Rob Tempio for their faith in this book and their many efforts in its support. My friend Walter Scheidel offered encouragement and advice on several points. Two anonymous readers for Princeton University Press made helpful suggestions. I also express here my deep appreciation of my husband, Sean Hribal, without whose patience and help this book could not have been written.

THE
SLOW
MOON
CLIMBS

•

The long day wanes; the slow moon climbs; the deep
Moans round with many voices. Come, my friends,
'Tis not too late to seek a newer world.

—ALFRED, LORD TENNYSON, "ULYSSES"

Prologue

THE GRANDMOTHER OF US ALL

•

GENGHIS KHAN'S MOTHER

One cold spring day, Genghis Khan's mother found herself abandoned on the banks of the Onon River in Mongolia. With her were seven children. The oldest, the future conqueror of much of Asia, then called Temüjin, was nine. The two youngest children belonged to her husband's other wife (the latter a very shadowy figure in the story). Her husband, Yisügei Ba'atur, was dead. Hoelun was still a young woman, but she would never remarry and would have no more children. But though her reproductive career was over and her future looked bleak on that morning in 1170, Hoelun was one of history's most biologically successful women—the progenitor of an astonishingly large fraction of the people alive today.[1]

Her story comes down to us in *The Secret History of the Mongols*, the anonymous thirteenth-century document that records Genghis Khan's life and conquests. At the time of her exile, Hoelun had been living among her husband's people, the Borjigin clan. When her husband died, a rival branch seized the opportunity to shut out her sons from power while they were still small. The two widows of the former leader Ambatai, long dead, humiliated Hoelun by offering her family no food at a feast celebrating their joint ancestors. Hoelun protested, demanding her family's share, but instead of providing a grudging handout, the widows responded with a flat rejection. Their people would move on without Hoelun and her children. They packed up their felt tents, herds, horses, and carts, and left. They most likely did not expect her to survive.

Hoelun had no people of her own to turn to. As was customary, she had married outside her own clan, and her blood relatives, other than

3

her children, were far away. She was supposed to have married someone quite other than Yisügei—originally she had been betrothed to a young man of the Merkit clan—but like other women in *The Secret History*, she had been married by capture. Descending on the couple as the groom escorted his bride to their new home, Yisügei and two of his brothers, on horseback, had stolen Hoelun from her luckless fiancé. The fiancé had fled, and survived; but Hoelun, despite her desperate protests, had become Yisügei's wife.

Now alone after Yisügei's death, Hoelun supported her five children and two stepchildren by foraging for wild roots and berries. The family had eight gelding horses that grazed on wild grasses and provided transportation but no milk. She plotted revenge. She was furious when, as her sons grew up, her oldest two—Temüjin and his brother Qasar— killed one of her stepchildren in a fight over a bird they had trapped for food. She raged at them—how were they supposed to punish Ambatai's family, the Tayichi'ud, when the young men of her own family were killing each other? They had to remember the big picture.

The family endured attacks from the Tayichi'ud and also from the Merkit, seeking revenge for Hoelun's capture long ago. They survived by cunning and deception. Slowly, they gathered support—Temüjin found and married the girl to whom he had been betrothed as a child. He won the loyalty of an old sworn friend of his father, Ong Khan, by regifting a wedding present. He acquired a few loyal companions. With this help Temüjin won a major victory against the Merkit, and from there, the chains of loyalty multiplied until he was leader of a super-coalition of equestrian nomadic tribes called for the first time, all together, the Mongols.

Even after Hoelun's sons grew up and were able to provide for themselves, her role in their lives did not diminish. She adopted and raised four prisoners captured as children, and they became loyal allies of her sons. When Temüjin, now khan, divided up the command of his subjects in 1207, he considered his mother first—she had, after all, as he said, done the most work—and gave her 10,000 subjects and a military guard of 4,000. These numbers she deemed too small, but she held her tongue.

She was by now an old woman, but she was still able to intimidate her son, the khan. When he became suspicious of his brother Qasar and ordered him seized, Hoelun was furious. She hitched her white camel to her cart and traveled all night, arriving at sunrise to find her son interrogating his bound and frightened brother. She berated him, exposing her breasts and recounting how she had nursed the two boys as babies. Genghis was terrified and ashamed. He let his brother go. But the conflict was not fully resolved, and it gnawed at Hoelun, whose health deteriorated; while *The Secret History* does not say so, it is implied that she soon died.

The Mongol conquests were among world history's most brutal events, and in their course, Hoelun's sons raped, captured, and married many women. Their descendants, who often held high status, did the same. In 2003, geneticists announced the discovery of markers on the Y chromosome that occur with high frequency in 16 populations across a swath of Asia, from northeastern China in the east to Uzbekistan in the west, and that most likely identify the direct male-line descendants of Genghis Khan and his brothers. About 8 percent of those 16 populations, and about 0.5 percent of the world's population overall, carry the gene. If about the same number of women as men share this lineage today, Hoelun has more than 35 million direct male-line descendants through her sons.[2]

Hoelun's life illustrates several of this book's main themes, including the adaptive theories of menopause described in part I. Women stop bearing children in midlife for good reasons. Though most animals reproduce through old age—and this is what evolutionary theory predicts—the stage that we now know as menopause evolved in humans because the value of women's contributions, once freed from childbearing, compensated for the lost fitness benefits of continuing to reproduce. Although Hoelun had many children by modern standards, she stopped reproducing long before her body began to fail. Her most important contributions came after all her children were born—her work to provide them with food and keep them alive; her leadership in her family's mission of revenge; her role in transmitting the values, brutal as some of them were, that made the Mongols a people. She cared for grand-children and adopted nonrelatives, fended off attacks and gave advice. In all this she was hugely successful, and although she stopped repro-ducing at a young age, she has millions of descendants today. Had she continued to bear children during her dark years in exile from her hus-band's people, that outcome seems much less likely. Her story, though unusual and, perhaps, partly apocryphal, makes evolutionary accounts of menopause seem quite plausible.

The Mongol peoples of Hoelun's time were an equestrian, nomadic, pastoral—that is, herding—population. *The Secret History* catalogs raids, feuds, and alliances, favors rendered and promises made, complex and sometimes contradictory clan genealogies. Its people attack with bow and arrow, herd sheep, make and move the portable felt structures usually called "yurts" in English, steal women and horses, and eventually unite and found the largest contiguous empire in world history. Hoelun lived, that is, in the agrarian-pastoralist era, in a world dominated by the high-stakes struggle for hereditary property on which life depended. Her isolation and vulnerability in a patriarchal and hierarchical social system, among a people she had not grown up with and to whom she had no blood ties, is typical of that era; but so is her influence in midlife and later over a family that could not have survived without her.

MYTHS OF MENOPAUSE

Today, most women I know think of menopause as a medical problem, to be endured stoically or managed with drugs. Thousands of books offering medical advice on menopause are available, their prescriptions confusing and conflicting at best, and sometimes downright misleading. This is not one of them. Menopause only became a subject of *medical* interest in Europe in the eighteenth century, and its place in modern medicine rests on that recent foundation. For most of human history, people have seen menopause for what, as I argue, it really is: a develop-mental transition to an important stage of life; not a problem, but a solu-tion. For the most part, they have had no word for menopause and have not paid too much attention to the end of menstruation, instead recognizing midlife as a transition to the status of elder, grandmother, or mother-in-law.

A great deal of scientific research has tried to solve the puzzle of menopause: Why do women stop reproducing in midlife when it seems obvious that natural selection should have favored those who continued to have children? The explanations that this research has produced are ingenious and fascinating, but mostly inaccessible to the public, and they have had little impact on the popular understanding of menopause—or on that of the medical community, for that matter. Another, smaller body of research, by anthropologists and historians, has explored how the idea of menopause developed over time, and the role of culture in the experience of menopause today. My purpose in this book is to address these larger questions about menopause; that is, to explore its evolutionary, historical, and cultural aspects.

My hope is that when we see menopause in this larger context, our understanding of it will be transformed. Menopause is part of a life cycle and reproductive strategy that is unique to humans and that may well account for our successful colonization of almost all parts of the Earth. Because women have a long "post-reproductive life stage," humans are able to combine qualities that don't usually go together—fast reproduction and intensive investment in offspring; rapid population growth and a large role for experience and technology. Because of a life cycle and reproductive strategy that included menopause, human populations could explode across a landscape in favorable circumstances, but also limit the number of dependent children competing for resources and maximize the number of adult providers to young consumers. Menopause is part and parcel of an extraordinary ability to cooperate that has been critical to humans' success in the past and seems likely to be critical to our future as well.

Later in our history, when most people had abandoned the foraging way of life and were living on peasant farms, whole economic systems depended on a family structure in which households were managed by post-reproductive women and reproduction was controlled according to the constraints of the resources available. Most of these societies were deeply patriarchal and hierarchical, and I discuss the origins of those trends in chapters on agrarian societies. Finally, in the last section I will describe how we came to think about menopause the way we do in modernized, Western cultures—where the new idea of medical menopause came from and how it spread around a world that had no equivalent concept before.

I wrote this book to answer my own questions about menopause, and to share those answers with others who might have the same questions but cannot devote the necessary years to research because they are not, as I am, paid to do that sort of thing. Because I write and teach about the history of the premodern world, I knew that menopause did not have the same significance in the cultures I studied that it does today in my own. Some years ago, in one of my graduate seminars, I was discussing with my students the phenomenon of green sickness—a disease of adolescent girls—in Renaissance Europe. Why, I wondered, was puberty believed to be so difficult and dangerous then, in the way that menopause is now? At the time, in 2013, I was 47 years old, my children were becoming teenagers, and I had recently married my second husband, who is much younger than I am. I was trying to figure out what it meant to be middle-aged. (For the record, the latter is no longer something I worry about.)

As I became involved in the subject, I realized how deeply menopause is implicated in the human condition and how questions about menopause touch on, and are part of, even deeper questions about the nature of humanity, the trajectory of our history, the structure of our society, and the relationship between men and women. I have not tried to exclude these larger questions from the book, and so it has a broader focus than what readers might be expecting, although I have tried to show how all of its interlocking theses fit together. Its message can't be reduced to a simple formula or a 10-minute TED Talk. Instead I offer a scientific (yes, there will be some science), historical, and cultural tour of a phenomenon that has played an essential role in the development of civilizations, and also impacts the everyday lives of countless women (and men, in ways they may not realize). I have learned more from this project than from any other that I have undertaken, with the possible exception of raising children. Its subject sustained my unflagging attention over several years, mostly in the early hours of the morning before the demands of family, students, and colleagues took over. I am excited to share my findings with my readers, and I hope that they will find this tour as fascinating, illuminating, and transformative as I have.

Because menopause is a modern notion with negative connotations, it was hard to choose a title for this book. Tennyson's "Ulysses" is one of very few works of art with midlife as its subject—I interpret the climbing moon as a metaphor for this phase of life—and because it interprets this stage as one of vigor and expansiveness, it fits well with the book's message. There is a good case, outlined in chapter 3, that men as well as women have an evolved post-reproductive life stage during which their productivity is at its peak. For this reason, although my book's title might surprise readers accustomed, as we are, to thinking of menopause as a women's medical problem, it seemed like a good choice to me.

PLAN OF THE BOOK

I should define a few terms and concepts before continuing. For our purposes there have been three great eras of human history: the Paleolithic or "Old Stone Age," the agrarian-pastoralist era, and the modern era. Each of the book's three sections focuses mostly on one of these periods, although its organization is only loosely chronological. Part I, "Evolution," explains current theories of how menopause evolved during our long prehistoric past. Part II, "History," explores the role that menopause and reproductive strategy more generally have played in human society in all three periods, but focuses mainly on the agrarian era. Part III, "Culture," explores how modernization has changed our ideas and experiences of menopause.

The Paleolithic era began with the appearance of the first stone tools about 2.5 million years ago, and its main characteristic was that human life depended entirely on foraging for wild foods. This period is not only the longest in human history by far, but also the most obscure and least known—we must reconstruct it with no documents, depending almost entirely on whatever has survived many thousands of years in the ground and can be located and dug up. Today, we can supplement this material with increasingly sophisticated genetic testing on another survival of the Paleolithic—ourselves, the descendants of our Paleolithic ancestors and carriers of their genes. These tests have revealed a great deal, though they are not as good at telling a coherent story as one might wish, especially if we care about chronology—*when* things happened. Genetic science can try to assign dates to things but is not good at it, whereas archaeologists are very proficient. Finally, a few foraging populations survive today and are the source of our most valuable insights into Paleolithic social organization.

The second great era of human history, and the next longest, is what I call in this book the agrarian era. Archaeologists refer to the period after the invention of agriculture and the domestication of animals as the Neolithic, or "New Stone Age." By convention other eras follow the Neolithic—in Mediterranean archaeology, it ends with the onset of what scholars call the Bronze Age—but for our purposes this era extends to the eighteenth century CE. To avoid confusion, however, I will mostly use the term "agrarian" to describe this era and will use "Neolithic" only for its earliest segment, following the normal practice of archaeologists. Agrarian and pastoralist societies have been extremely diverse, but it is possible to make some generalizations about the history of population, reproduction, and family structure. Most of my discussion in part II will be about agriculturalists rather than pastoralists, because agriculture was the dominant system in this period.

In writing about the agrarian period, I use the word "peasant" to mean someone who worked on a farm owned, leased, or rented by his or her family in the era before mechanization, when the margins that farms produced above subsistence were not very great. I use this word in its sociological sense, and no pejorative meaning is attached to it.

The last, very brief era is the modern period, during which economies have been industrialized, and are based on manufacturing and wage labor rather than peasant agriculture. For our purposes, the most important characteristic of this era has been the Demographic Transition. It is possible to imagine the Demographic Transition happening

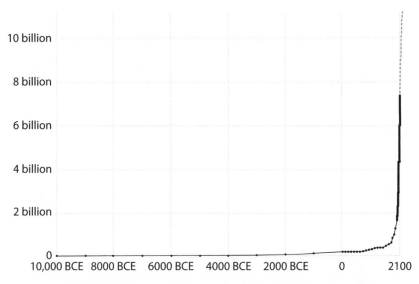

FIGURE 1. World Population over the Last 12,000 Years and UN Projection until 2100.
Modified from Roser and Ortiz-Ospina 2018.

without industrialization and vice versa, but historically, the two have been closely linked. In the modern era, rates of infant and child mortality have plummeted; adult mortality has also fallen, but changes in juvenile mortality have been much larger and have had more impact. The global population has exploded from perhaps 600 million people in 1700 to over 7 billion people today, in a function that is clearly exponential when plotted on a graph (figure 1). Most deaths before the Demographic Transition were caused by infectious disease; in modern societies, most deaths are caused by chronic, degenerative conditions like heart disease, and average life expectancy at birth is much higher. Other features of modernization, besides industrialized economies and the Demographic Transition, are high rates of education and literacy, urbanization (more and more people living in cities rather than in villages or on farms), and a tendency toward more democratic governments. This is the age in which an increasingly interconnected world has become a meaningful socioeconomic unit, with a shared fate, for the first time. Finally, this is the era when the medical system that some researchers confusingly call "biomedicine," and that I will call "modern medicine," became dominant. How all of these things are connected is debatable, and it is not necessarily the case that they all had to happen together, but such is their relationship in history, as it actually worked out, that it is reasonable to speak of a single phenomenon called modernity and of "modernization."

So defined, we see that these eras (Paleolithic, agrarian, modern) do not have firm chronological boundaries. Some people live mostly foraging lives today; a greater number live agrarian lives. But few in the twenty-first century are entirely untouched by modernization, as almost all economies, for better or for worse, have become at least partly industrialized, almost all the world has been organized into nation-states based on these economies, and the Demographic Transition has happened or is underway almost everywhere. Still, I have avoided using the term "premodern" unless I mean it literally; it is not necessarily the case that all agrarian societies are destined to modernize, although this has been the overwhelming trend. I especially dislike the terms "developed" and "developing" and have avoided them as much as possible.

The world before industrialization and the Demographic Transition is sometimes called "traditional," and I use the term "traditional societies" to refer to both foraging and agrarian peoples. The constraints of language force me to write of foraging and agrarian societies in the past tense, as though they no longer existed, or in the present tense, as though they have persisted unchanged; readers should bear in mind that neither of these things is true. Many societies have only partially transitioned from peasant economies, though hardly any society today is totally unaffected by modernization.

I make no value judgment about which of these systems is better, although many of the things I will say about the agrarian era may give the impression that I believe this was an especially dark turn in human history. The agrarian era was fantastically diverse and therefore fascinating, and for this reason, and because much of it is well documented and accessible to scholars, I have spent most of my life studying it. The changes of the most recent era have brought tremendous relief from the burdens of high mortality and reproduction to much of the world, and most beneficiaries of modernization live out a natural lifespan of

70 years or more, and never experience famine or the loss of a child. These changes have also unleashed a great many other potentialities. But modernization is at present unsustainable, its benefits have been very unequally distributed, and the horrors it is capable of perpetrating, in a world still ruled by older habits of territoriality, patriarchy, and hierarchy, are immense. Still, it is a feature of both the transition from the Paleolithic to the agrarian era and that from the agrarian to the modern one that there is no going back without terrible suffering. Humans, like other animals, tend to be conservative in their behavior, and values have lagged behind economic and technological change. But it is critically important to acknowledge the ways in which the world has changed, to let go of the past, to embrace the potential of the industrialized age, and to use the gigantic reserves of energy it has liberated to solve the problems it has created—that is, to make this a world of equality, sustainability, and resilience to external catastrophes. Although it is likely hard for the reader to see how menopause fits into this vision, I hope it will be clear by the end of the book that menopause is not just an artifact of our evolutionary past; it is critical to our present and future.

PART I

•

Evolution

CHAPTER 1

•

Why Menopause?

TO A HUMAN, it seems natural to stop reproducing in midlife. The very thought of becoming pregnant, giving birth, and caring for an infant through, say, age 70 is exhausting, even perverse. But this is what most other animals do. Only in rare circumstances does nature select for lifespans much longer than an organism's reproductive life; most female animals, that is, continue to reproduce in old age. Human menopause is one of science's profound puzzles, the hinge on which much discussion of our evolution turns: one of the most unique features of our species, it must be explained, or explained away.

Menopause is probably adaptive. That is, it's not a mistake or an artifact of modern life whereby women live past some natural test of usefulness. This conclusion has important consequences for how we should think about it and how we should research and treat it. But first, let's talk about the puzzle of menopause, before discussing in the next chapters some of its solutions.

The discipline with the potential to answer the question "Why menopause?" is evolutionary biology—a field that can seem more abstract and more speculative than other natural sciences. Its hypotheses can be hard to test. But only evolutionary biology can answer the big questions about how humans came to have their unique life course, defined by long childhoods, long lifespans, short intervals between births, and, for women, long post-reproductive lives. Most evolutionary biologists and anthropologists agree that all of these factors are related. I am going to describe different ideas about why menopause exists and how it arose— theories that do not always agree, but that are all compelling in their own way. I think that several of these theories, and not just one of them, are probably right, and I will try to convey how they might work together.

DOES MENOPAUSE OCCUR IN OTHER ANIMALS?

This question has proved surprisingly difficult to answer. Any study of life cycles of large mammals takes many years, and even then the results can be muddled. How long, for instance, do chimpanzees live? We can't determine whether they have a post-reproductive lifespan without knowing the answer to that question, but that task is not as simple as it seems. Chimpanzees live longer in captivity than in the wild, some groups of wild chimpanzees live longer than others, and some individuals in both groups live much longer than average.[1]

Because chimpanzees are humans' closest living relatives, evolutionary biologists often study similarities and differences between the two species to determine when, and whether, a trait might have evolved. If we share a trait with chimpanzees, it is possible (though not certain) that this trait evolved sometime in our common history. For example, many biologists believe that the tendency of both humans and chimpanzees (as well as other great apes) to exchange females among groups is a behavior that evolved before the divergence of the human and great ape lineages.[2] This "male-philopatric" (meaning "male-father-friendly") dispersal pattern is less common among most mammals than the practice of exchanging males. On the other hand, some scientists have argued that human male philopatry is a result of socioeconomic developments in the agricultural period and not typical of our Paleolithic ancestors.[3]

In a similar way, biologists have tried to determine whether chimpanzees experience an equivalent of human menopause and whether female chimpanzees commonly live past their reproductive lives. Depending on one's definitions, these may be two different questions. In humans, fertility ends some years before the ovaries stop ovulating and producing sex hormones. While human menopause as defined by most researchers—that is, as the last menstrual period—occurs around age 50 in most populations, historically only a small percentage of women have given birth after age 45. In a collection of data from 31 populations with "natural fertility," average ages at last birth cluster around 39 and 40.[4] Among the Hutterites, an Anabaptist sect in North America often studied by demographers because of their very high fertility rates, the average age at last birth in the mid-twentieth century was 41.[5] Some animal studies test their subjects' hormone levels or dissect their ovaries, but most wild animal studies rely on the observed ages of females at the birth of their last offspring. Because evolutionary fitness is measured in terms of reproduction, the end of fertility, rather than menopause per se, is usually the more relevant factor when thinking about evolution and natural selection.

The evidence suggests that humans' long post-reproductive lifespan emerged or evolved sometime after the divergence between humans and chimpanzees around 6 to 10 million years ago. But because menopause occurs in all known human populations, it probably emerged before our species divided into groups with little contact with one other; that is to say, probably before about 130,000 years ago.

The most spectacular documented example of an animal that undergoes menopause is the Japanese aphid *Quadrartus yoshinomiyai*, famous among insect researchers for its "glue-bomb" stage of life. Older adult aphids stop reproducing and instead secrete a sticky substance in their abdomens. When predators attack the colony, they selflessly fling themselves into the fray, sticking to the predators and defending the colony at the cost of their lives. These aphids reproduce parthenogenically—all females, they clone themselves in a series of "virgin births"—so the phenomenon called "kin selection" is an especially powerful force among them. A sacrifice by one aphid might save several with identical genes.[6]

What about animals more closely related to us? Do mammals, including our close relatives the chimpanzees, have post-reproductive lifespans? And what counts as a post-reproductive lifespan? Neither of these questions is easy to answer, but based on the research now available, it appears that humans share this trait with very few other mammals, and not with our closest relatives.

In the past, an obstacle to understanding whether menopause is unique to humans has been the problem of how to measure post-reproductive lifespan. This challenge has been overcome recently by Daniel Levitis of the University of South Denmark and his colleagues, who introduced two measures in 2011 and 2013. First, a simple measure called "Post-Reproductive Viability" solves the problem of how to define a maximum reproductive lifespan and a maximum natural lifespan for a species; another, more complex calculation called "Post-Reproductive Representation" describes the proportion in a given population of adult years lived after fertility ends.[7] Both of these calculations require information that we don't always have: detailed statistics on fertility and a demographic life table, which tabulates mortality, survivorship, and life expectancy at different ages.

Post-Reproductive Viability is the age at which 95 percent of a cohort's years have been lived, minus the age at which 95 percent of its children have been born. (A "cohort" is a group within a population whose members are the same age.) For the women of the !Kung, a foraging population of the Kalahari Desert in southern Africa, this number is 25 years. It is possible for the number to be negative, in which case the animal has no Post-Reproductive Viability.

Post-Reproductive Representation is a little more complicated. Let's imagine a cohort of 1,000 women, all born in the same year (figure 2). Imagine that 5 percent of this group's babies are born by the time its members reach age 20—we can call this the age of adulthood for that group. At that age, 600 of the original cohort are still alive, and they have an average life expectancy of 40 more years—that is, the group at age 20 has a combined total of 24,000 years of adult life ahead. By the time the women are 40 years old, they have given birth to 95 percent of all the babies they will ever have. Four hundred are still alive, and they have an average remaining life expectancy of 25 years; as a group, they will live about 10,000 more years past the age at which they will produce very few more children. To find the *proportion* of adult years lived post-reproductively, we divide 10,000 by 24,000 to get about 0.42 (or 42 percent), which is close to the value that Levitis and his team calculated for !Kung foragers. In a "stationary" population that is neither growing nor

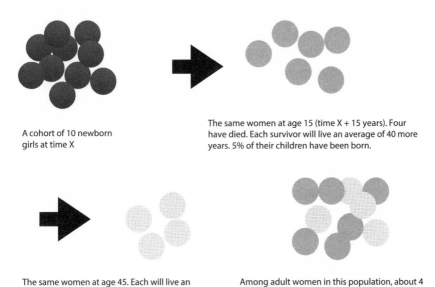

A cohort of 10 newborn girls at time X

The same women at age 15 (time X + 15 years). Four have died. Each survivor will live an average of 40 more years. 5% of their children have been born.

The same women at age 45. Each will live an average of 25 more years. 95% of their children have been born.

Among adult women in this population, about 4 in 10 are post-reproductive (age 45 or older). PrR is 0.42.

FIGURE 2. Calculating Post-Reproductive Representation in an Imaginary Population.

shrinking, this is also *the proportion of adult women in the population who are past reproductive age.*

Because of the way it is calculated, Post-Reproductive Representation, or PrR, is always a positive value between 0 and 1, so there are further complicated tests to determine whether it is significant (a huge value of 0.42 is obviously significant, however). Animals with significant PrR can be said to have a *post-reproductive life stage* that suggests some force of natural selection. Human populations have very large values for PrR that are hard to explain except as an adaptation of some kind.

Post-Reproductive Representation has become the gold standard in menopause research and is better than other methods of calculating post-reproductive lifespan—including Post-Reproductive Viability—because it considers the percentage of the population that lives to post-reproductive age. If some long-lived animals in a species have extended post-reproductive lives, but only a few individual animals live that long, then post-reproductive life has probably not been important in the evolutionary history of that animal. Many animals have some

Post-Reproductive Viability, but very few can claim PrR comparable to that of humans.

Another problem in menopause research has already been noted: zoo and laboratory animals can have very different life histories than populations in the wild. A few chimpanzees in captivity have lived lives much longer than average without continuing to reproduce; for example, Fifi at the Taronga Park Zoo in Sydney, Australia, died in 2007 at age 60, 20 years after she had her last baby. But after all, some humans survive past the "normal" maximum lifespan of around 75–80 years, to reach age 100 or more.

Protected zoo populations are like humans living in industrialized countries with low mortality, for whom PrR is much higher than for foragers. In order to understand how animals have evolved, it is important to use data from wild animals living in the environments that shaped their natural histories; likewise, we must use data from traditional human societies without industrialization or modern medicine to understand how humans have evolved. Only in the last few decades have researchers begun publishing the results of labor-intensive, long-term studies of animal populations in the wild. For large, relatively long-lived animals, these "demographic" studies—inquiries into questions about population size, fertility, longevity, and mortality—take a long time; researchers must observe groups of animals over decades, in conditions in which even catching sight of them can be difficult.

Thankfully, several research teams have studied wild chimpanzees over the very long term, beginning with the famous work of Jane Goodall, who has studied the Kasakela chimpanzee community of Gombe National Park in Tanzania since 1960. Researchers have published demographic studies of other wild populations in Tanzania, Guinea, and the Ivory Coast, as well as analyses that combine all of this information.[8] The demography of captive chimpanzees has also been studied, based on the records kept by zoos and primate laboratories.[9]

Among most populations studied, the natural lifespan for a wild chimpanzee is around 40 years. Only 7 percent of wild chimpanzees live past this age, though a few individuals have lived to age 50, and this is more common in captivity. Chimpanzee fertility peaks around age

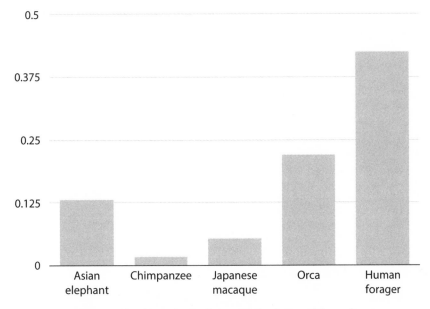

FIGURE 3. Post-Reproductive Representation in Some Mammals.

25–30 and declines after that. But about half of the small number of wild chimps who live past age 40 have at least one more baby. Chimpanzee fertility reaches zero around age 50, about the same age as in humans, but this is close to the normal limit of a chimpanzee's natural lifespan.[10] By the calculations of Levitis and his team, a typical cohort of wild chimpanzees has lived 95 percent of its years by age 37 but has not had 95 percent of its babies until age 45. That is to say, the reproductive lifespan in chimpanzees is actually longer, by this measure, than the "somatic" lifespan of the body itself (*soma* means "body" in Greek); that is, they do not have Post-Reproductive Viability. Post-Reproductive Representation among most groups of wild chimpanzees is only 0.018 (figure 3).[11]

As we learn more about wild chimpanzees, their demography becomes more complicated. The published studies of chimpanzee mortality cited previously are based on populations that are mostly declining, from disease and the catastrophic effects of humans on their environment; consequently, mortality is probably higher in these populations than has been historically true for most chimpanzees. Newly published

studies of chimpanzees in healthier environments have shown much
lower rates of mortality, especially for the Ngogo chimpanzees of Kibale
National Park in Uganda, which are thriving.[12] This population is sur-
rounded by other groups of chimpanzees and has little contact with
humans, no large predators, no epidemic diseases, no history of logging
in their forest, and an abundance of wild foods. As a result, its members
live much longer on average than other groups of chimpanzees that have
been observed, and perhaps longer than has been historically normal or
average for chimpanzees. Average life expectancy at birth for Ngogo
chimpanzees is 35.8 years for females and 29.6 years for males—similar
to the range for human foragers (though the difference between male
and female mortality is strikingly high and much greater than for humans). Early in life, mortality for this group of chimpanzees is actually
lower than for human foragers.

However, even in this healthy and long-lived population, the contrast
with humans at older ages is strong. Human foragers show much higher
survival at ages past 40. Mortality begins to escalate earlier in the chimpanzee population than it does among humans, at about age 20. More
chimpanzees in the Ngogo population outlive their reproductive lives
than do members of other groups, but they apparently do not have a
PrR comparable to that of humans.[13]

What, then, is the extent of the post-reproductive life stage among
humans? Among the mid-twentieth-century !Kung, 95 percent of cohort years were lived by age 67—a full 25 years past the age at which 95
percent of fertility ended, at 42. Among foragers generally, the average
age at which women last give birth is about 39, and life expectancy for
those who reach that age is about 25 to 30 additional years. Typical Post-
Reproductive Representation for foragers ranges between 0.35 and 0.5,
though an unusually low figure of 0.256 has been calculated for the hor-
ticulturalist Yanomamo of Brazil. Among plantation slaves in eighteenth-
century Trinidad, a very high-mortality population in which mortality
exceeded fertility so that its numbers could only be maintained by im-
porting more slaves, PrR was still 0.315. Post-Reproductive Representation among modern human societies is of course much higher, reaching
about 0.76 in Japan today.[14]

Thus, while humans and chimpanzees stop reproducing at about the same age, a 40-year-old chimpanzee—who still has a nearly 50 percent chance of giving birth again—is quite old. If humans had similar reproductive patterns, women would continue to bear children well into their 70s.

Besides chimpanzees, the long-suffering Rhesus macaque commonly used in laboratory experiments is among the most thoroughly studied of all primates. In 1938 researchers established a colony of Rhesus macaques imported from India on the Puerto Rican island of Cayo Santiago; originally intended as a source of laboratory animals, the colony, managed by the University of Puerto Rico, is now used to research the natural behavior and life cycle of macaques. Other research colonies of macaques have lived since the 1970s on Key Lois and Raccoon Key in Florida, and since the 1950s in Japan, where many wild monkey parks have conserved populations of Japanese macaques (a different species) for both tourism and research. Particularly well studied among these are the monkeys of Arashiyama ("Stormy Mountain").[15] All of these macaque communities are "provisioned"—that is, supplied with food by people. For this reason the demographic patterns observed may be somewhat different from what is typical in the wild—though it is hard to say how much, and the difference may be slight.

Although there is more evidence for a post-reproductive life stage in macaques than in chimpanzees, it is still small. An analysis of 50 years of data from the Arashiyama population of Japanese macaques found that fertility declined steeply after age 22 in this animal, reaching zero by age 26. Nineteen percent of female macaques who reached maturity survived to age 26—past the upper limit of fertility—but only about 8 percent lived beyond age 30. A few, mostly from one female lineage, survived to age 33.[16]

The average lifespan after last birth for the Arashiyama macaques was about 4.5 years. This is three times the average interval between births of about 1.5 years, but results were skewed because a few females lived much longer than the rest; in comparison, the median lifespan after last birth was fewer than three years. The researchers concluded that although females who survive long enough will outlive their reproductive capacity, and some individuals can live much longer than that, post-reproductive lifespan is not in general an important part of the life history of Japanese macaques.[17] Post-Reproductive Representation for Japanese macaques is 0.054, higher than that of chimpanzees (and statistically significant, meaning that the researchers are 95 percent confident that it reflects a real tendency for macaques to outlive their reproductive lives), but it is a tiny fraction of any value found for humans.[18]

How unusual is the long human post-reproductive life stage? Very, as it turns out. It is true that many or most mammals—and possibly other animals and organisms (though mammals have been studied the most)—can outlive their reproductive lives, as Fifi or the long-lived lineage of macaques at Arashiyama did. This is also the conclusion of an exhaustive survey by Alan A. Cohen published in 2004[19] and based on older methods of calculation from before the invention of PrR. For the purposes of his study, Cohen defined "post-reproductive lifespan," or PRLS, as the period of time between the average age at the birth of an animal's last offspring and the average age at death for animals living past that age. That is, if we call the first number x and the second number y, PRLS is equal to $y - x$. By this definition all animals that survive their last birth have some post-reproductive lifespan; Cohen defined a significant PRLS as a figure greater than the average interval between births for that animal, plus one standard deviation.

Cohen's study identified a pattern across many species, from lions and baboons to polar bears, ground squirrels, and several other mammals: fertility tends to cease before the end of the natural lifespan, and the oldest females may live significantly beyond the age at which they give birth for the last time.[20] Cohen counted 35 mammals that met his criteria for post-reproductive lifespan, out of 42 studied in the papers he surveyed. These studies included captive animals, laboratory animals, and domestic animals as well as wild animals, and some of their conclusions have been challenged by more recent evidence. Still, it is fair to say that reproductive lifespan and somatic lifespan can evolve independently, and that, in many mammals, fertility ends significantly before death. For most of these animals, though, post-reproductive lifespans are short and PrR, where this has been calculated, is small. For example, a recent study comparing seven primates that have been observed over long periods of time in the wild—including chimpanzees, baboons, gorillas, three kinds of monkey, and one lemur—calculated small PrRs of between 0.01 (for baboons) and 0.06 (for spider monkeys).[21]

To sum up, it is not unusual for individual nonhuman animals to outlive their reproductive lives, and reproductive lifespan does not always exactly match somatic lifespan, suggesting that the two can evolve separately. But the more we learn about other animals, and the more we develop good methods of calculation, the more divergent humans seem to be, with Post-Reproductive Representation that is much higher than that of our nearest relatives and of almost all other animals. Old arguments that menopause is only an artifact uncovered by recent medical advances that have lowered mortality and that few premodern humans lived past menopause can be relegated to the garbage heap of scholarship. Humans have a very significant, naturally occurring post-reproductive life stage.

As far as we know, among undomesticated mammals, only two species of whales have a post-reproductive life stage comparable to that of human females.[22] Several demographic studies on whales date to the 1980s, before commercial whaling was banned by the International Whaling Commission in 1986. In this period, whale drive fisheries corralled and killed whole pods at once, allowing researchers to study the demography of the populations that were hunted. More humane studies based on photographic surveys of pods over years or decades have also been published.

One of these photographic-survey studies analyzed data about orcas, or killer whales, living off the coast of British Columbia and Washington state from 1973 through 1987.[23] Two separate ("northern" and "southern") communities of orcas lived in the area year-round; a "transient" community also visited the region but was not studied. Each community was made up of several pods of orcas; the northern resident community had a larger number of pods of smaller average size than the southern community.[24] Observations of the "southern resident" whale population are still ongoing, and the Center for Whale Research in Friday

Harbor, Washington, has trained many amateur volunteers to recognize individual whales by sight.

On average, female orcas give birth to calves about five years apart, beginning around age 15. The average age at last birth for females is 39, similar to the age at last birth for humans and chimpanzees. But mortality rates remain very low for females of this age. Several females reached ages beyond 60 during the study, and the researchers estimated the ages of the two oldest females as 76.5 and 77; maximum lifespans for females, they concluded, might be around 80 or even 90. Female orcas, then, have a post-reproductive life stage comparable to that of humans. One team of researchers has calculated orca PrR at 0.22, lower than that of most human foragers but higher than that of any other wild animal except the short-finned pilot whale.[25] The ages of male orcas were harder to estimate, but researchers did determine that they had much shorter lifespans, with maximums of about 50 or 60 years.

Orcas live in complex matrilineal societies in which each pod is composed of several families and pods cooperate in communities. Calves of both sexes continue to live with their mothers in adulthood, so families may be large, extending to as many as four generations.[26] It is possible that this social structure is connected to the evolution of a post-reproductive life stage in orcas. Having a living mother is highly beneficial even to adult orcas: adult females are 2.7 times more likely to die, and males 8 times more likely to die, in the year after their mother's death.[27] Post-reproductive females are more likely than others to lead salmon hunts, especially when salmon are scarce—perhaps because of their accumulated knowledge and experience. In this way, their skills may enhance survival for their descendants in the group.[28] In one charming story broadcast on National Public Radio, a grandmother orca seems to have helped deliver her daughter's baby by pulling on its dorsal fin, and she continued to swim with the baby afterward.[29] It is possible that post-reproductive female orcas help their descendants survive in other ways that may be hard to observe or understand.

It is less clear why female orcas do not continue to reproduce as they age. One theory is that when a mother and daughter are reproducing at the same time in the same group, there is too much competition for resources and higher mortality for juvenile orcas. Researchers have observed that when calves are born into groups with this kind of reproductive competition, the offspring of the older female has a greater chance of dying; the offspring of the younger female actually has stronger chances of survival than when no competition is present, probably because of the benefits of having a grandmother in the group. Also, because of orcas' social structure, older females are more related to the others in their group, who are all likely to be these females' own descendants, than are younger females, whose fathers are likely to be males outside the group. For these reasons, researchers argue, kin selection favors suppressed reproduction in older females rather than younger ones.[30]

Short-finned pilot whales (*Globicephala macrorhyncus*) were hunted at Taiji in Japan through the early 1980s. Studies published by Toshio Kasuya and Helene Marsh in 1984 analyzed data from the carcasses of more than 800 whales stranded or killed in the drive fishery from 1965 through 1981. Because the fishery captured whole groups, including pregnant females and juveniles, the researchers were able to study the demography and reproductive life cycle of this species. They dissected the reproductive organs of both males and females and also recorded length, weight, age (determined by layers of growth in the teeth), and other characteristics.[31]

Like orcas, short-finned pilot whales live in matrilineal groups in which females spend their entire lives; some males migrate to other groups at adulthood. Females become sexually mature much earlier than males, at around age nine (compared to the late teens for males). The oldest pregnant female in Kasuya and Marsh's sample was 34, and the oldest female still ovulating was about 40. The youngest post-reproductive female was 29, and all the females over 40 were post-reproductive. But females of this species, the researchers found, often lived decades past this age. The oldest female in the sample was 63, and they calculated the average post-reproductive lifespan for females at about 14 years. Some 25 percent of adult females in the sample, in fact, were post-reproductive.[32]

While most calves of young females were weaned by age three, post-reproductive females sometimes nursed their last offspring much longer, up to 15 years. The average interval between births was about seven years but was much lower (around five years) for young females under age 24. Males of this species lived much shorter lives. The oldest male in the sample was 46, and Kasuya and Marsh estimated that the male lifespan was about 15 years shorter than that of females for this species.

Because studies on whales in the wild are so difficult to conduct, we don't know whether these results showing long post-reproductive life stages for two species would hold true over time and in different ecological conditions. But they suggest that in rare cases, other animals have developed a female reproductive life cycle similar to that of humans. In both of these whale species, it is noteworthy that males die much earlier than females and continue to reproduce throughout their shorter lives. This suggests that females do not stop reproduction early; rather, some selection pressure has caused female lifespans to lengthen *beyond* reproduction. Whatever this pressure is, it has not affected the lifespans of male orcas or pilot whales.

Like humans, these two whale species are long-lived—at least the females are. So do all animals with long lifespans have a post-reproductive life stage? Is there some hard age limit for reproduction in mammals—say, age 40 or 45—that evolution cannot get around? Few animals live longer than humans, and the life cycles of long-lived animals are especially difficult to study because of the vast spans of time necessary to track changes. But we do know of at least two mammals that reproduce long past the human maximum of about age 45.

Cynthia J. Moss has studied African elephants in the Amboseli National Park in Kenya since 1972.[33] In 2001, based on observations and records of that population (1,778 individuals in total, including those who died in the course of the research), she published the most comprehensive and reliable study of wild elephant demography available today. On average, African elephants at Amboseli gave birth for the first time around age 14. Infant mortality (death in the first year after birth) was relatively low—almost 90 percent of calves born to mothers over 20 survived their first year. Calves were usually born about 4.5 years apart.

Elephant fertility began to decline around age 40, and this decline accelerated late in life, after age 55 or so. But only 9 of the 38 elephants who survived to age 50 stopped reproducing (that is, only these 9 females survived longer than seven years past their last birth). Of 12 females who survived past age 60, 5 gave birth. Maximum life expectancy for female elephants was around 65, and for males it was a little less, around 60. Males faced much higher mortality rates throughout the lifespan, and only a minority survived to reproductive age, which is late for male elephants—only mature bulls, usually over age 30, have a good chance of reproducing.

Another study has focused on a population of Asian elephants used for logging in Myanmar, based on records of the Myanma Timber Enterprise dating as far back as 1900. While these elephants worked for humans during the day until the retirement age of 54, they were otherwise unsupervised and allowed to forage and breed naturally. Results were similar to those published by Moss. Fertility in these elephants decreased after age 50 but did not end abruptly. Ninety-nine percent of elephant births were not complete until age 57, 10 years later than in the nonindustrialized human population to which researchers compared them (a dataset representing over 5,000 Finnish women born between 1595 and 1839), and the latest elephant birth on record occurred at age 65. Furthermore, mortality was higher in older elephants than in humans, and elephants were less likely to outlive their reproductive lives for that reason. Elephants over 40 years old had somewhat shorter intervals between births—less than five years—compared to younger elephants, even when researchers controlled for every confounding factor they could think of. Among the Asian elephants of Myanmar, PrR was 0.13—relatively high for a nonhuman mammal—but in the human population used for comparison, it was almost four times greater, at 0.51.[34]

Studies of wild African elephants have shown that very old matriarchs (over age 55 or 60) are better at protecting their herd from lions, at discriminating the signal calls of friends from those of strangers, and possibly (though this is harder to prove) at finding food and water during periods of drought. In one study of elephant behavior during a 1993 drought in Kenya, the oldest elephants seemed to remember water sources from a drought that had occurred 40 years earlier! In Myanmar, elephants born to young, inexperienced mothers were eight times more likely to survive to age five if their grandmother lived in the same group; in fact, 93 percent of newborn elephants survived to age five when their grandmother lived with them. Although elephants do not have long post-reproductive lifespans, these studies remind us that it is important to include the value of experience when calculating the benefits of longer lives in foraging populations.[35]

Another long-lived species that reproduces late in life is the fin whale (*Balaenoptera physalus*), which is the second-largest animal in the world, after the blue whale. The fin whale was hunted commercially until 1987, and studies of this species' life cycle have been based on carcasses; no long-term observational studies have been attempted, and these would be difficult, because fin whales live a very long time. In a publication from 1981, Sally Mizroch calculated the age of the whales by counting growth layers in their earplugs, which, along with the ovaries, had been preserved by the Japanese expeditions that killed them. The whalers also recorded other information about their catch, such as length and whether the females were pregnant. Twelve years of whaling data produced information on 1,556 female whales. Using these figures, Mizroch estimated the age of the oldest whale in her dataset to be 111. Four of the next oldest were in their 80s.

Fin whales reach sexual maturity early, around age six or seven. Calves are born around 2.5 years apart on average. Although Mizroch's examination of the ovaries suggested that ovulation rates declined with age, pregnancy rates told a different story: some pregnant females were more than 70 years old, and rates of pregnancy did not seem to change with age.[36]

Reproduction in long-lived mammals, then, can continue to very advanced ages. Furthermore, in some mammals with much shorter lifespans than humans—laboratory rodents, for instance—fertility declines and rates of abnormal oocytes, stillbirth, and genetic abnormalities in offspring increase in old age.[37] That is, mammalian eggs do not have a standard shelf life; "old eggs" alone cannot explain why reproductive life ends so early for women.

HOW DOES MENOPAUSE HAPPEN?

To understand how it is possible for animals—even closely related ones, like humans and chimpanzees, or certain species of whales—to have different reproductive life cycles and post-reproductive lifespans, we must consider the physiology of reproduction. I should caution, though, that at this point—when we try to explain *why* certain things happen— answers become more complex and debatable, and it will be impossible to avoid oversimplifying. Most theories of reproductive senescence (a scientific term for "aging") rely on the idea that, in mammals, ovarian follicles containing "oocytes," immature eggs, become depleted over time. Because individual animals, including humans, vary widely in the number of follicles their ovaries contain, and because follicles can only be counted by dissection and cannot be counted for any individual more than once, it is difficult to get a clear picture of this process of depletion over time. Nevertheless, the physiology of reproduction in mammals offers an obvious way for nature to select for fertile lifespan separately from somatic lifespan, at least in females.

It is still the scientific consensus that in mammals and birds, males continue to produce sperm cells throughout life, but females produce all their oocytes early in embryonic life. In humans, the number of follicles containing oocytes reaches a maximum of about 600,000 at around five months of fetal development—around 295,000 per ovary on average, but with wide variation—and declines from this peak until menopause. There is some debate about the best model for this decline, but both leading candidates postulate an exponential decline that accelerates with age, even if they reject the more old-fashioned "broken stick" theory that saw a sharp increase in the rate at which oocytes are lost around age 38. Because of the exponential nature of the decline, however, even though it speeds up with age *as a percentage of remaining follicles*, the *absolute number* of follicles lost is much higher per month in youth, with the rate of loss peaking at age 14 or 19 depending on the model used.[38]

A maximum of only about 400 follicles mature for ovulation over a woman's lifetime (fewer in women who have many pregnancies). The remaining follicles degenerate at different points in the process of development; this degeneration is sometimes called "follicular atresia," and sometimes "apoptosis," a more general name for the kind of programmed cell death that is presumed to cause this process. These latter, doomed follicles produce estrogen, progestin, and other hormones necessary for reproductive cycling. Menopause occurs in humans when about 1,000 follicles are left. Most researchers believe that variation among individuals in the number of follicles they are born with explains variation in the age at menopause—women born with fewer follicles reach menopause sooner.

There is growing evidence from several independent studies that some female mammals—including laboratory rodents, some primates, and humans—may produce new oocytes throughout their reproductive lives.[39] Though this "eggs forever" hypothesis is still debated, scientists are working on new infertility treatments that may one day use stem cells from adult human ovaries ("oogonial stem cells") to generate new oocytes that can be fertilized.

In the traditional model of mammalian reproductive physiology, natural selection can act on the reproductive lifespan by increasing the number of oocytes produced before birth or by the relatively less costly method of slowing down the rate of atresia. The evidence that natural selection has, in fact, acted this way is strong. Larger mammals with longer lifespans have more follicles and lose them more slowly than smaller mammals, and mammals with both short and long lifespans show a similar pattern of declining fertility that eventually ends as the number of follicles decreases and the quality of oocytes declines.[40] If the creation of new follicles during reproductive life is an important factor in some mammals' fertility, presumably nature could also select for the production of more follicles over a longer period by delaying or turning off whatever changes are responsible for the decline of this process. We do not yet know what those changes are, but the new "eggs forever" model certainly begs the question: If both men and women produce sex cells throughout life, why do women stop, while men do not? If ovaries really do renew their eggs during adulthood, it should have been even easier for nature to select for a longer reproductive life, were it advantageous to do so.

Whatever causes reproductive senescence in women, age at menopause is both variable among individuals and heritable. Menopause can happen as early as age 40 (if not earlier; 40 is the arbitrary cutoff for "premature" menopause) or as late as 60. Furthermore, we tend to reach menopause at an age similar to that at which our mothers experienced it. Seventeen genes have so far been identified that relate to age at menopause, although they explain only a small part of its heritability, which is estimated at 40 to 60 percent (that is, inherited genes are thought to account for about half of the individual variation in age at menopause, whereas other influences account for the other half).[41] Both of these conditions—variability and heritability—are important for natural selection to happen, and they are clearly present. We begin to see the evolutionary puzzle of menopause: it is hard to imagine a greater and more straightforward fitness benefit than a longer reproductive life. But although our lifespans have lengthened, age at menopause apparently has not changed since before our ancestors diverged from the lineage we share with chimpanzees.

As we have seen, there is no reason to believe that nature cannot select for shorter or longer fertile lifespans, and, more importantly, *nature has in fact done this*: when we compare the reproductive and post-fertile lifespans of other mammals, we find broad variation. "Pleiotropic" arguments about menopause—including the Patriarch Hypothesis discussed later—rely on the idea that nature *cannot* modify or extend reproductive life, but this is clearly not the case.[42] Furthermore, reproductive aging can evolve independently from general somatic aging (aging of the body). Natural selection can favor ending reproduction at a certain age, while the aging of the body is controlled mostly by other mechanisms and can be selected for separately. In most animals, the pressures of selection have, *on average*, placed age at reproductive cessation close to the maximum lifespan or a little before that, when somatic senescence is advanced—that is, when the animal is old. But in every population, some individual animals outlive their reproductive lifespans for a longer time than average (like Fifi the chimpanzee or the four female macaques at Arashiyama who lived to age 33), just as some reproduce longer than average. Those individuals provide an opportunity for nature to favor longevity past the reproductive lifespan, if they have a fitness advantage. This has happened at least a few times, in humans and in two species of whales. But it is unusual.

MENOPAUSE AND THE EVOLUTION OF AGING

It is not surprising that most animals do not outlive their reproductive lifespans for very long, because theories of senescence—aging—strongly predict this. Modern theories of senescence trace back to 1951, when Sir Peter Medawar delivered a famous lecture at University College, London, called "An Unsolved Problem in Biology," in which he addressed the question: Why does aging occur, given that nature should select for longevity and against mortality? His answer, although based on a much simpler model of genetics than the one that prevails today, remains foundational to evolutionary theory. Even without aging or mortality of the body, Medawar postulated, old animals are rare because organisms tend to die as a result of external causes (predation, accident, disease). Given this condition, harmful genes that act on the organism late in life can only be selected against weakly, and it is the accumulation of these genes over time that causes aging.[43]

This idea was strengthened and developed further in a well-known 1957 article in the journal *Evolution* called "Pleiotropy, Natural Selection, and the Evolution of Senescence," by George C. Williams. "Pleiotropy" refers to a gene's ability to cause "many turnings" or outcomes, some of which may be "antagonistic"; that is, they have opposing effects under different conditions and, specifically in Williams's theory, at different ages. If a gene is beneficial early in an organism's lifespan (when its reproductive potential is high) *but also* harmful later (when its reproductive future is shorter), nature will select for that gene. Because such genes will inevitably arise, they will inevitably accumulate, causing aging. Nature will continue to select against senescence in other ways, but the effects of this selection decrease with the organism's age as its reproductive future declines and it has fewer opportunities to pass on the gene.[44]

Variations in longevity and aging among different species of animals reflect different balance points between these two types of selection, and organisms with high "extrinsic mortality," such as insects and rodents, tend to age faster and live for much shorter periods than organisms with low extrinsic mortality. This is because it is important for them to reproduce earlier and faster, and because there is little selection pressure against aging when few animals in the species survive for very long. Williams's theory predicts that large animals and those with unique protection from predators should have longer natural lifespans than smaller, more vulnerable animals; real-life examples include elephants, whales, some birds and bats (which can fly away from predators), and tortoises. Animals that reproduce early should age and die faster than animals with long periods of sexual immaturity. Where there are differences in extrinsic mortality between the sexes, the sex with higher mortality (usually males, who often compete for mates) should age and die faster. Most of these predictions hold up pretty well, and some of them are important for the hypotheses discussed later.

In 1966, an influential paper by William D. Hamilton built on Williams's work by developing mathematical formulae to express the effects of natural selection across the reproductive lifespan and the relationship between reproduction and mortality. Because Hamilton's model agrees with Williams's in predicting that animals will not long outlive their reproductive lives, some scholars call it the "Wall of Death" model.[45]

Finally, in 1977, Thomas B. L. Kirkwood added the concept of the "disposable soma" to senescence theory.[46] Previous theories had not considered that maintaining the body's cells is costly and requires energy and resources that could be used for other things, such as growing or reproducing. Because of these costs, nature tends to select against maintenance longer than the animal can reasonably survive and reproduce in the wild. This theory emphasizes that the body is only a vehicle for reproducing the germ cells (egg and sperm cells) that carry our genes to future generations and will tend to become disposable. The body, that is, is manufactured cheaply and designed to be used for a relatively short time. Again, if extrinsic mortality is high, there will be little payoff to investing in the maintenance of a body that probably won't survive for long anyway, and nature should select against maintaining the body beyond the point at which it can reproduce.

Indeed, for the most part, powerful selective pressures seem to have kept reproductive lifespans close to the natural lifespan in most animals. Where differences arise, they can result from two main causes. Pleiotropy, as described by Williams, and "kin selection," a concept developed largely by Hamilton, are common sources of such counterintuitive effects in evolution.

With kin selection, a trait that may be disadvantageous to an individual—an early end to reproduction, for instance—might still be selected for (or at least not selected against) if it is advantageous to close relatives of that individual who have a good chance of sharing the gene. In groups in which relatedness is high, some traits or behaviors that benefit the group more generally, but not the individual, may be selected for. Kin selection is one reason that humans and other animals often behave altruistically; that is, they help others at a cost to themselves.

Integral to kin selection is the concept of "inclusive fitness." This is an extension of the idea of evolutionary "fitness," which refers mainly to survival and reproduction—how many copies of its genes an organism passes on to offspring. An individual's genes are shared not only by offspring, but also by grandchildren, siblings, nieces, nephews, and so forth. Some groups with high "relatedness" share many genes, even among individuals who are not close family members. Hamilton coined the term "inclusive fitness" to describe this factor in natural selection. "Hamilton's rule" models mathematically how nature will select for a trait or behavior if its cost to the organism is outweighed by its benefit to related organisms, according to the likelihood that they share the gene for the trait. In an often-quoted reference to earlier versions of this idea, the geneticist J.B.S. Haldane is supposed to have quipped that he would lay down his life for two brothers or eight cousins.

Sometimes a gene that is selected for because it is advantageous also controls other traits that are not advantageous, or are even disadvantageous; Williams's theory of the evolution of aging depends on this concept of pleiotropy. One well-known example of pleiotropy is the case of sickle-cell anemia, in which a single mutation to a gene controlling the production of hemoglobin causes a cascade of physiological effects. Even with modern medicine, this mutation is highly lethal and devastating for people with two copies of the gene, but those with one copy are resistant to malaria; thus the gene persists despite conferring severe disadvantages on some who inherit it. Disadvantageous or neutral byproducts of natural selection for advantageous traits are sometimes called "epiphenomena." Some researchers have argued that menopause is such an epiphenomenon,[47] but most current theories see it as adaptive in some way.

Whether we see menopause as an epiphenomenon or as an adaptation carries important implications. If menopause is an epiphenomenon, the "menopause as disease" approach that pervades much of modern medicine might make some sense. If it is adaptive, however, then it is more appropriate to see menopause as a normal, even healthy development that calls for little, if any, intervention.

The amount of dispute, investigation, testing, and theorizing that has gone into the question of the origin of menopause in recent decades is staggering. Emotions run high; camps are large and entrenched. Whatever one's conclusion, it is clear that the issue is an important one currently exercising some of evolutionary biology's best minds. It is fair to say that the Grandmother Hypothesis, described in the next section, is the dominant theory: even those who disagree with it must address it. Thus, even though I am going to propose something a little different myself, I will describe the Grandmother Hypothesis first.

The Grandmother Hypothesis challenged influential theories about the importance of hunting and monogamous marriage in human evolution, and debate on it continues to rage today. Whatever one concludes, the debate is significant in itself: menopause is obviously a defining feature of our species—a feature demanding explanation. My own conclusion is that, regardless of whether we accept all of the Grandmother Hypothesis, menopause is not only adaptive; it is related to other unique features of human life history and to our extraordinary success in almost every terrestrial environment.

CHAPTER 2

•

"Thank You, Grandma, for Human Nature"

THE GRANDMOTHER HYPOTHESIS

BOTH WILLIAMS AND Hamilton thought that because women's long post-reproductive lifespan defied their predictions about how natural selection should work, it needed special explanation and was probably adaptive. Williams proposed an early version of what is now called the "Mother Hypothesis"—the idea that as a woman's reproductive future declines, the risks to her life and health of continuing childbirth outweigh the benefits of investing more care in the children she already has. Hamilton, a founding father of kin selection theory, was among the first to propose that post-reproductive women's care for both children and grandchildren caused nature to select for post-reproductive lifespans.[1] Records of a Chinese farming population in Taiwan around 1900, in which the low average life expectancy at birth suggested minimal influence of industrialization on health and survival, but in which many women nevertheless survived past age 50, inspired him to observe:

> The 15 or so years of comparatively healthy life of the post-reproductive woman is so long in itself and so conspicuously better than the performance of the male that it suggests a special value of the old woman as mother or grandmother during a long ancestral period, a value which was for some reason comparatively little shared by the old male.[2]

(Average post-reproductive lifespan for women, as described in chapter 1, is actually closer to 25 years in traditional populations.)

Mothers can help their remaining dependent children and their grandchildren at the same time, and there is no obvious reason to separate Williams's "mother" effects from Hamilton's "grandmother" effects when talking about why menopause arose, though some researchers describe these as separate theories.

It was not until 1998 that the Grandmother Hypothesis was fully formed and articulated, appearing in a body of work, still ongoing, led by Kristen Hawkes of the University of Utah.[3] In developing this hypothesis Hawkes and her colleagues have relied partly on a set of mathematical models for the evolution of life history developed by Eric Charnov.[4] We have seen that in classical senescence theory, species with low extrinsic mortality have longer lifespans because there is more opportunity for nature to select against genes that are damaging late in life and more payoff to an investment in maintaining the body. Also, in Charnov's theory, organisms that do not have to rush to reproduce quickly before they die can grow for a longer period of time and reach larger sizes. Humans and other primates—especially our closest relatives, the other great apes—have slow life histories and relatively low adult mortality. Chimpanzees, for example, nurse from their mothers until about age five and do not begin reproducing until their early teens. Humans reach sexual maturity even later than the other great apes; average age at first birth among foragers is 19 or 20.[5]

Charnov's model predicts that adult mortality and age at maturity will co-vary in a specific and predictable way. It also predicts that slow-growing animals will have longer intervals between births (lower "annual fecundity"), as the "production" that is spent on growing during childhood is channeled to offspring once the animal reaches its maximum size; the amount of production itself remains more or less constant. His model works well for most mammals, including our closest relatives, but humans are outliers: we have shorter birth intervals and much longer post-reproductive lifespans than other apes.

Both of these features would be explained if older women past reproduction were transferring their quota of extra production to their daughters and grandchildren (and nieces and grandnieces), providing food and other support that allowed the younger women to wean their babies earlier and give birth more frequently. In this scenario, nature might select for longer lifespans because the younger relatives of older women would benefit from this longer period of vigorous productivity. There would not be the same pressure to select for longer reproductive lives, since women bearing more of their own children in old age would not necessarily have a fitness advantage over women investing the same energies in helping their grandchildren. Lifespans, then, would lengthen, while reproductive lifespans might not—and although reproduction in human females ends at the same age as in female chimpanzees, our total lifespans are much longer. Put another way: to longer *lives*, nature said, "Yes!"; but to longer *reproductive* lives, at least among women, nature said, "Meh."

Several other theories of human evolution combine well with parts of the Grandmother Hypothesis. One of these, proposed by Sarah Blaffer Hrdy, retired from the University of California at Davis, focuses on how overlapping childcare may have influenced social development. Humans are the only apes that care for more than one dependent child at a time, something we can accomplish only with help. Like many species of birds and some other mammals, humans are "cooperative breeders" who share childcare—behavior that contrasts strikingly with that of female chimpanzees, who may kill one another's children if any are left unguarded for a moment. And because human children have had to engage the attention of multiple caregivers, we have evolved the social skills necessary for complex forms of cooperation to develop.[6]

Although the Grandmother Hypothesis has never directly addressed what anthropologists call "encephalization"—the evolution of larger brains and the intelligence that expands with them—it is also compatible with theories that emphasize the role of skill and learning in human life history, such as the Embodied Capital Hypothesis.[7] According to the latter hypothesis, the proficiency of older people in the many skills necessary for survival as a forager increased the value of their help.

Finally, although advocates of the Grandmother Hypothesis have, like most other scientists of human evolution, argued that a change toward a cooler and drier climate at the onset of the current Ice Age about 2.5 million years ago was the stressor that caused humans to diverge

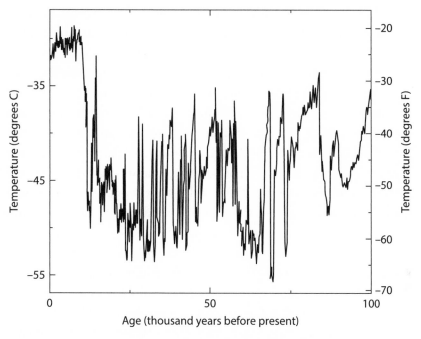

FIGURE 4. Temperature in Central Greenland. From Alley 2014, 119.

from other apes and turn to new food sources and reproductive strategies, this hypothesis has also been invoked by researchers who emphasize that the last 3 to 10 million years—and especially the last million years—have been an era of unusual and increasingly volatile climate changes.[8] In this view, the climate of the Pleistocene—the geologic era that began about 2.5 million years ago, in which *Homo erectus* and their descendants, including us, evolved—cycled dramatically between glaciations and sudden warmings, and volatility within these periods was also intense (figure 4). The climate of Paleolithic East Africa swung like a wrecking ball between extremes of moisture and drought, the deep lakes of one era becoming arid craters in the next. Because we have been living in a brief interval of warm, eerily stable climate conditions for the last 10,000 years—the era scientists call the Holocene—and because this interval encompasses our entire recorded history since the invention of agriculture, scientists until recently have not appreciated the volatility of the climate in which humans evolved.[9]

A fascinating theory of Jonathan C. K. Wells (a professor of anthropology and pediatric nutrition at University College London, not the "intelligent design" advocate whose name is similar) finds many similarities between human life history traits and those of seabird species that have adapted to the severe climate oscillations that *El Niño* causes in some regions. He proposes that climate volatility has shaped the main features of human reproductive strategy in several particulars. Long lives, slow growth, and long childhoods allow us to wait out periods of scarcity; stacking offspring (that is, caring for more than one child at a time) and cooperative breeding (with the help of post-reproductive women) allow us to ramp up reproduction quickly in favorable conditions, so that we may better survive the next ecological hammer blow that will pound our numbers down again.[10] Wells also argues that humans' unusual fatness, compared to our closest relatives, is another flexible adaptation to a shifting environment. Fat not only allows us to store energy as a buffer against hard times; it also communicates with the immune system, the reproductive system, and other systems about how to allocate whatever energy is available.[11]

The findings of researchers like Kim Hill and A. Magdalena Hurtado, whose work is discussed in chapter 4, that most foraging populations today are growing at rapid rates of more than one or even two percentage points per year, tend to support the "boom and bust" scenario proposed by Wells.[12] If Paleolithic humans reproduced at rates similar to those of modern foragers, then only decimations and extinctions followed by recolonizations could have kept population growth near zero over the course of that era. Comparisons with the natural history of other animals suggest that a typical "crash" phase would reduce the population by about 70 to 90 percent.[13]

In theory—and this is one form of the boom-and-bust hypothesis—even without external factors such as pestilence or climate change, a foraging economy might produce cycles of population growth and rapid decline as food sources are used up and then recover.[14] Because foragers can temporarily escape this type of cycle by moving to a new area with undepleted resources, there is a lot of incentive to do so where possible, and this would also explain the rapid spread of humans over the world outside of Africa in the last 50,000 years.

The boom-and-bust theory is harder to match up with the recorded history of the last 5,000 years, in which the pattern has been slow growth punctuated by occasional setbacks—catastrophic events that reduce the population by more than a few percentage points are quite rare, although an event of this type seems likely to occur in our near future. But human economies and societies changed radically with the transition to agricultural economies dependent on domesticated plants and animals in ways that may have mitigated the boom-and-bust pattern (more on this in chapter 6), and climate has been much more stable in this period.

All forms of the boom-and-bust theory characterize humans as a *colonizing* species—one capable of expanding rapidly in favorable conditions by reproducing at a fast rate. The idea that menopause is part of a reproductive strategy that is, above all, opportunistic and adaptable is appealing in light of our history. Humans spread quickly around Eurasia 80,000 to 50,000 years ago. Most scientists agree that humans spread the length of the two Americas in less than 2,000 years, from a founder population that left Beringia—extreme northwest Canada—only 16,000 years ago.[15] The ability to colonize large areas quickly when conditions are favorable is part of our heritage.

While the proponents of these different theories have different guesses about what *drove* the evolution of human life history, and while some disagree with Hawkes that this driving force was, in fact, grandmothering, all of these theories are compatible with aspects of the Grandmother Hypothesis, and especially its view that menopause is useful and adaptive.

The Grandmother Hypothesis emphasizes provisioning—in the classic form of this theory, older women help mainly by collecting food to share with their daughters and grandchildren—but grandmothers may help in other ways too. Anthropologists studying the Tsimane, a foraging-horticultural people of Bolivia, have tried to quantify a large range of helping across generations and among kin, such as provisioning, childcare, child adoption after the death of a parent, care for the sick, and conflict resolution, and have shown that older adults are more productive donors than those under age 40, even through quite advanced ages.[16] A mathematical model by Chu and Lee explores outcomes when older, post-reproductive women specialize in childcare to support younger generations; it finds that not only might nature select for post-reproductive life, but once menopause is established, it might pay to support older women with food so that they can provide more childcare, even if they can no longer forage enough to feed themselves.[17] This model sees the origin of menopause and, later, of filial piety—support for the elderly—in a division of labor: reproductive-age women forage and care for children; post-reproductive women forage more and care for children less as their fertility ends; elderly women care for children more and forage less as their physical abilities decline. Some contributions of older adults, such as the information and experience they can transmit to others (a contribution most important in societies without writing), are very hard to measure, and no one has tried to do this yet, with the result that no study captures all of the advantages of long life, and anthropologists may tend to underestimate the value of the elderly.

A strength of the Grandmother Hypothesis is that it does not postulate that humans once reproduced through old age and then stopped doing this, an idea for which there is so far no evidence. We only need to explain why lifespans lengthened while women's reproductive lives failed to catch up. Another strength of the Grandmother Hypothesis is the elegance with which it explains several unusual features of the human lifespan, including long post-reproductive lives in women, as related variables in a single equation.

It is easy to imagine how the changes predicted by the Grandmother Hypothesis might have arisen. Perhaps we, like chimpanzees and many other mammals, used to outlive our reproductive lives by only a short time on average. But because of the normal variation in lifespan that we see in any population, or perhaps because some other factor changed the rate of extrinsic mortality in adults, some individuals lived much longer, and a few females outlived their reproductive lives—as has happened among the macaques of Arashiyama or the chimpanzees of Ngogo. Some of these post-reproductive women, with no dependent children of their own, helped their older children rear grandchildren, perhaps by foraging for foods that required strength and skill to collect. Selection might well have favored the multiplication of whatever genes helped them to live longer, and lower adult mortality in turn would have favored longer periods of immaturity, because without having to hurry to reproduce, humans could afford to grow larger and stronger.[18]

Hawkes's theory is not just hypothetical. It was inspired by a year she spent in 1985–1986 studying a group of Hadza, a foraging population of northern Tanzania. Along with James O'Connell and Nicholas Blurton Jones, she lived with the group and collected very extensive records of their activities.[19] At the time, the dominant hypothesis of human evolution claimed that hunting and male provisioning had shaped human reproductive strategy, but Hawkes noticed that hardworking, post-reproductive women played a major role in supporting their younger relatives.

Of the 90 subjects studied, eight were post-reproductive women who helped to feed and care for the children of younger women. These included four grandmothers (two on the mother's side and two on the father's), a maternal great-grandmother, two maternal aunts, and one more distant relative; for the sake of simplicity, Hawkes' team called all these women "grandmothers," but not all of the women were grandmothers in a strict sense. There were 18 women of childbearing age in the sample and a total of 9 post-reproductive women, virtually all of whom helped younger women; furthermore, all nursing women had a post-reproductive helper.

Among the Hadza, older women, reproductive-age women, and also children all foraged extensively for plant foods. Tubers—the underground storage organs of the plant *Vigna frutescens*, called //ekwa by the Hadza—were the main staple of their diet. But older women spent more time foraging than did females of any other age group, and more than any group except for unmarried teenage boys. Men hunted or scavenged for big game, alone or in groups, and sometimes helped with plant foraging. Both sexes collected honey. Nursing mothers with new infants spent less time foraging, while "grandmothers" spent more time foraging when helping to support new infants. Adult women concentrated on foraging for foods that children could not procure for themselves, especially tubers that were hard to collect and were usually cooked before eating.

Humans are not the only animals that help care for children other than their own. But unlike other animals, weaned human children do not collect enough food to support themselves for many years, even among the Hadza, whose children collect a great deal. By continuing to provide food to weaned children, Hawkes theorized, humans have been able to survive in a wide variety of ecological conditions and adapt to changes in climate over time. When someone other than the mother helps provide food, mothers can wean children earlier and have more children sooner. And if grandmothers are providing much of the food, this could explain why post-reproductive life is so long in women.

Teenage girls were also important helpers among the Hadza, and children over the age of five were able to gather about half of their own daily food requirement. But the weight of young children with nursing siblings correlated significantly with the foraging time of "grandmothers," suggesting that the latter's help was critically important to those children.

TESTING THE GRANDMOTHER HYPOTHESIS, PART 1: DO GRANDMOTHERS HELP GRANDCHILDREN SURVIVE?

Like most evolutionary hypotheses, the Grandmother Hypothesis is difficult to test, but there have been a great many efforts to do just this. Hawkes was not able to show direct positive effects of Hadza grandmothers on their grandchildren's survival, but Nicholas Blurton-Jones's recent results, based on more data, showed a strong effect of grandmothering—Hadza children without living grandmothers were much more likely to die, especially when the mother was young.[20] Furthermore, he found that about 90 percent of children had a living grandmother when they were born, and, more surprisingly, about 70 percent still had a living grandmother at age 25.[21]

Most studies that examine whether children with living grandmothers have better chances of survival use information from nonindustrialized farming populations with good written records; foraging populations that lack writing are much harder to study in this way. This choice of study populations is not unreasonable, because patterns of fertility and mortality are broadly similar among foragers and agriculturalists who have not modernized. In particular, child mortality is high among both populations; typically, about 25 to 50 percent of children die before age 15. Even small differences in child survivorship, then, can have large effects on evolutionary fitness.

On the other hand, our history as foragers is much longer than our history since the invention of agriculture. Our genus *Homo* evolved about 2 million years ago in Africa, and our species, *Homo sapiens*, evolved about 200,000 years ago, also in Africa. But humans began practicing agriculture about 11,000 years ago at the earliest. That has been enough time for some evolutionary changes, but because most features of our life course—such as age at menopause and maximum lifespan— are about the same everywhere, researchers believe that they evolved during our long pre-agricultural past. One weakness of survivorship studies, therefore, is that they reflect important social conditions— farming, settled societies, mostly "philopatric" residence, and patriarchal patterns of property inheritance—that may change behavior greatly, and that probably did not prevail at the time when longer lifespans evolved. It is nevertheless important that results of these studies tend to support the Grandmother Hypothesis.

Most of the studies show that having a living grandmother increases a child's chances of survival, and the effects of this difference can be strong. In rural Gambia in the mid-twentieth century, children with a living maternal grandmother were about 10 percent more likely to survive to age 5, and in rural Finland and Canada in the eighteenth and nineteenth centuries, women gained an average of two extra grandchildren for every 10 years they survived beyond age 50.[22] Some studies show this effect only for maternal grandmothers, a few show it only for paternal grandmothers, and a few find that having a living grandmother actually decreased a child's chances of surviving.

A 2008 paper by Rebecca Sear and Ruth Mace reviewed 45 papers that investigated the role of kin—mothers, fathers, older siblings, aunts and uncles, and grandparents—in infant and child survival to determine what sort of help has benefited women in raising their children. The authors separated the studies into a group of 31 that were more rigorous and tried to control for important confounding variables like wealth, and a second group of 13 "supplementary" studies that did not.

Not surprisingly, all studies showed that the death of a mother reduces a child's ability to survive, especially at younger ages. Historically, only a small percentage of children have survived their mother's death in childbirth, but in several of these studies, the effect disappeared entirely after age two, suggesting that other kin may take over the mother's functions quite well once children are weaned.

Evidence about other relatives was more ambiguous. Living fathers enhanced their children's chances of survival in less than half (7 out of 15) of the more rigorous analyses.[23] Maternal grandmothers increased survivorship in about two-thirds and paternal grandmothers in about 60 percent of the more rigorous studies, and older siblings had a positive effect in almost all (5 out of 6) studies that looked for this. Grandfathers had a beneficial effect in only 2 of 10 studies. These researchers, like others before them, remarked on a pattern: paternal grandmothers are most beneficial around the time of birth or shortly thereafter, while maternal grandmothers are most beneficial for children around ages two or three. This could reflect different kinds of benefits: whereas maternal grandmothers may help with childcare directly, especially around the time of weaning, paternal grandmothers, who often live with their daughters-in-law, might affect the mother's environment, for better or (in those studies that showed negative effects) for worse.

Finally, one recent study tested a theory that surviving grandmothers might help some grandchildren more than others. Because only girls inherit the X chromosome, grandmothers are slightly more related to their sons' daughters (who inherit about half of the genetic material on their two X chromosomes from her) and to their daughters' children (who inherit about one-quarter of this material from her) than to her sons' sons, who inherit none of her X chromosome. The XX or XY pair of chromosomes is only one of 23 chromosome pairs that we inherit, so this difference in relatedness is not very great. Still, it inspired Molly Fox and colleagues to investigate whether distinguishing grandmother effects according to the sex and paternity of the grandchild might make a difference. When they analyzed the results of seven child survivorship studies that distinguished results according to the sex of the child, they found support for the hypothesis that grandmothers invest in grandchildren according to their degree of relatedness.[24]

Studies based on statistics can be frustratingly shallow—they don't tell us much about how or why the effects they describe might occur. For more insight it can be helpful to look to exploratory, "qualitative" studies that offer more depth about a small sample of subjects. In one qualitative study, researchers in Accra, the capital and largest city of Ghana, interviewed 24 adult survivors of childhood protein deficiency malnutrition (*kwashiorkor*, which causes swelling of the belly, diarrhea, and other symptoms and is fatal in most cases, although it can be cured with a rigorous 36-month treatment program). They also interviewed the survivors' main caregivers, who included 17 mothers, 10 grandmothers, one great-grandmother aged 100 at the time of the interview, two fathers, three siblings, and six "aunts" of various relationship to the survivors. The original intent of the study was to find out whether survivors of childhood malnutrition went on to live normal lives; it was quickly established that they did. The researchers then became interested in the role of grandmothers, and especially maternal grandmothers and "aunts," who were frequently the first to identify the illness. They concluded that because the grandmothers' long memories included times of civil conflict when malnutrition was especially widespread, they were attuned to the problem, and were highly motivated to seek medical care and to help their families follow the Ministry of Health's recommended program; to this end they provided all types of support, from childcare to money to transportation. Also, because older women were considered society's experts on medical issues—and not unreasonably, as most of them had borne and raised many children—mothers turned to them for advice about whether their children needed help. The researchers concluded that grandmothers and other older women involved in the children's care had been a critical factor in their survival, and had perhaps made the difference between those who survived and those who did not.[25]

In another example, a time-budget study of rural Oromo families in Ethiopia showed that maternal grandmothers visited their daughters frequently even after they married and moved to other villages. They often helped with heavy domestic chores like hauling water, freeing mothers from hard work and allowing them to spend more time on childcare. Children with grandmothers had better chances of surviving the dangerous first five years of childhood, during which mortality was 15 percent overall; maternal grandmothers were more beneficial than paternal grandmothers, and girls benefited more than boys. Even though matrilocal marriage was practiced by the poorest families and nutrition was worse in these families, grandchildren who lived with their maternal grandparents survived more often than those whose mothers had married outside the village and lived with their in-laws.[26]

TESTING THE GRANDMOTHER HYPOTHESIS, PART 2: MATH

Mathematical modeling is one way to test evolutionary theories. It might seem plausible that women could increase their inclusive fitness by caring for their relatives' children after they stop reproducing, but would that really lead to selection for longer lives? Researchers have created models that incorporate the likelihood that a gene will be passed on to a relative, maternal age at first birth, interbirth intervals, age at last birth, sex of the child, length of lifespan, the effects of behavior, and other variables. Values are best guesses based on ethnographic, biological, or demographic research. Sometimes, where very complete genealogical data on a population are available, models are based closely on those data to determine whether, given the patterns we see in that specific population, grandmothering has enough effect to account for the evolution or maintenance of menopause. The Grandmother Hypothesis has done fairly well in these tests, though results have been mixed.

Especially important in discussions of the evolutionary basis for menopause have been data collected by the Japanese on the farming population of Taiwan in the early years after the island came under

Japanese rule in 1895. In his 1966 article on senescence, Hamilton used these data to generate a life table and to make other calculations. He found that there was no sign of the influence of modern medicine and public health on this population; average life expectancy at birth was 26, typical for nonindustrialized societies; fertility was very high; and the population was growing at a rate of about 1 percent per year. It was these data that impressed Hamilton with the need to explain menopause, because even though mortality was high, a large number of women lived well past the age of last reproduction, challenging parts of his theory.

In 2001, Daryl P. Shanley and Thomas B. L. Kirkwood used these data, including rates of fertility and mortality, to calculate the effects of mothering and grandmothering, assuming that living grandmothers might aid the survival of both daughters and grandchildren. They also assumed that no child under the age of two would survive its mother's death, and that the harmful effect of a mother's death continued, though it declined, through age 15. They included a risk that the mother would die in childbirth, which in humans is usually small but does escalate with age.

Shanley and Kirkwood then varied the age at which fertility declines and stops to see what effect this would have on fitness. Even though they assumed that a mother's death would result in the death of her newborn and reduce survival rates for her other children, they found that the increasing risks of childbirth later in life had little dampening effect on the benefits of continuing to reproduce. That the "mother" effect described in Williams's original hypothesis—the idea that menopause balances the risk of dying and losing one's investment in existing children against the benefits of having more children—is not as important as one might think had already been demonstrated and was not a surprise. In 1993, Alan Rogers had used these same data from Taiwan to demonstrate that the risk of death in childbirth could not explain menopause even if it were 10 times the rate of 1 in 100 live births observed in the early twentieth-century United States. (Comparisons with foraging populations show that Rogers's estimate of maternal mortality is reasonable: among the Ache of Paraguay, the rate of death in childbirth before peaceful contact with Europeans was about 1 in 150 live births;

among the Hiwi of Bolivia, it was about 1 in 55 to 75 live births; and among the Hadza at the end of the twentieth century, it was about 1 in 100 live births.[27])

Shanley and Kirkwood also found that grandmothering in itself was probably not enough to explain menopause: grandmothers would have to increase their daughters' total fertility by more than 40 percent to replace the fitness lost by continuing to reproduce themselves. When they combined the two "mother" and "grandmother" effects, however, menopause enhanced fitness under conditions that seemed reasonable, although this model predicted that menopause would happen somewhat later, between ages 50 and 60.

Rebecca Sear produced another test of the Grandmother Hypothesis based on an exceptionally complete set of data from four farming villages in a rural area of The Gambia. These data were collected by Ian McGregor, a British physician researching malaria for the Medical Research Council (MRC) in what was, until 1965, a British colony. (McGregor died in 2007.) Until 1974 modern medicine had little impact on this population, and fertility and child mortality were both very high. In 1975, the MRC established a permanent clinic in the area, and mortality declined dramatically, and for this reason Sear excluded data from after that year. Married women usually lived with their in-laws, but sometimes not until a few years after marriage, and most women continued to live near their maternal family in the same village. Children often went to live with their maternal grandmothers for a time when they were weaned. Most property was inherited and passed on by men, but rice fields were inherited by women through their mothers, and women did most of the farming work. Polygyny was preferred, and many women had co-wives. Divorce, widowhood, and remarriage were common for both men and women.

In a 2002 study using these data, Sear and colleagues found that maternal grandmothers and older sisters, but not fathers or paternal kin in any category, improved a child's chances of survival. Mothers were essential to a child's survival up to age two, but the death of a mother had no effect after that.[28] A living paternal grandmother slightly reduced birth intervals, however; the researchers speculated that the most likely explanation was increased social pressure on the daughter-in-law to have more children.[29] Whether a woman had co-wives did not affect birth intervals or the mortality of her children, somewhat contrary to expectations.

In 2007, Shanley and Kirkwood, together with Sear and Mace, published a mathematical test of the Grandmother Hypothesis based on McGregor's data, using assumptions similar to those of their previous model based on the farmers of Taiwan, but with the more precise figures for child mortality in the presence or absence of kin that were available for the Gambian population. They got the same result as in their previous analysis: menopause is adaptive when "mother" and "grandmother" effects are combined, but it should occur 5 to 10 years later than it does.

It was not a slam dunk; the numbers worked, but barely. Because only a small effect is necessary to cause evolutionary changes over time, however, and because it is likely that no mathematical model takes account of all the factors actually in play, these models suggest that the Grandmother Hypothesis is *plausible*, even if they cannot be conclusive.

One of the earliest mathematical challenges to the Grandmother Hypothesis came in 1991, when Hawkes and colleagues were just starting to publish their data on the Hadza and ideas about grandmothering. Anthropologists Kim Hill and A. Magdalena Hurtado based their model on studies of the Ache of Paraguay, a population that had had little contact with the industrialized world before 1971.[30] In the 1980s, using structured interviews, the researchers produced a rigorous demographic study of this population that accounted for 955 people born between 1890 and 1970. To calculate the costs of menopause, the researchers used the fertility curve for macaques—if human lifespans had a similar curve, women would still have 43 percent of their peak fertility at age 65. Using their data on mortality, fertility, and child survivorship among the Ache, with and without grandparents, they calculated that although the benefits of grandmothering increase and the costs of reproductive cessation decrease over the course of the lifespan, natural selection would not have favored menopause in this population, partly because older women would have too few adult offspring to help.

Mathematical models based on single datasets have the advantage of being rooted firmly in reality, but they are constrained by a single set of conditions and may not be valid across populations. More general models can test a wider range of theories about how menopause evolved. Assuming an ancestor with a life history similar to that of today's chimpanzees—an ancestor from which humans diverged but chimpanzees mostly did not—these models test the conditions under which we might have evolved longer lifespans and long post-reproductive lives.

In 2003, Ronald D. Lee proposed a formal modification to Hamilton's theory of senescence by considering the effect of what he called "intergenerational transfers"—that is, any investment by older generations in the well-being of younger ones. These transfers of food and other resources are important in all animals that care for their offspring, and are most important in highly social species like humans, but had not been considered in previous theories of aging. When Lee introduced mutations affecting fertility and mortality into model populations with intergenerational transfers, his model showed high infant and child mortality for animals that invest heavily in a few surviving offspring, like mammals and birds. The reasons were that, first of all, the sunk cost of the parents' investment was lower at that stage; also, higher early mortality reduced competition among offspring for resources. The model, then, explained one mystery of evolution in humans and in some other animals: high juvenile mortality.

It also explained the mystery of post-reproductive life. In Lee's model, small changes in fertility were unlikely to make much difference to fitness in species with intergenerational transfers; changes in investment, on the other hand, had a stronger effect on fitness. A relatively high ratio of adults to children was most advantageous, and lower mortality in people of post-reproductive age—adults who would not be adding more children to the balance—was selected for.

By incorporating the idea of transfers, Lee's model was able to explain two traits observed in humans and some other animals that defy classical senescence theory: high infant and child mortality rates, and long post-reproductive lives. His model for humans, using a high value for transfers, produced a mortality schedule very close to what is observed in foraging and preindustrial societies.

A few years later, Lee tested this model on an imaginary population of small, genetically diverse bands sharing food, in which children are dependent through their late teens and adults produce a surplus (like human foragers today and, perhaps, in our ancestral past). In this version, he introduced random mutations that raised mortality at different ages, but no mutations affecting fertility. He showed that long post-reproductive lifespans would evolve under the set of conditions he used. This conclusion held true even when 50 percent of transferred resources went to distant kin or non-kin rather than to close kin; in fact, this last simulation led to lower mortality in middle age than the simulation with transfers to kin only, though late-life mortality was somewhat higher, because there were more substitutes for a parent or grandparent who had died. Lee's models only considered a one-sex population, without sexual reproduction. But in theory—and this idea will become important later, in chapter 3 of this book—transfers could cause nature to select for post-reproductive lifespans in both men and women.[31]

Lee's models included only one mechanism for evolutionary change—the accumulation of negative mutations—and did not try to model the effects of pleiotropy or of selection for positive traits. Other work that he published with Cyrus Chu and Hung-Ken Chien incorporated transfers into models that simulated positive selection according to the "disposable soma" theory—the idea that organisms have a fixed amount of energy that they allocate to growth, reproduction, maintenance, or (in Chu's model) transfers to younger generations.[32] It would have been too complicated to incorporate the complex social effects explored in Lee's 2008 study into this model, but in any case, Chu, Chien, and Lee again showed that high early-life mortality and post-reproductive life tend to evolve when transfers are added to classical senescence theory.[33]

Samuel Pavard and Frédéric Branger tested a model that considered the effects of several features of human reproductive strategy over a wide range of values observed in real, traditional (that is, nonindustrialized) societies. They considered the effects of a mother's death on her daughters' chances of survival, the degree of dependency of children on their mothers, and the possibility that grandmothers, if alive, could help grandchildren survive or even substitute for mothers entirely. They arrived at the interesting result that *helping* grandmothers (but not, oddly, *substitute* grandmothers) strongly drove evolution for longer lives if adult mortality was already relatively low; grandmothering also drove selection for higher fertility at young ages and lower fertility at older ages over the whole range of the matrix.[34]

The models described here all considered asexual reproduction—mothers and daughters only—which is much easier to model than scenarios that consider sexual reproduction with men. The first two-sex model to test the hypothesis that grandmothering could drive selection for menopause was designed by Friederike Kachel and colleagues at the Max Planck Institute for Evolutionary Anthropology at Leipzig,[35] with much-publicized results that seemed to challenge the Grandmother Hypothesis. In a model population of about 1,000, with both fertile and natural lifespan initially set at age 50, this model allowed surviving grandmothers without dependent offspring to help their daughters reproduce by allowing them to wean children earlier and by reducing mortality in those daughters' children. Random mutations were introduced that varied the length of the natural lifespan in both men and women, or of the reproductive lifespan in women only, but maximum potential fertility remained constant, so that more births earlier in life meant fewer later on. In this model, selection against aging and mortality at ages past reproduction turned out to be too low for kin selection to result in longer lifespans for older women. Kachel and her team did not consider their results inconsistent with the broader theories described earlier, which explained how post-reproductive life could evolve by incorporating the idea of transfers into classical theories of senescence. But they did reject the idea that grandmothering specifically could have driven the evolution of a longer human lifespan.

This model did not consider "mother" effects—according to which if a mother dies, young offspring should die or have increased mortality—although Shanley and Kirkwood had suggested that menopause could evolve if mother and grandmother effects were combined.[36] It did include constraints on reproduction, because high fertility at young ages decreased fertility at older ages, but perhaps did not capture the physiological costs, or costs related to competition for resources, that some researchers believe are important in shaping human life history. Including a tradeoff between fertility and mortality—as postulated in classical senescence theories that invoke the constraints of the "disposable soma," in which a finite store of energy can be allocated to growth, reproduction, or maintenance—might have changed the results.

Kim and colleagues raised this last point in their response to Kachel's model and developed their own model in response. In this model, they defined a "grandmother" as any post-reproductive female, and she could care for any weaned child in her group, freeing the child's mother to have more children. They included tradeoffs between fertility and longevity for both women and men: men were less likely to father children as they got older, and women who lived longer had children who remained dependent longer.

In the more sophisticated and realistic of the two versions of the model that the team developed, longer lifespans took a long time to evolve (about 275,000 years) and did not always happen. Some model populations remained stuck at chimpanzee-like lifespans, while others reached equilibrium at a human-like lifespan. Without grandmothering, some populations escaped low lifespans because of selection for longevity in men, who continued to reproduce throughout life in this model, but these populations went extinct as lifespans increased, because women could not produce enough children to avoid negative growth rates. When the researchers experimented by setting ages of sexual maturity and intervals between births at low limits, long lifespans did not evolve, suggesting that grandmothering is only adaptive in species that are relatively slow-growing and long-lived in the first place—a possible reason why menopause is not more common across species.[37]

One weakness of this model is that it fixes the end of reproduction at a specific age (45 years)—that is, it is strictly a test of whether grandmothering could have driven longer lifespans, not of whether it would maintain menopause in a population in which lifespans increased for some other reason. More recently, a new version of Kim and colleagues' model, this time published by Matthew H. Chan, Hawkes, and Kim, incorporated age at last birth as another variable in the model.[38] This study found that the transition from an ape-like life history to a human-forager-like life history only occurred with grandmothering and did not happen in all, or even most, populations with grandmothering, but only in a small subset. When it did occur, the new equilibrium took 30,000 years to arise and resulted in human-type lifespans and an average age at last birth of about 48, somewhat higher than what is observed in historical populations. The main factor determining the length of lifespans, however, was not grandmothering but the level of competition among males. Because grandmothering reduced the number of reproductive women in the population and increased competition among men for paternity, high levels of competition among men reduced lifespans when the researchers assumed a strong tradeoff between longevity and what they called "vigor in youth" for men. In this case, men needed a lot of "vigor" for competition, and the resulting shorter male lifespans muted the effects of selection for longer lifespans in women. Lifespans increased only when the researchers assumed that the vigor/longevity tradeoff for men was weaker (even a small change in this parameter had a big effect). The effects of male competition on human lifespans had not been considered in quite this way before—although mating strategies have been an important part of the grandmother debate in other ways—and they need more exploration.[39] That most foraging husbands remain with their wives after menopause—a pattern discussed later— has likely buffered the effects of menopause on male competition.

Mathematical modeling is difficult for many reasons already mentioned—it can't take account of all relevant factors; it relies on data from societies that may be different from our ancestral condition; and it requires researchers to make assumptions that, if incorrect by even a little bit, can yield misleading or anomalous results. It is also not surprising when the hypotheses of researchers are confirmed by the equations and parameters that they themselves have created. Modeling is better at showing what is plausible than at proving what really happened, and in that limited sense can be said to support the Grandmother Hypothesis.

A further complication of mathematical modeling is that, for the sake of simplicity, models almost always assume a stationary population that is neither growing nor shrinking. But if our species was normally expanding and sometimes crashing—as suggested earlier in this chapter— the results of the model might not be accurate. In growing populations the timing of reproduction is more important than it is in stationary ones. Births earlier in life have more impact on fitness than later births do, and a point of diminishing returns for additional births is reached more quickly. Mathematically, this means that menopause is less costly than it would be in a stationary population.[40]

What stands out strongly among these models, however, is Lee's formulation incorporating "transfers" into the theory of how aging evolved, which, unlike classical senescence theory, can explain high child mortality and post-reproductive lifespans. This idea that reproduction is not complete at birth but may involve a continuing investment by parents, grandparents, and other kin is a fundamental addition to senescence theory and is especially important for human life history, because in humans this investment is very large.

I myself am not convinced that grandmothering drove the evolution of longer lifespans. I think it is easier to suppose that lifespans increased when something happened to reduce adult mortality—perhaps when humans started cooperating to share food. But grandmothering does not have to be the only force that drove longevity in humans in order for post-reproductive lifespans to be adaptive. If grandmothering, perhaps combined with other factors, has acted to maintain long post-reproductive lifespans that arose for other reasons, it is adaptive in that sense. Still, Hawkes and others maintain, and they make a strong argument, that grandmothering is in fact the force that drove the evolution of humans' unique life history.

COMPLICATING THE GRANDMOTHER HYPOTHESIS: REPRODUCTIVE COMPETITION

While human families can be cooperative, as demonstrated by many studies of child survivorship, competition is also a universal factor among humans. Some of this competition is fundamental to our reproductive strategy: we invest a great deal in offspring, and this means that children may compete with each other for finite resources like food or a mother's attention. In agricultural societies, inherited wealth creates new kinds of competition.

If grandmothers are young enough to continue to have children, their own children may compete with those of their offspring for resources, and mortality rates for both children may be higher. Researchers observing orcas have been especially influential in developing the theory that "reproductive competition" might have played a role in the evolution of menopause. Orcas, like human foragers, live in co-residential bands, share food, and nurse and care for offspring for many years. When reproduction overlaps between generations in the same pod, the survival chances of the older whale's children suffer.[41]

Where researchers have looked for damaging effects of reproductive competition in human societies, they have found these effects to be negligible in some but very strong in others—that is, the evidence so far is mixed.[42] There is, however, substantial indirect evidence that reproductive competition has been a factor in the evolution of menopause. In humans, fertility ends at about the same time that women may become grandmothers, and compared to other animals, there is little reproductive overlap between generations. Social customs often reinforce this avoidance of competition. For example, cross-culturally, men tend to marry later than women and—in agricultural societies, if not so much in foraging groups—women tend to live with their in-laws. This means that women are unlikely to live in the same house as a reproductive-age woman of the previous generation. In The Gambia, Ruth Mace and Alexandra Alvergne found that women and their mothers-in-law hardly ever overlapped in reproduction for these reasons.[43] Customs in which women retire from sex and reproduction after becoming mothers-in-law or grandmothers are also common across cultures (but not universal).[44] One study of women in Poland even found that women who had daughters earlier in life reached menopause earlier—if this effect is widespread among human populations, it is intriguing evidence that nature has selected against reproductive competition with our daughters.[45]

The disposable soma theory of senescence implies a different kind of competition: a "quantity versus quality" tradeoff in human fertility.

Recall that according to this theory, organisms trade off fertility against maintenance, with the implication that higher fertility comes at the cost of higher mortality and poorer health for the mother.[46] These tradeoff effects are widely assumed by evolutionary theorists, and the *belief* that childbearing is depleting or that women have a limited number of potential births in their lifetime is common in traditional societies. But for a number of reasons tradeoffs can be hard to observe in real populations and hard to distinguish from one another, and the results of studies looking for them have been mixed.[47] Among the Tsimane of Bolivia, researchers found no long-term effects of many or frequent births on women's health.[48] On the other hand, a study of the records of more than 20,000 couples who married in Utah between 1860 and 1895 found that women with many children died earlier in life, and that their children also suffered higher mortality as a result. Mortality was higher for children in large families, especially later-born children, than for children in smaller families.[49] Similarly, a recent demographic survey of 27 modern countries in sub-Saharan Africa found that the odds of child survival declined as family sizes grew, by about 14 percent for each child born after the first two.[50]

Factors like these could reduce the costs of menopause, because producing the maximum possible number of children may not be the best fitness strategy anyway: *optimal fitness*, in other words, may not be the same as *maximum fertility*.

In agricultural societies, there is the further problem that a family's survival depends on its heritable property—its land—and this may affect reproductive strategies and child survivorship profoundly. Monique Borgerhoff Mulder's study of the Kipsigis people, a farming and herding population in Kenya, showed how the ferocious, sometimes violent competition for property among brothers, co-wives, and in-laws can affect children's chances of survival.[51] Among the Kipsigis, all land is inherited through the father's line, and sons by tradition inherit equally. Only registered tracts of land owned by the family can be inherited, or new land has to be bought (a twentieth-century development resulting from colonialism and modernization; before that, it was easier for sons who inherited a father's herd to expand onto unused land). In this study, competition among sons and cousins for inheritance was intense, and because sons could only marry if they could make large bridewealth payments in livestock, marriage could be difficult if there were not enough sisters to bring in the necessary payments.

In Borgerhoff Mulder's study, child mortality among the Kipsigis, at about 18 percent, was higher than in the industrialized world, but lower than in most agrarian societies of the past. Children were less likely to die if paternal grandfathers were alive and if their father had living brothers, but this effect was strongest if the father was rich. These relatives improved a child's chances even if they lived far away, most likely by providing help in times of crisis (cash for a doctor, food in a drought, blood money in a feud). In poorer families, paternal male relatives had less effect—perhaps they were less motivated to help because they had to worry about dividing an inheritance among many grandsons or sharing their sons' patrimony with nephews.

Mothers' fathers and mothers' brothers increased survival in the opposite case, for poorer families. They stepped in with aid when the mother's new family lacked resources, buffering the effects of wealth on child mortality. They were most helpful when they lived far away, a counterintuitive but not unique pattern, for which the most likely explanation is that they provided refuge for a woman and her children when they were abused or neglected by her husband or in-laws. Paternal grandmothers also helped grandchildren survive in these poorer families, perhaps shielding and buffering them from the neglect or hostility of male relatives.

Children of mothers with co-wives and children with more siblings had higher mortality than other children, especially in poorer families; here the likely explanation is competition for resources and inheritance. Borgerhoff Mulder observed many examples of this latter effect firsthand: one woman tried to poison her husband's brother's children, a man burned down his brother's house with his family inside, and accusations of witchcraft and lawsuits over land were common. Nevertheless, Borgerhoff Mulder also witnessed impressive acts of cooperation and altruism, and children with paternal grandparents did benefit from lower mortality among the Kipsigis.[52]

It may be significant that the only population in which having a living *maternal* grandmother is known to have a negative effect on survival, the Chewa of Malawi, is also unusual in another way—property in this culture is inherited and controlled by women, and most women live with their mothers' families after marriage.[53] A study of the Chewa by Rebecca Sear found that the harmful effect of having a living maternal grandmother was strongest on girls. She hypothesized that grandmothers in control of property might wish to avoid dividing it among too many grandchildren—a child had a better chance of thriving if the household was headed by her own mother rather than by a grandmother, who might prefer to concentrate resources on other grandchildren. Maternal aunts were also detrimental to child survival in this society, an effect that disappeared in families in which men controlled property, lending more weight to Sear's hypothesis. But because this study of the Chewa relied only on a demographic survey, and not in-depth field research, it offers little insight into what caused the results that Sear observed.

Finally, besides competition for resources among dependent children in the same group or family, competition within the individual over the allocation of resources to reproduction or maintenance, and competition among siblings or cousins for inheritance, another form of competition that has been important in our history is the tension between mothers-in-law and daughters-in-law that is so common in societies where brides move to their husbands' houses. The bride is not genetically related to any of her in-laws, and their reproductive and economic interests may conflict with her own, even though she is the means by which the family will produce new children. This conflict may affect grandchildren through their mother, which could explain why some studies have found that paternal grandmothers are either not helpful or even harmful to their grandchildren.[54] The mother-in-law problem was a near-universal feature of societies in the agrarian period and will be discussed further in chapter 7; it is unclear how much it applies to the foraging period, when family structures were different.

Inheritance is far more important in agricultural societies than in foraging societies, and, given its great power to influence reproductive strategies, studies of agricultural peoples must be used cautiously in arguments about evolution in the Paleolithic. Only one researcher has looked for the effects of reproductive competition among foragers— Nicholas Blurton Jones in a recent study of the Hadza.[55] Surprisingly, he found no evidence that children's chances of survival were lower if their grandmothers had newborns or young children; in fact, the opposite seemed to be true. This result is important because if grandmothers are able to care for their own young children without harming their ability to provide for grandchildren, this would undermine not only the reproductive competition theory, but also most adaptive theories of menopause. Looking more closely, however, Blurton Jones found that almost all women who gave birth after age 40 *had a living mother or mother-in-law*, as well as a childless teenage daughter, and these women were in marriages of longer duration than those of women who did not give birth after 40. This is the kind of correlation that makes theories of reproductive competition so hard to test, and it could be indirect evidence in support of them if the Hadza (for whatever reason), or humans in general, tend to have children near the end of their reproductive lives only when circumstances are very favorable. Among the Hadza studied by Blurton Jones, only women with many available helpers gave birth after 40; women past this age with fewer helpers did not reproduce at all, so that the mortality of their children could not be measured. That is, some mothers were able to give birth while still helping their daughters because they themselves had lots of help; other mothers stopped reproducing earlier.

I think it is likely that reproductive competition was a factor in the evolution of menopause, but I also think that the advantages of menopause in limiting reproductive competition became much more important in the agrarian period. That is, menopause not only helped us to survive as foragers; it is one of the reasons that humans were able to organize successful agricultural economies. I will discuss agrarian societies in more detail in part II.

To sum up: The Grandmother Hypothesis postulates that nature selected for longer lifespans in women because post-reproductive women helped their daughters, daughters-in-law, younger sisters, nieces, and granddaughters to feed and care for their infants and young children. This allowed the younger women to have more babies closer together, which in turn increased the inclusive fitness of the older women. Once weaned children could be supplied with hard-to-acquire foods, it also allowed humans to live in new environments and to colonize the world. Once adult lifespans lengthened, longer childhoods evolved as a result. Humans took advantage of these longer childhoods to develop higher levels of foraging skill; social skills also developed rapidly as cooperation became more important at all ages.

As Alison Gopnik put it in an AARP blog post reporting on Hawkes's theory, "Thank you, Grandma, for human nature."[56]

•

CHAPTER 3

Putting the "Men" in Menopause

MALE-CENTERED THEORIES
OF HUMAN EVOLUTION

BEFORE THE GRANDMOTHER Hypothesis, the best-known theory explaining human life history was nicknamed "Man the Hunter," after the title of a published collection of proceedings from a 1966 conference, edited by anthropologists Richard B. Lee and Irven Devore.[1] In this theory, as our ancestors switched to a new set of strategies for survival and reproduction, several changes happened at close to the same time. Humans became bipedal, leaving their hands free to hunt, carry food, and provision a family. They began to use stone tools for hunting and defense. Brains expanded as new skills—especially hunting—became important. But bipedalism placed a hard upper limit on the size of a newborn's head by limiting the width of the female pelvis, which meant that women had to give birth to less developed, more dependent babies, and care for them longer. This they accomplished with the help of men, who, in return for their investment in children, needed more assurance that the children were their own; thus humans began to form monogamous pair bonds. The Man the Hunter Hypothesis postulated that the nuclear family, bipedalism, tool use, "encephalization" (the development of big brains), and hunting were all closely linked factors in the evolution, about 2 million years ago (although it was thought to be much later at the time), of *Homo erectus*, so named because this hominin was supposedly the first one who walked upright.[2] This is the theory that I learned in school, and probably many of the people reading this book learned it too.

The "Man" part of the Man the Hunter Hypothesis is important—among foraging peoples today, it is mostly men who hunt big game, and anthropologists have assumed that this was also true in the Paleolithic. If hunting was a driving force behind human evolution, it was also a male activity, and this accounts for some of the weight and emotion behind the debate about human evolution—why the Grandmother Hypothesis was so controversial when it was first proposed and why the arguments can seem so strident today. Questioning Man the Hunter meant questioning whether characteristics of human society long assumed to be fundamental—male dominance and the nuclear family— are really essential to human nature. The two most influential theories of the evolution of menopause—the Grandmother Hypothesis and the Embodied Capital Hypothesis, discussed later—began as opposing factions in this debate over the relative importance of men and women in shaping the characteristics of our species. Over time, the Embodied Capital Hypothesis has come to look more like the Grandmother Hypothesis, but questions in the history of gender are central to understanding both of these hypotheses and the debate itself, and will be discussed in this chapter. A few hypotheses—most notably the Patriarch Hypothesis, which I will also discuss—still argue that menopause is a pleiotropic result of selection for other traits, driven by men.

Man the Hunter made a lot of sense at the time it was proposed, especially in the context of Louis and Mary Leakey's exciting discoveries in the 1950s of early stone tools and hominin bones in the Olduvai Gorge in Tanzania. Before that, anthropologists had assumed that only Neanderthals and "modern" humans used tools, but the older tools found at Olduvai raised the possibility that they might have been part of a suite of traits that evolved together.

Over the decades, however, new finds and interpretations undermined the Man the Hunter Hypothesis.[3] Perhaps most importantly, anthropologists now agree that *Australopithecus afarensis* was fully bipedal, almost 2 million years before the first use of stone tools and 3 million years before the first reliably attested big-game hunting. On the other hand, chimpanzees and other primates have been observed hunting vertebrate animals, and it is likely that our ancestors hunted small game long before tools were invented. This makes it less likely that there is a critical causal connection between hunting, bipedality, tool use, and social development. It no longer seems clear that being bipedal constrains the human pelvis, which theoretically could grow wider without cost to efficient locomotion; that pair bonding and the sexual division of labor date as early as *Homo erectus*; or that *Homo erectus* brains, or even human brains in general, take a great deal longer to grow after birth than chimpanzee brains do.[4] All of these assertions are, at the very least, debatable. Most importantly for our purposes, Man the Hunter did nothing to explain menopause.

It was the role of hunting in human evolution, however, that struck the deepest and most persistent chord with anthropologists and the general public. Hunting is overwhelmingly (though not exclusively) a male activity; besides that, the idea of hunting added a thrilling, heroic element to the story of our development. By focusing on the contributions of grandmothers and the tubers they collected, Hawkes's theory challenged the centrality of hunting—and of men—to this story. Much of her team's work has focused on reexamining the evidence for big-game hunting early in our history and on analyzing the contribution of hunting to the diet of today's foragers.

Today, anthropologists continue to argue fiercely about the fossil evidence for hunting by early toolmaking hominins. With the emergence of *Homo erectus*, hominin brains began to grow bigger, and some late *Homo erectus* skeletons have brain cases overlapping the range of modern humans in size. For a long time, researchers argued or assumed that *Homo erectus* was a hunter, and that more meat was one reason that nutritionally expensive brains began to grow.[5] But evidence for hunting this early is very difficult to interpret—it can be hard to distinguish hunting, scavenging, "power scavenging" (scaring predators away from a kill), and the natural accumulation of human and animal remains near locations like water sources in the record. Several assemblages of animal bones and stone tools once thought evidence of human hunting have been found, on close examination, to show no support for hunting and little even for scavenging. The first unequivocal evidence of hunting is probably the wooden spears recovered from Schöningen, Germany, that have been dated to about 400,000 years ago. A little later in the record, tools and bones from Qesem Cave in Israel, dating to between 380,000 and 200,000 years ago, show that humans were certainly butchering bones and may have hunted animals (mostly deer) as well, but evidence for hunting and meat eating is much more abundant after about 130,000 years ago.

Today, researchers who question the role of big-game hunting in early human evolution emphasize the high costs of hunting—the need to find, track, and catch food; transport it; and process it, all of which would have been even more difficult before projectile weapons were invented. They point out that humans and human ancestors are distinguished not so much by meat eating as by the *versatility* of our diet: we are able to procure and digest an unusually wide range of foods. *Homo erectus* certainly ate meat, and perhaps ate more meat than their ancestors, but other food sources were probably more important.[6]

A recent challenge to, or complication of, the thesis that the addition of meat to human diets was a critical factor allowing the development of bigger brains has come from Richard W. Wrangham of Harvard University, who argues that cooking, rather than hunting, was the fundamental change in nutrition that enabled *Homo erectus* to evolve.[7] Another recent study argues that the glucose released by cooking tubers nurtured human brains—supporting a faction, including proponents of the Grandmother Hypothesis, that has long argued that tubers, rather than meat, were crucial to human evolution.[8] Arguments about cooking are compelling and would explain many features of *Homo erectus'* anatomy: this species had smaller back teeth and jaws than their ancestors (uncooked meat and plants are much harder to chew) and a barrel-shaped thorax similar to that of modern humans, rather than the flared thorax and larger guts of earlier hominins, suggesting a more efficient diet. However, although we know that *Homo erectus* did eventually use fire, there is no evidence yet that fire dates as early as the origins of the species. The oldest archaeological evidence for controlled fire comes from a site at Zhoukoudian, China, dating to about 780,000 years ago, and our ability to produce enzymes that quickly break down the sugars in cooked starchy foods appears to date to about 1 million years ago.

The thesis that hunting was an important force in human evolution becomes more plausible for later periods, when weapons were more advanced and evidence for hunting is more abundant. In the case of *Homo sapiens* (anatomically modern humans), hunting is clearly one of the strategies that helped us adapt to every environment, including Arctic environments with few edible plants at all. Furthermore, if long lives and long childhoods are a recent phenomenon specific to *Homo sapiens* only—that is, if modern life histories developed only long after brains had become big and not, as many researchers argue, in close connection with big brains—hunting may be part of a pattern of behavior that shaped modern human lifespans. But Hawkes and other researchers have painted a new, quite different picture of this behavior than the classical Man the Hunter Hypothesis proposed. Why, they ask, do humans, and especially men, hunt? Who gets the meat when an animal is killed— does it go to wives and children, as the Man the Hunter Hypothesis suggests that it should?

What Hawkes and her colleagues observed about hunting among the Hadza did not seem to support the Man the Hunter Hypothesis, and this disconnect inspired the Grandmother Hypothesis. Blurton Jones's thoughts on these findings in a recent publication set the stage for the complicated debate that follows:

> We began to wonder why Hadza women kept husbands. Women toiled in the hot sun all day digging tubers, then came home and hammered baobab seeds until dark, while men sat in the shade talking, smoking, and fiddling with their arrows. . . . Seldom did men, avidly professed hunters, materialize with meat. Why did the women put up with this? Occasionally a man would arrive with a small bag of honey, or a game bird, or even a bundle of baobab pods and leave them at his wife's house for her to do the lengthy processing. Only rarely, a man would arrive with meat over his shoulder, park it on the roof of his house, report that people could carry more meat if they cared to go and get it, and then retire to the men's place to smoke. . . . Every house, even the house of a single old woman, ended up with some meat. If waiting around for a hunter to succeed was a reason to keep a husband, then why not just wait around for someone else's kept man to announce a kill?
>
> We still puzzle about this.[9]

Most foraging societies divide labor, including food acquisition, along gender lines—women do certain work and men do other work, usually with some overlap. In general, men do most big-game hunting, while women collect most plants, insects, shellfish, and other "gathered" foods. (Thus many scholars refer to foragers as "hunter-gatherers"; I prefer "foragers" because it is more succinct.) The amount of food contributed by men and women varies widely; women contribute from just 10 percent (in some extreme climates) to about 65 percent of the calories consumed. On average, men provide about 65 percent of calories among foragers, though the percentage is lower—about 50 percent— among mobile, "low-latitude" foragers living in tropical climates.[10]

A factor sometimes overlooked by anthropologists in making these calculations is that, on average, men consume many more calories than women. That is, men are expensive to maintain.[11] Nancy Howell calculated that the average !Kung woman aged 25–29 required 2,169 calories per day to maintain her body weight, whereas the average man in the same age group required 2,613 calories, or about 20 percent more; these numbers include 352 calories for reproduction (pregnancy and lactation) charged to the woman. Post-reproductive women aged 55–59, by her calculation, required only 1,883 calories, while men the same age needed 36 percent more food, or 2,568 calories, to make it through the day.[12] Conversely, because of physiological changes with age in both men and women, including lower metabolic rates, older people are cheaper to maintain than those of reproductive age. The "Thrifty Aged Hypothesis" proposes that this is an adaptation that allows us to help our younger relatives more effectively—we can consume less and give away more.[13]

Women typically do most of the food processing in foraging societies, and other tasks are also divided between men and women, with most childcare usually falling to women. In societies where men procure most of the food, women may do almost all the other work, specializing in technology rather than foraging (making tools, making clothing from skins, building shelters, cooking, and so forth). Men range more widely when foraging, while women stay closer to home. Men pursue foods that are riskier because harder to acquire, but that offer a huge payoff if successfully obtained; these foods are also usually more dangerous to pursue. Large game is the main example—individual hunters may go many days or weeks with no success, but a single kill can feed a large group. Women, on the other hand, tend to target foods that they can reliably acquire every day, such as plants, small game, and shellfish.

There are exceptions: research among the Agta of the Philippines in the 1970s and early 1980s showed that, at least in some groups, most women hunted, and men and women targeted similar game; older women and teenage girls hunted more than women with small children, and children past weaning age were often left with grandmothers, co-wives, or older siblings while their mothers hunted.[14] In some societies women hunt more often as they reach post-reproductive age. Among the Martu of Australia's western desert, for example, older women are highly effective hunters of sand lizards,[15] but here too, Martu grand-mothers continue to focus on the more reliable foods normally pro-cured by women rather than switching to the high-risk, big-game hunt-ing of men. (Some researchers have speculated that it would be impractical to take up big-game hunting in middle age, but we have no good evidence one way or the other.)

Men's foods tend to be shared with the whole band, often according to strict egalitarian rules meant to ensure that the hunter and his family get no more meat than anyone else. Women's foods are also often shared, not only within the family but also outside it, though less widely than men's foods.

While the sexual division of labor seems to be a unique and pervasive feature of human foraging societies, how and why it arose is less clear. Many anthropologists have pointed out that tasks assigned to women tend to be compatible with nursing and the care of small children, but there are probably other factors involved as well. Men's activities are often riskier to health and less reliably productive but more conducive to building broad alliances and winning prestige. Women's activities aim more at providing for their families. Clearly, by specializing, men and women are cooperating in some way. But how?[16]

Important background for much of the debate about the nature of cooperation between men and women is the idea that the sexes have different reproductive interests and strategies that account for the physical differences between them. Classical evolutionary theory, beginning with Darwin, starts with the observation that it is much less costly to produce sperm than eggs—for most sexually reproducing species, females must invest more in reproduction than males. This is especially true of mammals, which nurse their young after birth, but less true of species in which fathers invest a great deal in their offspring, such as many bird species. Certain traits tend to go together. Species that mate (mostly) monogamously tend to have higher paternal investment and low sexual dimorphism (that is, in monogamous species, males and females are similar in size and other attributes). In these species, most males mate and reproduce, and there is little variation in male reproductive success. By contrast, in species that compete for mating opportunities, males' *potential* reproductive rate is much higher than that of females. Because they are constrained by the rate of female reproduction—a population can only produce as many children as the females can bear—some males may have many more offspring than others. Among these species, males gain more fitness by competing for mates than by providing for offspring. There is also more sexual dimorphism, because males evolve traits, such as large body size, that help them in this competition, a process that tends to escalate like an arms race.[17]

Recently, Cyrus Chu and Ronald Lee developed an elegant model of the evolution of sexual dimorphism in which males and females allocate varying amounts of energy to growth, reproduction, maintenance of the body, and transfers to offspring at different ages.[18] A complement to their models of aging and longevity discussed earlier, this model shows that in general males should make fewer transfers and that menopause should arise in females, whose costs for reproduction are high, but not in males, who benefit less from being released from these costs. This model interprets the difference between what it costs males and females to reproduce as an *energy surplus for males*, similar to the energetic surplus that menopause provides to females in the Grandmother Hypothesis. Males can use this surplus to provide for their offspring, to grow bigger or to grow weapons like antlers or canine teeth, to defend their offspring from infanticide, to compete for mates in any of a huge number of ways, and potentially, in some circumstances, to do other things.

On the continuum between, on the one hand, species with low sexual dimorphism, monogamous mating, and high investment by fathers in their offspring and, on the other hand, those with high sexual dimorphism, polygynous mating (that is, mating with many females), and low paternal investment, humans seem to lie somewhere in the middle, which is one reason there is so much debate about human reproductive strategy. If it is the case that humans mate monogamously, that human fathers provide for their children rather than expending most of their energy competing for mates, and that there is little physical difference in size between the sexes, then there is more support for traditional theories of human evolution and even a theory that men, like women, experience post-reproductive life (more on this later). If we think that men compete for mates with varying reproductive success, invest little in their offspring, and are significantly bigger than women, then there is less support for the Man the Hunter Hypothesis, with its emphasis on male provisioning and monogamous mating. It does not help that men's mating and parenting behavior varies a great deal across cultures and even among individuals. As anthropologists sometimes say, there are "dads" and "cads," and the same person might pursue both strategies at different points in life or even at the same time.[19]

Many observations about modern foraging economies do not support the classic Man the Hunter Hypothesis of human evolution. If an individual hunter's chances of success on any given day are not good—and this is true even of highly skilled hunters in most foraging societies—can the thesis that pair bonds formed so that men could provide for their children hold up? Moreover, if men's foods are shared with the whole band—and sometimes, if the kill is large, with other bands of the same ethnic group—how do women benefit from being married to a hunter? Is the division of labor really related to the "nuclear family," and did Paleolithic humans form nuclear families anyway?

Anthropologists have made enormous efforts to answer these questions, but none of the answers they have offered is very simple. In particular, they have had to go beyond kin selection and parental investment to explain the high level of cooperation—especially sharing food—that is found in foraging societies; humans appear to cooperate more than any other species except for some "eusocial" insects like honeybees (though we are learning more about cooperation among whales and other vertebrate animals), and much of our cooperation is with people to whom we are not related.[20]

Some of the concepts deployed by evolutionary biologists to explain cooperation among non-kin derive from, or overlap with, ideas in game theory, which was invented to explain economic behavior in a world in which our decisions partly depend on what we think others will decide to do. One of the most important of these ideas is reciprocity, introduced to evolutionary biology in an influential 1971 article by Robert Trivers: animals may exchange help over time, so that an act that is temporarily costly to the actor is repaid later. This cooperation could provide fitness benefits and be selected for, especially if the animal is long-lived, and if it lives in stable social groups without too much inequality, so that all members have more or less equal power to reciprocate benefits. Reciprocity is one reason humans form bonds of friendship, which are important in all societies.

When a person or animal cooperates by performing an act that benefits both itself and others at the same time, instead of reciprocity, this is called "mutualism," a concept often invoked to explain group cooperation. If the benefits of living in a group (or the costs of being ostracized or punished by the group) are high, performing an action that promotes one's acceptance by the group can be mutualistic (this strategy is sometimes called "pay to stay"). "Tolerated scrounging," discussed later, is an example of mutualism.

Cheating is a problem in these cooperative strategies: reciprocity partners can defect, and group members can potentially get a "free ride" by contributing less than others. All human groups have complicated social and psychological methods of dealing with such "collective action" problems. Reputation, for example, is a tool to discourage cheating—because humans have language, people can interact with others based on what they know of their partner's history of reciprocating fairly and contributing to the group, and not only on the basis of their own experience with that person. Humans have highly developed social emotions, such as gratitude and guilt, that control cheating and allow unusual levels of cooperation.

With all this in mind, four (or more) main hypotheses to explain hunting and food sharing among foragers have been advanced. The first, the classic Man the Hunter Hypothesis, postulates that men hunt to provide for their children and, because of monogamous pair bonding, usually know who their children are; therefore, selection has favored men who help their offspring to survive. The second suggests that men practice reciprocity in hunting and food sharing: they share food with the expectation that others will share food with them in the future ("delayed reciprocity") or because they get other, nonfood benefits from sharing meat ("indirect reciprocity"). The third proposes that men hunt and share food to win admiration and prestige. This is sometimes called the "Show-Off Hypothesis" (a designation that is a little misleading, because foragers are typically humble and egalitarian, but they may still be motivated by these concerns). When linked to signal selection theory—a branch of evolutionary theory that postulates that some traits, like a peacock's tail, evolve because they convey valuable information, rather than offering some direct physical advantage[21]—this idea is called the Costly Signaling Hypothesis. Hunting skill may signal physical fitness (attractive to potential mates whose offspring may inherit advantageous traits); sharing may signal generosity and prosocial intention (creating a reputation attractive to potential allies). Finally, the fourth hypothesis suggests that men share meat because it is more costly to defend it from others than it is to give it away. This is often called "tolerated scrounging" or "tolerated theft."

Studies that have attempted to distinguish which motives for sharing food are most prominent among foraging groups have found support for all four of these hypotheses. I think the evidence shows that fathers are more relevant to childrearing than the Grandmother Hypothesis suggests—at the very least, whatever their reasons for doing so, men in foraging societies produce large surpluses of food that they share with their residential group, which is likely to include some of their own children or other relatives. But it is too simple to say that men hunt to provision their families; nor is it obvious that humans form pair bonds in order to cooperate in raising offspring, though this is probably part of the benefit.

If humans do not pair up so that fathers can provide for their families, why do they do this? Other reasons might be that men are better off monopolizing one woman if competition for women is high,[22] men might find it convenient to ally with women who can process food for them, and women may benefit from protection against harassment by other men or against those who might kill their children. Again, researchers have found support for all of these hypotheses.[23]

THE PATRIARCH HYPOTHESIS

The Grandmother Hypothesis was inspired by research among the Hadza, among whom the role of hardworking grandmothers is very obvious. Even Frank Marlowe, a Hadza researcher who opposes it, agrees that "with the Hadza as an example, the grandmother hypothesis is not only plausible but compelling."[24] His own theory, however, is based on a different phenomenon that he observed among the Hadza: some men, especially those with reputations as good hunters, left their post-reproductive wives to start second families with younger women. An extreme version of a similar pattern has been documented among some of the aboriginal peoples of northern Australia, where gerontocratic—meaning "rule of the old"—traditions were very pronounced: only older men advanced in the ritual hierarchy had much chance of marrying fertile wives and could accumulate large polygynous households, sometimes by marrying all the sisters in a family.[25] Clearly, among at least some foraging peoples, male reproduction can continue until late in life.

Marlowe hypothesizes that as culture developed, giving rise to technology that allowed men to hunt past their physical prime and values that granted status to older age, some men competed successfully for mates and fathered children even late in life. These men passed their reproductive advantage to their offspring, both sons and daughters.

Sons benefited because their longer lives allowed them to reproduce more than their shorter-lived rivals. Daughters received no fitness benefit because they became sterile once their oocytes were depleted at menopause. Physiological constraints meant they could not evolve longer reproductive lives: if nature accomplished this by increasing the number of oocytes produced before birth, ovaries would have to be much bigger. But because they shared genes with their brothers, they lived long too; in fact, they lived even longer, because the harmful effects of testosterone lowered men's life expectancy. In this hypothesis, which Marlowe named the "Patriarch Hypothesis" because it postulates that a group of relatively high-status men drove the evolution of the life course, menopause is not adaptive; it is a pleiotropic byproduct of longevity in men.[26]

Although this theory has the benefit of simplicity—it is easier to explain how longer lives could be adaptive for men, who continue to reproduce, than for women, who do not—I do not find it convincing. First, it depends on the assumption that women could not have evolved longer reproductive lives, which seems unlikely for several reasons discussed in the first chapter of this book. It also does not offer a very convincing explanation of why women live longer than men. Despite a common perception in agricultural societies that women age more quickly than men—an idea attested in the literature written by the elites of ancient Greece and China, and of medieval England—in those modern and historical populations whose mortality rates can be calculated, women usually have lower mortality at all ages.[27] In some foraging and agricultural populations girls die more often than boys because of infanticide or neglect, and in some, women's mortality in youth is slightly higher than that of men for similar reasons (social disadvantage) and also because of deaths related to childbirth; most researchers agree that women's "natural" advantage over men is smallest in young adulthood. But even in those populations, women over age 45 or 50 almost always experience lower mortality than men, and their "natural" advantage is greatest in this stage of life. That is, women outlive men most robustly at exactly the same period of life when, according to the Patriarch Hypothesis, selection for longer lives in women should be nonexistent. Marlowe explains men's shorter lifespans by pointing to their higher testosterone levels, but there is no consensus in the research on the proximate causes of mortality differences, which are likely very complicated. The Patriarch Hypothesis also assumes that nature cannot select for different lifespans in men and women. But males and females have different lifespans in many other species (although this has been a neglected area of research)—notably in two whale species that, like humans, have menopause. In both of those species, male lifespans are much shorter: selection has acted to prolong female lifespans, but not those of males.[28]

The most interesting challenge to the Patriarch Hypothesis, however, comes from those who have questioned its basic premise that men reproduce late in life. Some men do, of course, but is it enough to have driven the evolution of longevity? In a 2007 article titled, perhaps somewhat defensively, "Why Men Matter," Shripad Tuljapurkar and colleagues at Stanford University and the University of California at Santa Barbara answered "Yes." Furthermore, by combining evidence from several cultures that some men continued reproduction past the age of female menopause, they also showed mathematically that any late-life reproduction in men could lead to longer lifespans in women too. This was important support for Marlowe's Patriarch Hypothesis, although it did not address the weaknesses described earlier.[29]

Other scholars have questioned whether, among our Paleolithic ancestors, men's reproductive lives really continued later than those of women. Tuljapurkar's paper combined evidence from foraging and agricultural/pastoral societies, but a lot more male late-life reproduction has been found in the latter than in the former.[30] This is because polygyny—marriages with more than one wife (the word means "many women" in Greek)—is more common in societies with property, as in those societies women are economically dependent on men, and some men accumulate more resources than others. For example, the salmon-fishing foragers of the Pacific Northwest, whose economy was based on stored resources (dried salmon) and thus had much in common with agrarian economies, also had wealth inequality; permanent residences; social stratification, including slavery; and more polygyny than is typical for foragers.[31] It is likely that our ancestral condition was closer to that of mobile foragers like the Hadza or the !Kung, among whom rates of late-life reproduction in men are low. This is especially true of the !Kung, among whom men father only 3.6 percent of their children after the average age at which women last give birth, and men's Post-Reproductive Representation is almost as high as women's.[32] Among the Tsimane, a people studied exhaustively by Hillard Kaplan and Michael Gurven, whose Embodied Capital Hypothesis is discussed later in this chapter, 90 percent of men whose wives reached menopause did not have more children; among those who did, most of their younger wives also reached menopause after the next child.[33] It seems likely that among our ancestors, as among foragers today, only a few men were able to acquire younger wives as they reached middle age, either in addition to, or as replacements for, their post-reproductive wives. Instead, they stopped reproducing when their wives reached menopause.

A study by Richard A. Morton and colleagues attempts to address one of the problems with the Patriarch Hypothesis, namely, that it does not explain why reproductive life did not expand to match somatic life in women.[34] They hypothesized that male mating preferences could cause menopause to evolve, if men preferred to mate with younger women. In their model, they began with the assumption that women remained fertile through old age and introduced a male preference for younger women. When genes affecting fertility in men and women were introduced to the model, genes damaging to fertility in women accumulated after the age at which men ceased to prefer them, though women would continue to live long lives.

While this model addresses one weakness of the Patriarch Hypothesis, it shares the same assumption that nature could not select for different lifespans in men and women, and thus does not consider sex-specific mortality mutations that might affect female mortality more than male mortality or vice versa. Had these mutations been allowed, women's lifespans would have shrunk to the age at which they stopped reproducing. Also, and perhaps more importantly, the model assumes that male mating preferences for younger women are exclusive—the chances of a man of any age mating with a woman older than 30 (!) are zero in the model, rather than the much higher probability that is observed in reality.[35] Among the Hadza, for example, one-third of women over age 55 are married, and among the !Kung, well over one-half of women over 55 are married—a situation that should lead to substantial selection pressure for longer reproductive life.[36]

But the weakness in Morton's study most frequently cited by other researchers is its assumption that men prefer younger women.[37] This preference is a favorite topic of evolutionary psychologists and has been exhaustively studied in modern populations. However, critics of Morton's model have argued that any innate preference among men for younger women is more likely a result of menopause than a cause of it. That is, in an imaginary world in which women remain fully fertile through age 80—the world of Morton's model—there is no obvious fitness benefit to deserting a 40-year-old woman to raise more children with a 20-year-old woman. In fact, researchers found that among chimpanzees, a species in which fertility remains high late in life, males prefer to mate with old females—and the older, the better![38] For chimpanzees, old age may be a signal of "good genes" or of parenting experience that will help the next offspring to survive.

Besides menopause, another reason commonly cited to explain human males' preference for youth is that humans form long-term pair bonds, and young women have more of a reproductive future. But it is too simplistic to say that humans form long-term pair bonds. Among foraging societies, short, experimental marriages and frequent divorce are common in youth, but marriages become longer and more stable as the parties age—and not the reverse, as theories like Morton's would predict. Altogether, for these reasons, I am skeptical of the "male mating preference" hypothesis.

Finally, this may be the place to mention a further complication to theories that depend on male mating preferences, and specifically men's preference for younger women. Such theories are weakened if it is the case that women also prefer younger men, a contingency that lowers the chances that the "patriarch" will succeed in reproducing, especially in mobile foraging societies in which women choose whom they marry and accumulated resources are not a factor. In these societies, there are no rich older men whose money compensates for the disadvantages of age. It is true that cross-culturally, men tend to be older than their wives by a few years—among the Hadza, the median age difference is six years, and among the Ache, five or six years is typical. The average age difference between spouses among modernized populations is usually only two or three years, but still, men are older on average.[39] This tendency for women to marry older men may have several explanations—one is that men reach maturity later than women; another is that their pre-reproductive work may be a useful source of food for the group; still another is, to be sure, that men's fertility lasts longer than women's, and in agrarian and industrialized societies, older men are usually richer. There have been societies in which older men monopolize reproduction, accumulating wives while younger men are forced to wait—these include certain Australian foraging peoples, East African pastoralists, and North American religious sects.

However, we have no evidence of a general tendency among young women to prefer *much* older men, and there are several reasons why they might not. Among foragers, to the degree that men provide for their families, women should prefer a man with no previous children of his own, with a longer term of peak hunting productivity in front of him, and with a lower chance of dying while her children are still small—that is, the argument that long-term pair bonding causes men to prefer younger women cuts both ways.[40] Also, male germ cells, like those of females, decline in quality with age, so that the children of older males have a higher risk of genetic diseases and problems. (Research beginning in the 1990s has shown this effect to be strong in humans, although it does not get as much attention as the better-publicized risks of advanced maternal age.[41]) Finally, many or most older men among the Hadza and other foragers do, in fact, stay married to their post-reproductive wives, suggesting that although they might want to marry young women and continue reproducing, the opportunities are not always there, perhaps because young women prefer men closer to their own age.[42] Conversely, in Australian polygynous societies, in which older men monopolized many younger women, and among the forest-era Ache, among whom men outnumbered reproductive-age women by a wide margin, it was not unusual for young men to take older wives while they waited for the chance to marry a young woman—that is, exactly in those societies in which older men reproduced most, older women still had mating opportunities.[43] These factors are all challenging to any theory that relies on male mating preferences for younger women to explain menopause (or anything else).

EMBODIED CAPITAL, OR PUTTING THE MENOPAUSE IN MEN

The main rival to the Grandmother Hypothesis as an explanation for the evolution of menopause emphasizes the role of learned skills in human foraging strategies. This theory, developed by Hillard Kaplan of the University of New Mexico and Michael Gurven of the University of California at Santa Barbara, is sometimes called the Embodied Capital Hypothesis. The phrase refers to humans' investment in big brains, in skills that take a long time to acquire, in knowledge and experience, in highly trained "muscle memory," and in other things that require large up-front costs but pay off over time—that is, in types of "capital."[44] Gurven and Kaplan are co-directors of the Tsimane Health and Life History Project, and their theory is informed by their long-term, intensive study of the Tsimane, though both researchers have also studied other peoples in South America and elsewhere. The Tsimane are a forager-horticulturalist people, which means that besides foraging, they also cultivate some foods with hand tools, but without using a plow or animal traction. Although they acquire much of their food from foraging, their economy is different from that of mobile foragers like the Hadza—they live in villages, cultivate some crops, and do not share food in the same way. But because their mortality patterns are those of a non-industrialized population, they have been an important source for the scientific study of human health and life history. Today mortality is changing among the Tsimane as the research project has brought basic medical care to their villages.

The Embodied Capital Hypothesis begins with the observations that foragers do not begin to acquire enough food to support themselves until their late teens or early 20s; that they reach peak productivity in middle age, typically after the age of 40; that younger families operate in "caloric deficit" (mothers and fathers together do not acquire as much food as the family consumes); and that many or most men stop reproducing when their wives reach menopause. Kaplan and Gurven often refer to a "three-generational" pattern of organization in human families, in which grandparents, parents, and children are all interdependent and transfers are mostly downward from older generations to younger ones. In the Embodied Capital Hypothesis, the brain is an expensive organ that requires a long period of child development and a lot of provisioning to grow. Thus, it has been the driver of some of our most uniquely human qualities—long lives, extended childhoods, and high levels of social organization and cooperation.

Originally, the Embodied Capital Hypothesis focused mostly on hunting skills, on the transfer of resources among generations, and on explaining longevity rather than menopause; the theme that men provide for their children has always been an important part of it. It is thus closer to the traditional Man the Hunter theory than the more radical Grandmother Hypothesis, and is subject to some of the challenges raised by Hawkes and others, such as: How much do men really provide for their families? Is male provisioning the reason for pair bonds? How long does it take to learn to hunt? Even questions such as the number of calories in certain foraged foods have been hotly disputed.

In 2002 and 2003, Hillard Kaplan and Arthur Robson published mathematical models in which they borrowed from the economic theory of capital to add a dimension to the classical theory of aging—the idea that organisms may invest in features that do not pay off right away but will compensate for the investment over time. The most important example of this kind of capital is a big brain. In Kaplan and Robson's model, brains are costly to grow, but the investment is repaid in adulthood when they are used to acquire skills and experience that make the organism more productive and more likely to survive. In this way the authors link the physiological investment in large brains to observed rates of forager productivity over the life cycle—foragers remain dependent for a long time but eventually begin to produce a surplus, which peaks in middle age; downward transfers from adults to children even out the balance. Incorporating these ideas, they show how human mortality could have its characteristic U-shaped curve—decreasing in childhood, increasing in late adulthood—and argue that brains must have become bigger in tandem with longer lifespans and vice versa (that is, these things "co-evolved"). Thus, although their model is different from that of Lee (discussed in chapter 2) on important points—Lee's model focuses on the transfer of resources between generations, whereas Kaplan and Robson's model emphasizes up-front investment in expensive physiology—they both add nuance to senescence theory and arrive at some of the same results.[45]

The original Embodied Capital Hypothesis argued that longevity evolved because humans relied on their big brains and on skills that took a long time to learn; the investment paid off if, and only if, adult life was long. The Grandmother Hypothesis, in contrast, suggested that because older women helped younger women to reproduce, nature selected for longer adult lives. Thus, the Grandmother Hypothesis, strictly speaking, argues that grandmothering *caused* human lifespans to lengthen, while the Embodied Capital Hypothesis maintains that investment in brains and skills caused lifespans to lengthen.

To some researchers it seems simpler and easier, however, to suppose that it was a reduction in extrinsic adult mortality (for example, from predators or illness) that caused longer natural lifespans to evolve, an effect that is easy to explain and consistent with classical senescence theory and with observations across many species—recall that in species with low extrinsic mortality, long natural lifespans evolve because the benefits of investing in growth and development outweigh the need to rush to reproduce. More recent versions of the Embodied Capital Hypothesis have proposed exactly this scenario—a reduction in extrinsic mortality came first, longer natural lifespans evolved as a result, and humans developed big brains and elaborate skills in conjunction with their longer lives. What, then, caused adult mortality to drop in the first place? In this theory, it is cooperation that came first. Humans started sharing food—possibly at the same time that they began inventing the egalitarian social systems described in the next chapter—and this reduced mortality from disease, as they were able to survive bouts of debilitating illness in which they otherwise would have starved or become too weak to recover.[46] Researchers have found that hunters in traditional societies are disabled by illness on about 10 to 20 percent of all days, and long bouts of illness are not unusual—among the Ache, more than one-third of men were debilitated for 30 days or more in a period of three years.[47] Cooperation could have set up selection for longer lifespans and good conditions for the emergence of other distinctly human adaptations—menopause, big brains, long childhoods, and high levels of skill in foraging.

Cooperation is also critical in the theory of Sara Blaffer Hrdy that shared childcare drove the evolution of longer lifespans as part of a highly adaptive reproductive strategy. Humans who cooperated to care for children were able to "stack" offspring—that is, to care for more than one dependent child at a time—and thus to reproduce faster without sacrificing quality of care. In response, children evolved slow growth patterns and complex social skills and emotions; because childhoods were already long, big brains could evolve without too much cost.[48]

Several different hypotheses of human evolution, then, invoke cooperation as a fundamental driver—rather than just a result—of the suite of traits that distinguishes us from other animals, including menopause. The Grandmother Hypothesis, newer versions of the Embodied Capital Hypothesis, and Hrdy's Cooperative Breeding Hypothesis all place forms of cooperation at their center. It is possible that forager social organization is integrally related to the development of the biological traits that distinguish us, such as long life, menopause, and big brains.

A challenge to some of these theories—in particular, the Embodied Capital Hypothesis and the Cooperative Breeding Hypothesis—arises if both long childhoods and longevity (and therefore menopause) are features of *Homo sapiens* only; that is, if they came about after brains were already big, rather than evolving before or along with big brains. If Neanderthals had brains as large as those of *Homo sapiens*, but shorter lives and shorter childhoods, we must admit that the expensive brain was not part of the engine that drove the evolution of those features of human life history that we are interested in—what we did with the brains (behavior) was more important. The Grandmother Hypothesis, which does not invoke big brains as a driver of longevity, is more flexible on this point.

I discuss the timing of all these developments further in chapter 5. For now, it is enough to say that menopause fits into a human strategy that ultimately came to include long life, cooperative breeding, egalitarian social organization with food sharing, highly skilled foraging strategies, and a long period of dependency in childhood. It is part of a reproductive strategy that allows us to breed quickly while also investing in long-lived, high-quality offspring—a system that is adaptable to many environments; that is resilient in periods of stress; that can "boom" in good ecological conditions; and that, with its high adult-to-child ratios and longer natural lifespans, supports the technologies, skills, and cultures on which humans have increasingly relied to adapt to, and change, their environment.

A main difference between the Embodied Capital Hypothesis and the Grandmother Hypothesis is that in the former, both men and women help younger generations—grandfathers are as important as grandmothers, and selection has acted to prolong life in both men and women for similar reasons. Originally, the Embodied Capital Hypothesis did not directly try to model the emergence of post-reproductive life, but more recently its authors have incorporated some elements of the Grandmother Hypothesis into their theory, extending it to include men. In this newer version of the hypothesis, not only does male fertility decline with age for physiological reasons, but this effect is increased because men often remain with post-reproductive wives. That is, men have a post-reproductive life stage—menopause in that sense—just like women do.[49]

There are reasons why we might expect selection pressures to be different on men—males and females can have conflicting reproductive interests and strategies, and there is plenty of evidence that this is true in humans. Because men invest less in pregnancy, lactation, and childcare, the burden of reproduction is much lighter on them, and they benefit less from being relieved of these burdens—they can help their kin while continuing to reproduce, and some men do reproduce late in life. Still, some of the logic of adaptive menopause applies to men too: whether they provide for their children directly or only share resources with a group, men's children, like those of women, have to compete for the resources acquired by adults, meaning that there may be diminishing returns to fathering more children after some point. Also, it is hard to understand why men do not compete more vigorously for fertile mates throughout life unless they are reproducing in some other way, such as by helping their kin. Perhaps by withdrawing from the competition for young women, in which they are unlikely to be successful, and investing their energy surplus in their families instead, older men have received enough payoff in reproductive fitness that most outlive their reproductive lives by a long time.

It seems probable that a combination of factors has favored long life in men: some are similar to the pressures selecting for post-reproductive life in women, but on the other hand, the fact that men can continue to reproduce throughout life shows that classical selection pressures have also been influential. It is less of a mystery why men live long. Most animals will live for as long as they have a decent chance of reproducing.

WHY AREN'T WE MORE LIKE MOLE RATS?

So how did menopause evolve? It emerged and endured in a context of sharing and cooperation—and in this context, it was very useful. The ways in which it proved adaptive might be hard to model mathematically or trace with survivorship studies or by other means, but this does not bother me too much: nature has done the math for us. By helping their daughters to reproduce, grandmothers were part of a system that allowed humans to compress 50 years of reproductive life into half that time, and this factor has helped maintain the balance point of fertility's end in our 40s, rather than selecting for late-life reproduction in women, which would surely have happened otherwise.

Among humans, because children remain dependent for a long time and are born in rapid succession, the ratio of adults to children in any residential band of foragers is a matter of critical importance to the group's survival. If families with many small children run caloric deficits, too many dependent children can put the whole group in caloric deficit. Groups can be rebalanced by exchanging members, but a family with many dependent children and few adults might not be welcome in any group. In such conditions, longer weaning periods and much greater spacing between births might be advantageous, and ecology does influence these factors: the !Kung, who inhabit an environment that barely supports them, have longer birth intervals and fewer children than foragers in environments with more abundant food (although a direct relationship between these factors is hard to prove). But in general, it is a distinctive feature of humans that birth intervals are shorter than those of our nearest relatives; women are more fertile in youth, a quality that has allowed for rapid expansion of the population in favorable environments.

In this situation, it is easy to imagine how a woman who stops reproducing in midlife might not have much of a fitness disadvantage over one who keeps bearing children. A woman who kept reproducing throughout her lifespan would produce competitors to her other children and grandchildren for the share of food brought in by adults, while her own productivity would be lower as long as she was nursing infants. Her family would remain in caloric deficit for a much longer time, probably until the end of her life. Older children might help raise her young ones—siblings provide childcare in many forager societies today—but soon after they were able to feed themselves they would start producing dependent children of their own. The group might struggle to provide for her children and for the other children in the camp, and this family would not be able to provide for itself without the group.

One solution, not yet discussed, might be to delay or suppress reproduction in some of the woman's adult daughters, who could then help raise their younger siblings; nature has selected for this solution in several species. It works because in sexually reproducing species, we are about as related to our full siblings (if both parents are the same) as we would be to our own children—in each case, we share half our genes. In some circumstances, then, if reproductive success is unlikely, it can be advantageous to stay with a parent and be a "helper at the nest," as zoologists call this behavior (after a famous 1935 article by Alexander Skutch).[50] Among meerkats, African wild dogs, and some other animals, this type of "cooperative breeding" is obligate (these animals can't reproduce successfully without it), and reproduction is restricted to a few female "alphas." Naked mole rats are an extreme example—colonies of these underground rodents have reproductive "queens" that commandeer the labor of the rest of the colony, like queen bees.

But in humans, reproduction is suppressed in older women rather than in younger ones. Why? Some researchers have proposed that this is because residential patterns among humans are male-philopatric or "virilocal"—the wife lives with the husband's family after marriage—so that older women are more related to the others in their group than are younger women, with the result that kin selection works more powerfully through them.[51] Although I agree with these researchers that the effects of competition for resources were probably critical to the evolution of menopause in humans, I am not sure that a virilocal marriage pattern can explain why nature has mainly selected for repressed reproduction in older women. Modern foragers are not especially virilocal.[52] Alternatively, it may be possible to press the logic of the Embodied Capital Hypothesis into service here: because skill and experience are so important to human subsistence strategies and because foragers tend to become more efficient through late middle age, older women, rather than their daughters, could be more helpful to their kin when freed from the burdens of reproduction. Finally, it is likely that the selection pressures on humans are different from those that led to cooperative breeding in other species. Among other animals, according to the most influential theory, cooperative breeding arises when there is not enough habitat for offspring to occupy; chances of reproducing on one's own are small, so offspring stay behind and help their siblings ("delayed dispersal"). In humans, cooperative breeding seems less a response to ecological constraints; rather, it is part of a system that also includes long periods of dependency for children and a social organization in which groups share food and labor.[53] However, even among other animals, there are cases in which delayed dispersal does not explain cooperative breeding, and yet it is still more normal for daughters to suppress reproduction than for mothers to do so.

I am not sure we understand why it is older women, rather than younger ones, who suppress reproduction. But we are lucky that it turned out the way it did, because the human solution, menopause, is much more humane and fair than the alternative in which young animals delay or forgo childbirth. In our species, post-reproductive helpers for the most part *have already reproduced*, and indeed have probably given birth to all the children they care to have. In other species, dominant females may brutally murder the offspring of wayward subordinates, subordinates may plot violence against their oppressors, and "helpers" may cheat whenever they get the chance. But in humans, the transition to post-reproductive life and grandmotherhood does not cause these types of conflict.[54]

CHAPTER 4

•

Foragers Today

HUNTING, SHARING,
AND SUPER-UNCLES

OBSERVATIONS OF HUMAN foragers have provided crucial support for the ideas and theories of menopause discussed in the previous chapters. That is, a deep understanding of forager society, acquired in recent decades, has supplied much of the basis for newer theories of how we evolved the suite of life-history traits unique to us, including menopause. But I have not so far discussed any foraging societies in detail. Among them, two that have been exhaustively studied, that have deeply influenced theories of human evolution, and whose differences have been the source of much debate among researchers seeking to explain menopause are the Hadza of Tanzania (introduced in chapter 2) and the Ache of Paraguay. These examples will flesh out our understanding of the theories already described, illustrate the "human adaptive complex" at work, and add depth and complexity to our discussion of difficult questions about hunting, families, and the sexual division of labor.

Foraging societies around the world have been extremely diverse, as foragers have inhabited virtually every ecological niche from the Arctic, where meat and fish are almost the only available foods, to the Amazonian rain forest, to the deserts of western Australia. I emphasize here that while it is relatively easy to guess how the local ecology shaped individual foraging societies, and while *differences* among foragers are important to understand and explain, it is more important to understand the traits we share as a species that allow us to adapt so easily to different environments. Cooperation, longevity, the ability to acquire skills and develop technology, and a flexible reproductive strategy are interrelated and fundamental human qualities that make us very adaptable—and menopause is a part of this picture.

The Hadza have been discussed earlier as the population that inspired Hawkes, O'Connell, and Blurton Jones to develop the "Grandmother Hypothesis." In 2010 Frank W. Marlowe of the University of Cambridge, a former graduate student of Blurton Jones, published a book about the Hadza based on the cumulative history of Hadza research and his own studies beginning in 1995. Most of the description that follows is based on this book, called *The Hadza: Hunter-Gatherers of Tanzania*, and further references can be found there, though I have also cited Marlowe's sources where his conclusions depend on other research and where this seemed important.[1] (Marlowe himself does not support the Grandmother Hypothesis of menopause, and his competing Patriarch Hypothesis is discussed in chapter 3.) By the time he published *The Hadza*, Marlowe had spent a total of about four years among them, spread out over more than a decade, and had collected very detailed data on, among other things, food acquisition and food sharing.

The Hadza are among a very few societies that still survive almost entirely by foraging with little use of modern technology; most other foragers today also cultivate some crops or use manufactured goods and tools, and many foraging societies are known only from ethnographic descriptions by (sometimes obtuse and chauvinistic) Europeans, written in the age of colonialism. While many laypeople think of the Hadza and other foragers as primitive and inferior because they do not use complex technology—it was partly for this reason that several efforts were made by governments and missionaries over the course of the twentieth century to force the Hadza to take up a sedentary way of life—this attitude is less typical of today's anthropologists, including Marlowe, whose respect for the Hadza and their culture is evident in his work. It is impossible to overstate, and difficult even to imagine, the changes that agriculture and industrialization have brought to the human condition. For this reason the world's few remaining foragers, such as the Hadza, are an extremely valuable, and quickly vanishing, source of insight about our history as a species. For almost all of our history, humans were foragers, and it was a strategy that worked very well. The strategies that have replaced foraging have been in use for only a short time and are still experimental; we do not yet know how well they will work out.

Of course, it is important to appreciate that the Hadza are not stuck in time; their history is as long as that of any other people, and like all other peoples, they have changed in the last 10,000 years since most of the rest of the world took up agriculture and, eventually, industrialization. They have contact with the pastoral and agricultural peoples who increasingly surround and infiltrate Hadzaland, and whose activities (such as cutting down trees, grazing livestock, and killing lions) have changed its ecology. However, Marlowe makes the case that, among modern foragers, the Hadza are the best source of insights into how we came to be the way we are. As he argues, the Hadza are the "median low-latitude foragers"—that is, with regard to many factors such as camp population, fertility, and age at marriage, the Hadza fall roughly in the middle of the spectrum.

It can be difficult to know what verb tense to use when describing traditional societies. Historically, anthropologists have often used the "ethnographic present" even when the available sources are old, and in some cases even when the peoples they describe no longer exist. This can be very misleading, and also may convey the false impression that traditional societies never change. Life has changed a great deal for the Hadza in recent years because of the increasing encroachment of pastoralists, farmers, and tourists, and even contact with researchers has probably caused changes. But since Marlowe's fieldwork among the Hadza has continued through at least 2012 and his book is fairly recent, I have used the present tense, for the most part, in this section, although some information may be out of date where it is based on data collected decades ago.

Like most foragers, the Hadza live in fluid bands of about two or three dozen individuals, and although most people live with at least some relatives, camps are not strict kinship groups but loosely related associations of families. Camp population is highly variable, and large camps can have over 100 residents; on average, they are larger in the dry season, when groups are forced to converge around water holes. There are about 1,000 Hadza in total.

The Hadza revise much of what we in industrialized cultures may think we know about human developmental physiology. For example, how big is an "anatomically modern human"? Body size varies hugely among living populations, and not only because of genetics, but also because of nutrition, climate, disease burden, and other factors, as our genes respond to a large number of cues about our environment in the womb and early in life.[2] Like most foragers, the Hadza are small by Western standards. I am considered a small woman in the United States, but the average Hadza man is about my size, 5 feet, 4 inches, and 117 pounds; women are smaller, at around 4 feet, 11 inches, and 101 pounds. The Hadza grow more slowly than people in industrialized populations, reaching full height at about age 24 (for women) or 25 (for men), and the adolescent "growth spurt" so obvious in the United States, and thought by many researchers to be a distinctive feature of the human life course, is less pronounced among them.[3] Like many foragers, Hadza women and men reach their peak weight in youth, around age 30 or so, and gradually lose weight over the rest of their lifespan.

Most foragers today are smaller than most industrialized peoples, and Western researchers sometimes attribute the forager pattern of slower growth, later adolescence, and smaller stature to "nutritional stress." But despite the rigors of foraging life and a diet that many in industrialized cultures would consider meager, the Hadza are a healthy population with excellent physical fitness and high fertility.[4] Rather than seeing the Hadza as small because nutritionally stressed, it may be more accurate to see modern Westerners as large because unusually well-fed.

Hadza girls reach menarche—the age at which they first begin to menstruate—at about age 16.5, much later than girls in the industrialized world. Although they become sexually active and marry at about the same age as menarche or even before, Hadza women give birth for the first time at age 19 on average; researchers usually attribute this gap to "adolescent subfecundity," a period when young women are less than fully fertile. There is apparently not enough evidence to calculate the average age at last birth among the Hadza; some research suggests that the average age at hormonal menopause is about 43, but we lack good data, and this figure is probably too early, since Marlowe and others have observed that Hadza women commonly give birth over the age of 40.[5] Hadza women, like women in many other populations with "natural fertility," report no symptoms of menopause.

A demographic study based on data collected in 1966, 1967, 1977, and 1985 suggested that the average woman surviving to the end of reproductive life gives birth to about 6.2 children.[6] According to Marlowe's research, the average interval between births is about 3.4 years; mothers nurse their children for about 2.5 years and then become pregnant again. Of these children, a little more than half—54 percent—reach adulthood, and about 21 percent die in the first year of life, figures typical for foragers and also for agricultural populations before the modern period. Because infant and child mortality among the Hadza is high by modern standards, life expectancy at birth is low, about 32.5, but average life expectancy for women surviving to age 45 is an additional 21.3 years, and some Hadza live to age 80 or beyond. The Hadza population was growing throughout the 20-year period that data were being collected, at a brisk rate of about 1.3 percent per year.

Causes of mortality are hard to measure; when asked by researchers Blurton Jones, Hawkes, and O'Connell over the course of the 1990s to identify the reasons why people had died, the Hadza most often named tuberculosis and other respiratory diseases, measles (an epidemic of which killed several children during an attempted resettlement in 1986), other illnesses, and old age; other causes cited were childbirth (4 percent of the deaths described to Blurton Jones's team), falling from a tree (probably while collecting honey), poisoning or bewitching by a non-Hadza, and murder. A whopping 40 percent of deaths were not attributed to any cause.[7] In Marlowe's own research, the Hadza tend to attribute deaths to old age, falling from a tree, animal attack (for example, by a lion), magical attack, or violating taboos around meat eating.[8] Violent deaths are relatively rare among the Hadza, but Marlowe heard of two cases in which love triangles led to murder, and competition for women is the most common cause of violence among foragers more generally. Also, unfortunately, physical violence has become more common in recent years, as alcohol consumption has greatly increased in parts of the Hadza population.[9]

Marlowe did not see any evidence of infanticide in his research; even when they gave birth to twins, Hadza mothers tried to raise both of them, despite low probability of success. The Hadza also express no preference for sons over daughters or vice versa, though fathers tend to give more care to sons, and mothers to daughters.

As with other foragers and preindustrial peoples in general, mortality among the Hadza is highest in childhood and especially in infancy. Mortality in late adulthood is higher for men, which is also typical of most human populations. The ratio of men to women overall is about 0.97, but for people over the age of 60 it is much lower, at 0.73. By Marlowe's count about two-thirds (63 percent) of the 71 Hadza women over age 55 were single, because their last husbands had either died or left them to have families with younger women, a source of some bitterness; still, about one-third were married. These older women are the "hardworking Hadza grandmothers" of Hawkes's research, and Marlowe, like Hawkes, found them to be highly productive, bringing more food into camp than any other age and sex category. They usually live near one of their daughters and not only forage vigorously, but also help with childcare and other work.

Marlowe remarks on the striking independence of women and even children among the Hadza. Because women collect most of their own food and can also rely on support from their mothers and other kin, they seem not to need husbands very much (a point he notes repeatedly). There is, to be sure, gender inequality and male dominance among the Hadza: men are supposed to be bosses and decision-makers, and there are culturally important taboos that restrict the eating of certain parts of large kills to the group of full-fledged adult hunters. Men sometimes beat women, and unlike most foragers, the Hadza practice female circumcision at puberty, whereas boys are not circumcised. But compared to most agricultural or pastoral societies, and to most modern ones as well, there is much more equality between men and women. For this reason, Hadza women who marry pastoralists often divorce and return to camp with their children.

Mothers carry their nursing infants while foraging, but children over weaning age are left in camp under the supervision of older children, with one or more adults within earshot to scare off predators and intervene in emergencies (sometimes Marlowe and his team got stuck with this babysitting duty). Researchers have observed this type of communal care in several other foraging societies, though some ecological conditions seem to preclude it (such as rain forest, which is a more dangerous environment for children).[10] Hadza children over age eight or so sleep in groups together with other children of the same age and sex, and adolescent girls and boys also camp in sex-segregated groups. Children among the Hadza acquire much of the food they eat—about half, by age 8 or 10—which may explain their high degree of independence. Like other foragers, however, children, especially boys, do not produce as much food as they consume until their late teens.

Marriages are not arranged, and there is no legal formula or ceremony for them; couples who begin living together are considered married. As with some other foragers, the Hadza marriage system is best described as "serial monogamy." Early experimental relationships and first or second marriages often do not work out; for example, Marlowe and other researchers found that 11 of 28 marriages (almost 40 percent) that began between 1985 and 1990 ended within four years.[11] Divorce rates among Hadza couples married for more than five years were still high, even compared to other foragers, but some marriages were long-lasting, and older couples were more likely than new couples to stay together. Little stigma attaches to having children without being married, and remarriage is common. After divorce, children live with their mothers or sometimes with their grandmothers; in Marlowe's research, about 45 percent of children age eight and under (26 of 59 children total) were not living with their genetic fathers.[12] Men are usually older than their wives (by an average of about six years), and a few men have two wives, though these polygynous marriages tend not to be very stable; most Hadza women and men become jealous and angry if they think their spouse is pursuing someone else, and male infidelity is the most common cause of divorce. (Blurton Jones also notes that polyandry—in which a woman has two husbands at once—is not unknown among the Hadza, although it is rare; these arrangements last less than two years on average.[13]) According to Hadza custom, when a man dies, his widow sometimes marries his brother, a practice called the levirate, but this is not obligatory.

Men are supposed to keep their wives' mothers happy by providing meat, and in Hadza stories it is the groom's mother-in-law whose demands make her victim miserable, and not, as in many agrarian cultures, the bride's. Similar traditions of "bride service" are typical of other foraging populations and support the idea that women with young children are likely to live near their mothers, at least early in marriage, in contrast with the practice in most agrarian societies, in which women live with their in-laws. Research conducted in the 1990s showed that most Hadza couples did, in fact, live in the same camp as the wife's mother if she was alive, although the Hadza, like other foragers, are flexible and opportunistic in choosing whom to live with.[14] This point is important for our discussion, because the Grandmother Hypothesis is most plausible if young mothers live near their own mothers and can benefit from their help, although paternal grandmothers, aunts, and other female relatives can also act as "grandmothers."

The Hadza rely a great deal on "alloparenting" (this term signifies all conditions in which someone other than a parent cares for a child; for example, when small children are left in camp under the care of older children and teenagers). For children under age four, mothers provide the bulk of "direct" care, such as holding and feeding. Much of the direct care not provided by the mother comes from fathers and maternal grandmothers, who seem to trade off this duty: grandmothers do more when the mother is single or married to someone other than the child's genetic father.[15] There is a strong ethic among the Hadza that stepfathers should care for biological children and stepchildren equally, though Marlowe's research showed that in reality, men invested more effort in their own children. Still, Blurton Jones found that women with small children remarried faster after a divorce than women with no children; the reasons for this were not clear, but his finding argues against the idea that Hadza men find providing for stepchildren burdensome.[16]

The fact that Hadza stepfathers invest in stepchildren at all, or even profess to do so, illustrates that when men support children in pair bonds, it is not only for the fitness benefit of passing on their genes. This can also be what anthropologists call "mating effort"; men demonstrate that they will support (or at least not harm) a woman's children in order to win her sexual favor, so that future children may be their own.

Although fathers and maternal grandmothers are the most important alloparents of young children among the Hadza, other kin (such as sisters or aunts) and unrelated individuals together provide a great deal of direct care—and this proportion increases for children over age three or four, who are cared for not only by their siblings but also by unrelated older children in play groups. This habit of entrusting children to the care of others is not typical of our nearest primate relatives but is pervasive in all human societies—humans are talented and inventive alloparents. Many researchers cite the Efe, a short-stature (Pygmy) people of the Ituri Rainforest in the Democratic Republic of the Congo, as world champions of alloparenting. Well-known studies based on research conducted in the early 1980s showed that the average Efe one-year-old infant had 11 caregivers. Because adult mortality was very high for this population, much infant care was provided by other children, including the infant's older brothers and sisters. Grandmothers provided the most allocare when they were available, but few infants had surviving grandmothers. Because rates of sterility were high, non-reproductive aunts also filled this void. Families with infants sometimes recruited orphans or children with many siblings from other camps as helpers, and these foster children were more assiduous alloparents than natural siblings.[17] Sarah Blaffer Hrdy, who hypothesizes that sharing childcare was the behavior that drove much of human evolution, has discussed the subject in detail in her highly recommended 2009 book *Mothers and Others*.[18]

Humans use alloparenting adaptively in varying and changing circumstances, and this is one reason it can be so hard to find kin effects on children's well-being or survivorship—that is, it can be difficult to document tangible benefits to children of having a living grandmother or, even more difficult, a living father. Kin may step in to fill a void or help opportunistically—a maternal aunt if there is no maternal grandmother, a grandmother if the father has deserted, a father if a wife's kin are not nearby, an uncle if his brother has died, or a sister if two sisters are married to the same man (as was common among aboriginal foragers in parts of Australia through the twentieth century). In most foraging societies, as in agrarian societies, "artificial kinships"—ritualized relationships with nonrelatives—are also important, and unrelated helpers often contribute in raising children. Unrelated people may help because they expect the favor to be reciprocated; or because the costs of pitching in are lower than the social benefits of being seen to do one's share; or in some cases, such as that of the Efe foster children, because they are especially vulnerable.

Returning to the Hadza, much of Marlowe's research seeks to understand how this population acquire and share food. Hadza women occasionally collect tortoises, hunt small mammals or birds, or scare a predator away from its kill, but for the most part they bring in plant foods, especially tubers that they dig up with long tools made of stone. While men, women, and children eat some of what they collect in the field (with the result that men and women have somewhat different diets), all groups also bring food back to camp, which is where most food is shared. Men usually hunt alone, probably because it is easier to stalk animals that way, and use poisoned arrows with iron tips. Like Hawkes and other researchers, Marlowe observed that most hunts are unsuccessful—hunters bring down big game only a few times per year. Hadza men also forage for baobab and berries (when in season, everyone eats berries and little else), and they collect honey, a much-valued food, by climbing baobab trees with pegs and using smoke to calm stinging bees. Women also bring in honey but focus on that made by nonstinging bees with hives closer to the ground, which is much less dangerous to collect. Women usually forage in groups, bringing their nursing children with them; they often bring along an older boy to guard them, as they fear predators or attack and rape by pastoralists (not by other Hadza, as this is rare or unknown).[19] Hadza women do not use bows for their own protection, although it might seem useful to be able to do this.

Like other foragers, the Hadza share food, a phenomenon rare in nonhuman animals unless they are provisioning their own offspring. Both men and women share food while foraging in the field and not just at camp, but food brought to camp is especially hard to keep to oneself.

In Hadza camps there is little privacy: spouses and younger children share a hearth and a hut, but in general everyone knows what food is brought in, and everyone demands a share, so that it is quite difficult for researchers to keep track of where the food goes. In Marlowe's observation nobody saved food for anyone who was not in camp, not even a spouse or a child; any food brought to camp was eaten immediately until it was gone. People often tried to hide honey to share with their families, but with limited success. With some foods, such as small game, and with women's foods generally, there was less of an obligation to share outside the family, but this limitation was not strict, and people who dropped in at the hearth where it was being cooked would get a portion. When men killed a large animal, not only did everyone in camp demand a share, but other camps took portions as well; the group might try to keep the kill secret, but rumor spread fast. For the most part, researchers have found that the Hadza share easily, without begging or bickering. But Marlowe observed that in large camps, it was easier to hide food, and there was more bickering about unequal sharing, supporting theories that a high population density may generate social inequality even if no other factors are involved.

Large game is the food distributed most widely—acquired rarely and highly valued, it is a huge windfall of which everyone wants a share. The Hadza do not seem to have rules about how to distribute meat beyond the taboos already mentioned, though some foraging societies do have such rules. Among the Hadza, it seems that everyone simply takes a share of the meat until it is gone.

By careful observation, Marlowe and his graduate student Brian Wood were able to observe that hunters kept a larger share of the game they killed than other families received, and have used this in their argument for the importance of "male provisioning" in families. But this inequality in food sharing was apparent only on very close analysis. More obvious support for the hypothesis that fathers provision their families was Marlowe's finding that men with small children, especially infants, brought more food home to camp, provided that the children were their own and not stepchildren, whereas mothers of infants brought in less food, probably because it is harder to forage with a small nursing baby. (The extra food that genetic fathers provided was mostly honey, rather than more meat.) This provisioning did not, however, prevent the Hadza from having high divorce rates, and children whose mothers were married to their genetic fathers did not have lower mortality than other children,[20] probably because grandmothers or other kin could also support mothers. Women were clearly providing for their families, as they brought home food to share with their children every day, although they might also share with unrelated children and with others in camp.

Although they did not have more wives on average, good hunters had more genetic children (counting only children from wives; no one has tried to quantify extramarital affairs among the Hadza), an effect also found in other foraging cultures.[21] Good hunters of any age tend to remarry faster after a divorce, and their new wives tend to be younger than those of mediocre hunters.[22] Among humans in general, male fertility varies among individuals more than female fertility—some men reproduce more than others—and any factor that seems to correlate with male fertility could be an important fitness benefit. A reputation as a good hunter is one of these factors, and good hunters in some foraging societies have up to twice the fertility of mediocre hunters. Anthropologists have therefore devoted a great deal of effort to explaining this effect. Why do good hunters have more children—that is, why do they have higher reproductive success?

One reason could be that the extra food good hunters bring in increases their wives' fertility, and this is Marlowe's argument—hunters are providing for their families. But several other theories might explain the same effect. For example, everyone wants good hunters in their group because meat is such a highly valued food; if these men are treated better than mediocre hunters, their higher social prestige may make it easier to marry early, to keep a wife, or to marry a younger woman when a wife reaches menopause. Women may prefer to be married to good hunters because of the prestige that their husbands enjoy. This kind of effect is sometimes called "indirect reciprocity"—because hunters trade meat for something else, in this case intangible social benefits—and it is particularly hard to prove but seems likely to be important in societies like the Hadza.

In Marlowe's research both male and female Hadza ranked foraging skills highly when asked what qualities were important in a mate (both sexes ranked this along with character and good looks as the most important qualities). If good hunters marry good gatherers who bring in more food, this in turn could account for higher fertility, an effect called "assortive mating." Research by Hawkes, O'Connell, and Blurton Jones among the Hadza showed that good hunters did indeed tend to be married to good gatherers.[23]

According to another hypothesis, the Costly Signaling Hypothesis already mentioned, good hunters may be more attractive and father more children because hunting success signals traits that are valuable for children to inherit, such as strength and intelligence, or because it signals one's potential value as an ally or formidability as a competitor. Finally, hunting success could be related to fertility through "phenotypic correlation"—genetic qualities that contribute to hunting skill might also make the hunter more fertile (that is, good hunters are just in better shape generally). This hypothesis is almost impossible to test but might account for much of the observed difference in fertility between good hunters and so-so hunters.

Studying hunting and fertility across several cultures, Eric Alden Smith found the most support for the Costly Signaling Hypothesis, but also found some support for all of the effects just described.[24] In his work with the Hadza, Marlowe has sought to show that provisioning is, in fact, a part of the attraction of good hunters, against Hawkes and others who have challenged this more traditional view. But the case is a difficult one to make with the Hadza as subjects. Using all the same evidence as Marlowe, Blurton Jones continues to argue that male provisioning is not important among the Hadza, and that costly signaling, the Show-Off Hypothesis, best explains Hadza relationships.[25] This is, after all, the hypothesis that agrees best with what the Hadza themselves have said when anthropologists have asked them why they hunt—women like it!

What is most striking about hunting and food sharing among the Hadza, however, is not any of the things just discussed, but the pervasiveness of "tolerated scrounging" and a related value system that is highly egalitarian. Hadza hunters may wish to keep meat to themselves or reserve it for their families, but they do not do so; they accept that everyone who sees it gets a share of it. In this system, need and inequality are perfectly good reasons to share—ideally everyone ought to receive the same amount. And indeed there is little health inequality among the Hadza, because everyone in the group gets about the same amount of food. There is also little social inequality—the Hadza have no headmen, no shamans, no specialists of any kind. Although it is likely that good hunters enjoy more prestige than so-so hunters for the reasons just explained, such effects are subtle and hard to observe because of a strong ethic of humility that discourages arrogance, bragging, and telling others what to do.

A similar but perhaps even more extreme economy of tolerated scrounging has been much studied among the !Kung and most recently described by Nancy Howell in her book *Life Histories of the Dobe !Kung*, based on research she conducted in the late 1960s.[26] Probably because the ecology of their homeland is more dangerous and less productive of food than that of the Hadza, Howell found that the !Kung were thinner and had lower fertility, and their children acquired little food for themselves until their late teens. While the average age at last birth was only 35, older women remained vigorous foragers, reaching peak productivity in their 50s. Like the Hadza, the !Kung shared big game widely; they shared small game and gathered foods less widely. Households exchanged food every day, and most families with small children consumed more food than they produced. Nevertheless, like the Hadza, the !Kung showed little difference among households in health status (measured by variations in Body Mass Index in this case), because of food sharing.

The !Kung are famously "fierce egalitarians" with a great many ways of cutting people down to size or pointing out inequalities—while some Western researchers have been charmed by their adroit use of humor and other strategies to do this, others seem exasperated by what they have called "complaint discourse." In any case, Howell judged sharing and egalitarianism so fundamental to this thin and marginally nourished society that she proposed a "fatness hypothesis" explaining why humans, and especially children, lack hair and have large deposits of fat. Fatness, in her view, developed so that we could more easily estimate need and share food accordingly.

To be sure, studies of foraging societies—including the !Kung, the Hadza, and the Ache (discussed next)—often show that even in cultures with few formal status distinctions and widespread food sharing, inequalities exist in measures linked to well-being, such as grip strength, weight, hunting success, and reproductive success. The !Kung also recognize heritable foraging rights to certain waterholes, which translate into social connections, as people must negotiate and maintain gift-giving relationships with those families. Some of the other advantages just mentioned may also be inherited. In general, however, the degree of this heritability is not impressive. The measure most directly related to well-being—weight—is the most highly heritable, but it is also the measure that varies least.[27] The inequalities typical of foraging societies are therefore very different from the social stratifications that emerge whenever there is a shift to an economy that allows people to accumulate—and inherit—material goods such as livestock and land.

The egalitarian social system of the Hadza can make it hard to reach group decisions or take collective action. This may be one of several reasons that the Hadza, the !Kung, and most other mobile foragers, with a few exceptions, do not engage in warfare or other kinds of collective violence—it is hard to organize. Many researchers today view war, raiding, and feuding as Neolithic innovations typical of farmers and pastoralists but not of our Paleolithic ancestors.[28]

Research with the Hadza over the last few decades has mostly confirmed, rather than undermined, a view of foragers—and, by extension, of our foraging past—that has been influential in anthropology since the late 1960s: the view that foraging societies are egalitarian, with widespread food sharing, communal childcare, and equality between men and women.[29] While often tinged with an unfortunate romanticism, this image is not necessarily wrong. But not all foragers are like the Hadza—foragers have inhabited almost every ecological niche on the planet, and foraging societies vary widely. One of the most thoroughly studied foraging populations outside of Africa is the Ache people of Paraguay, and comparing them to the Hadza illustrates several points that can deepen our understanding of human reproductive strategies.[30]

The Northern Ache are the largest group of Ache in existence and the last of the native peoples of Paraguay to be contacted and pacified. The first peaceful contact between Ache and Paraguayans occurred in 1970. A turbulent period followed in which the Ache moved to reservations and nearly half died of infectious diseases (often from respiratory complications such as pneumonia), in squalid conditions and without adequate medical care, as the government sought to sell and develop the forest they once inhabited. (Kim Hill has argued from his own observations that with enough food and basic medical care, most Ache would have survived the illnesses that killed so many at first contact, and he himself helped many Ache recover.[31]) At first contact in 1970, there were 557 Northern Ache.

Hill and (starting just a few years later) Ana Magdalena Hurtado, both now at Arizona State University, began fieldwork among the Northern Ache in the late 1970s and have continued to work with them until recently. Although their research began after most or all of the Ache had moved to reservations, because contact was so recent, the adults with whom they spoke in the early years of their research remembered life in the forest quite well; also, because the Ache continued to make long foraging treks in the forest while living on the reservations, Hill and Hurtado were able to observe their forest life. Using an exhaustive process of structured interviews, they compiled demographic data on the Ache for most of the twentieth century and reconstructed their population history for the forest period, as well as for the contact and reservation periods. In 1996 they published a magisterial study of the Northern Ache called *Ache Life History: The Ecology and Demography of a Foraging People*, which is, in my view, some of the most fascinating reading in the field of anthropology.

The Ache of the forest period, like the Hadza, lived in fluid, mobile bands; the average band numbered about 50. Their rain forest environment was abundant in prey that was not too hard to capture, such as giant armadillo and coati, and Ache hunters had much higher success rates than Hadza hunters. On the other hand, hunting in the Paraguayan rain forest was more dangerous than it was in Hadzaland: snakebite was a frequent cause of death for hunters, and jaguar attacks were much feared. Dangers to infants and toddlers in the forest environment from snakes, insects, spiders, and accidents were extreme. Infants could not be placed on the forest floor even for a moment, and nursing mothers slept sitting up, bent over the babies in their lap. Children were three years old before they began to spend time more than one meter from their mother. Still, by age eight or so, children had considerable independence and often slept at night with other relatives or non-kin rather than with their parents.

Because hunting was relatively easy and childcare highly demanding in these conditions, men acquired much more food from foraging than women did—about 87 percent of total calories. Women spent less time foraging, and when they did forage, they acquired mainly palm starch. In the forest the Ache moved camp every day, and the women dismantled, transported, and rebuilt the camp while the men hunted. This extreme mobility was hard on the sick and frail, who might be left behind, sometimes dragging themselves into camp late on their hands and knees; some Ache who could no longer keep up would ask to be buried alive to avoid being eaten alive by vultures, a contingency rare in the forest period but more common with contact and the new diseases that it spread.

Some differences between the Hadza and the Ache probably trace to Ache men's greater role in supplying food.[32] A striking feature of forest life among the Ache was a high rate of infanticide and child homicide. In particular, infants and young children might be sacrificed when their fathers died, or sometimes when their parents divorced (Hill and Hurtado collected many blood-curdling stories about this). Girls were more likely to be killed than boys, with the result that the sex ratio among the Ache in the forest period was heavily skewed toward males through middle age (women eventually caught up because of their lower mortality in adulthood). Altogether, about 14 percent of boys and 23 percent of girls were killed in infancy and childhood. Culturally, the Ache perceived orphans as needy consumers of food, and the most likely reason for the population's high rates of infanticide and child homicide was that they ensured a favorable ratio of adult providers to dependent children, and also of males to females, in a society in which men's hunting provided most of the food. Researchers have long described similar high rates of infanticide and male-biased sex ratios among Arctic foragers, for whom hunting and fishing are almost the only sources of food.[33] Although Hill and Hurtado found no direct evidence for a cost of reproduction among the Ache—families that produced the most children did not appear to have lower fitness because of higher child or adult mortality—it is nevertheless hard to believe that infanticide and child homicide would have been so widespread among the forest Ache unless competition for resources, dependency loads, and costs of reproduction were compelling constraints on this population.

When Hill and Hurtado studied food sharing and food transfers among the Ache, they found that families with several dependent children consumed more calories than they produced, as Howell had found for the !Kung.[34] The extra food came from unmarried men in the same camp or married men with few children. Like the Hadza and the !Kung—but even more so—the Ache shared meat equally throughout the camp. Even though fathers did not provide more food for their own offspring than for other children, having the right number of hunters to support the group was important, and because of the high sex ratio of men to women, each group was likely to have some extra single men whose contributions were critical to its survival.[35] It may be relevant to recall here that among most or all foragers, men marry later than women—usually a few years after they have begun to acquire more food than they consume. One reason may be that it is helpful to have extra, non-reproductive hunters in the group to make up the caloric deficit of reproducing families. The Show-Off Hypothesis is especially plausible for this group: a likely motive for young unmarried men to hunt and share food is to impress young women.

Among Arctic peoples some widows and daughters without brothers simply learned to hunt.[36] It seems unfortunate that the Ache and other hunting-dependent foragers did not more widely resort to this solution, or even to the expedient of raising daughters as boys to adjust the sex ratio (transgender roles are a feature of many traditional societies, including the Ache, and are discussed later). Perhaps because infant and child mortality is high in foraging societies anyway, infanticide has been a simpler solution than adapting cultural ideals about gender and about men's and women's work.

Probably related to these factors is one of the most well-known features of Ache forest society, the practice of "partible paternity." That is, the Ache, like several other indigenous peoples of South America, recognized more than one father of a child. They distinguished the probable genetic father—the one who conceived the child—from other fathers, but considered all men who had sex with a woman during her pregnancy to be secondary fathers. The average person had about two fathers—this seems to have been the optimal number—and offspring were the best informants about extra fathers (that is, they were the ones most likely to remember and name all their fathers, probably because this was to their advantage). Children might refer to an uncle as a father if their father had died, and uncles were also frequently secondary fathers. Among other animals, polyandrous mating—that is, with more than one male—is a common strategy for a number of reasons. Females have an interest in choosing males with the most desirable genes (even if this is someone other than their own mate), securing parental investment from more than one father spreads risk, and a male in their group may be less likely to kill a female's offspring if he thinks he could be the father.[37] Some of these strategies may explain the tradition of partible paternity among some human foragers. Among the Ache, children with a secondary father were in fact less likely to be killed, though the effect was not statistically significant. Partible paternity is especially advantageous when men outnumber women and when men provide most of the food, as was the case among the Ache during the forest period.

Marriage among the Ache of the forest period was even more fluid than among the Hadza—by age 30, the average Ache woman had been married 10 times. This is despite the fact that divorce increased the chances of a child being killed—this fitness cost to both parents was not compelling enough to keep couples together. Although polygynous marriages were rare and unstable, and polyandrous marriages even more so, most men nevertheless reported having had at least one polyandrous marriage in their lives, and almost all women had had a co-wife at some point. Sexual experimentation before marriage and even before puberty was also common—part of the puberty ritual for girls involved washing all the men who had had sex with her so far.

As with the Hadza, a substantial number of Ache lived to advanced ages in the forest period. At contact in 1970, of 547 Ache, 30—about 5.5 percent—were over the age of 60. Although men greatly outnumbered women at younger ages, the sex ratio in this age group was about equal (16 men, 14 women) because adult mortality was lower for women. Some middle-aged men abandoned their wives to have children with other women, and a few married a second wife, but most commonly, they remained with their post-reproductive wives. Hill and Hurtado report that the Ache had a word for menopause but do not say what it meant to them—whether they understood menopause simply as the cessation of childbearing, or whether they experienced symptoms or attached significance to it. They noticed that women usually did not acknowledge being post-reproductive until they were quite old, and that last children were doted on and spoiled. Although Ache grandmothers did not produce as much food as Hadza grandmothers because of the different sexual division of labor among the Ache, they visited and helped their adult children however they could—gathering food and doing housework through middle age and switching to childcare after age 60 or so. Grandfathers, too, switched to foraging with the women and to caring for children as they became too old to hunt.

The Ache have become somewhat notorious in the enthnographic literature for geronticide—the killing of the elderly—and Hill and Hurtado report an interview they conducted in 1985 with one 75-year-old Ache man who described killing old women in his youth by stepping on them and breaking their necks. It is uncertain how much weight to give to this account, though, since Hill and Hurtado's own tabulations of causes of death before contact attribute only one death of a woman over 60 to "burial" (the Ache form of euthanasia); they report that only two people over 60 were "left behind," both of whom were men, and that no homicides occurred for this age group.[38] While I certainly do not want to trivialize the issue by suggesting that their interviewee was just trolling, there is not enough evidence to convict the Ache of widespread geronticide in their foraging past.

The lifestyle of moving camp every day was surely very hard on the sick, the elderly, and anyone else who could not keep up, and abandonment or euthanasia of the very old is not uncommon in the older ethnographic literature on nomadic populations,[39] but this extremely mobile lifestyle was perhaps not typical of the Ache over the long term. Among the !Kung, Howell observed that families tried to care for the elderly, but typically did not need to support them for a long time—most people died soon after they became too frail to forage. She emphasizes that the burden of caring for the elderly was trivial in this society compared to the burden of raising children, who remained dependent until their 20s. Among the !Kung, as among other foragers, transfers of resources overwhelmingly flowed downward through the generations.[40]

Adult mortality due to violence was much higher among the forest Ache than among the Hadza—violence caused about 36 percent of adult deaths, as well as a large number of child deaths (Hill and Hurtado treated cases in which children were kidnapped by Paraguayans and sold into slavery as deaths). Relationships between the Ache of the forest period, indigenous peasants, and Paraguayans were intensely hostile. The Ache sometimes raided farms for food and supplies, while ranchers, farmers, bounty hunters, reservation administrators, and even some early twentieth-century explorers and anthropologists kidnapped, killed, enslaved, or relocated the Ache in large numbers. Since the early 1970s, accusations of genocide have been made against the regime of Alfredo Stroessner, president of Paraguay from 1956 to 1989, although Hill and Hurtado have argued that the conflict between the Ache and the Paraguayans did not rise to the level of genocide but was similar to the conquest of many other indigenous peoples of South America.[41] In 2014, the Ache people filed a criminal complaint of genocide, which will be tried in Argentina because the Paraguayan judiciary has refused to consider it. Although Hill and Hurtado, and also some other scholars, have argued that the intervention of modern governments is the reason the Hadza and other African foragers are so peaceful and have fairly low mortality from violence, the case of the Ache suggests that the presence of a modern state and settled societies can cause more violent conflict, rather than less.

Though many adults died of violence among the Northern Ache, it was rare for them to kill or even fight with one another. Unlike the Hadza, however, they did have a ritualized form of within-group violence—what Hill and Hurtado call the "club fight." This was exactly what it sounds like: a fight with wooden clubs. These fights sometimes happened spontaneously, but on occasion groups of Ache would come together to clear a large space in the forest for a massive club fight in which all adult men would participate. Men would target their enemies, usually rivals for a woman or someone who had insulted them. Some men would die in these fights, but they were also social occasions at which people could find marriage partners, catch up with relatives, and change residential bands if they wanted to.

Ache fertility was higher than that of the Hadza, and higher than that of any other foraging people of whom we know. The average number of children borne by a woman surviving to menopause was more than eight. Although the Ache began reproduction at about the same age as the Hadza, intervals between births were shorter, and the median age at last birth was late, at 43 (that is, half of Ache women in the forest period gave birth after age 43). Hill and Hurtado calculated that during the forest period their population grew at a startling rate of 2.5 percent annually. This can't have been typical historically, because at that rate of expansion, their number would have reached 106 billion within 1,000 years! Hill and Hurtado propose that the Ache moved into the territory they now inhabit fairly recently, perhaps after the slave raids that depopulated the region in the seventeenth and eighteenth centuries, and found the food supply especially abundant. These researchers have become influential proponents of a theory that human population history, like that of many other animals, has been a series of booms and busts—periods of rapid growth punctuated by catastrophic declines every few generations—an idea discussed in chapter 2.[42]

The Ache, then, employ reproductive strategies of great depth and sophistication. But there are even more complexities to discuss. Hill and Hurtado observed that among this group, a few phenotypically male individuals opted out of hunting; they adopted female roles such as making handicrafts, gathering women's foods, cooking, and childcare, and generally, in modern parlance, they performed female gender. On one occasion, a *panegi*, as these transgender Ache were called, stayed behind to care for a number of sick Ache abandoned in the forest by their group, an episode later recounted with emotion by a hunter whose life he saved.[43] *Panegi* were accepted by other Ache but appeared to Hill and Hurtado to have low status. In 1970, the year of first contact, 3 of 150 Ache men over age 20 were *panegi*, or 2 percent.

The Ache provide support for a broader argument that transgender roles are an ancient feature of human society—perhaps one that is, like menopause, part of an evolved reproductive strategy. Transgender people are not unusual in traditional cultures (nor, of course, in modern ones). At least some researchers have proposed evolutionary explanations of transgender roles similar to those that explain menopause: by opting out of reproduction and focusing on other work instead, people like the *panegi* may help their siblings to thrive and reproduce.

Most researchers believe that there is at least some genetic component to both sexual orientation and gender presentation. This creates a paradox similar to the paradox of menopause, because people who form pair bonds with others of the same sex tend to reproduce at much lower rates than their straight counterparts. Do these traits persist because they are adaptive in some way that compensates for the reproductive disadvantages they confer? To answer this question researchers have focused on male same-sex pair bonds, which are usually more common, or at least more visible and better documented, than lesbian relationships. When studying male homosexuality, some researchers use the term "male androphilia"—meaning "man-loving" in Greek—because it is more specific and carries less baggage than "homosexual." (The opposite term, *gynephilia*, "woman-loving," is also sometimes used.)

Many traditional societies have accepted male androphilic behavior or at least practiced it widely. Among agrarian societies based on the conjugal unit of husband and wife, it is a common pattern for androphilic men to marry women and have families because social and economic pressures to do this are so strong. For example, in Qing dynasty China, gay sex, like heterosexual sex, played into dominance hierarchies that distinguished women and low-status men (passive sexual partners) from higher-status men (active partners). Being penetrated meant being stigmatized as subordinate and less masculine, and Qing legislation theoretically banned gay sex as part of a program that sought to abolish old hierarchies and enforce a uniform, minimum level of honor and status for all free Chinese. Nevertheless, gay sex continued to play a large role in society, and many examples of gay sexual relationships, including some long-term relationships, are attested in legal records from the prosecution of participants. Sometimes younger, subordinate, passive partners later married and graduated to the masculine role of householder and sexual penetrator.[44] All this is typical of what Stephen O. Murray, in his influential and highly recommended 2000 book *Homosexualities*, calls "age-stratified homosexualities."

However, there have also been groups of gender-nonconforming people who opt out of reproduction altogether, like the *panegi* among the Ache. Some researchers argue that this kind of homosexuality, which Murray calls "gender-stratified," may have been typical of our foraging Paleolithic ancestors, just as age-stratified homosexuality is typical of agrarian cultures. The partners of transfeminine people in traditional societies are usually "straight" men, not others of the same group. (Although I will use the term "transgender" in this discussion, following recent scholarship, most members of these groups identify not as either male or female, but, like the *panegi*, as a different category of person—"third gender"—or as a particular kind of male or female.)

On the other hand, "sex-gender congruent homosexuality" or "egalitarian homosexuality," in which both partners identify as male and do not play rigidly distinct social or gender roles, is more typical of modern cultures, and many scholars believe that it is a modern phenomenon. According to this view, male androphilia is a trait expressed differently in different cultural environments and probably, in the culture of our Paleolithic ancestors, expressed mainly as transgender presentation—the form attested among the Ache.

Transgender people in traditional societies may combine elements of male and female gender presentation, and of course individuals vary even within the same culture. Male-to-female role change has been studied more than the reverse. Female-to-male gender role changes often seem to serve purposes relating to property and legal status, such as allowing a daughter to inherit or to lead a family, as Ifi Amadiume described among the Igbo of Nigeria, or as in Albania, where Pepa Hristova has documented a last generation of female sworn virgins living as men in a series of piercing photographic portraits.[45] In agrarian societies there are usually many restrictions on what women can do, so these strategies are important. However, although some of Albania's sworn virgins, for example, were raised as boys from infancy—their parents made the decision for them—most of those interviewed by Hristova and by anthropologists in recent decades say that they made the choice at least partially themselves.[46] It is possible or likely that some traditional female-to-male role changes, like male-to-female role changes, have a biological element that makes this solution to cultural problems more viable.

Many other traditional cultures, like the Ache, acknowledge one or more types of transgender people. Readers may be aware, for example, that in a number of North American indigenous nations some people have adopted the social and economic functions of another gender and sometimes have married as that gender. Formerly called "berdache" by researchers—a term that originated with European ethnographers— these transgender people are now more commonly called "two-spirited," though the terminology varies across nations.

Scholars have long speculated that male androphilia may be an evolutionary adaptation favored by kin selection—men who do not reproduce help their kin, and thus their genes, to survive. Researchers have looked for evidence that gay men in modern cultures are more altruistic than straight men but have failed to find it. However, many aspects of modern Western culture could affect these results: modern societies have highly individualistic value systems; people often live far from their families; and whereas transgender people in traditional societies are usually respected or at least accepted, male androphilia in modern cultures often is not, meaning that gay men may be alienated from their families.

When Paul Vasey and Doug VanderLaan decided to study altruism in androphilic men of a non-Western culture—the third-gender *fa'afafine* of Samoa—they expected to find the same null results as other researchers had reported for Western populations. Instead, they found that the *fa'afafine* invested a great deal of care in their kin, especially their sisters' children, and were more altruistic and "avuncular" than their straight counterparts.[47] Samoan values are community-oriented, and the Samoans Vasey and VanderLaan interviewed perceived *fa'afafine* as especially committed to helping their families and communities. Many families depended on a *fa'afafine* son to help with domestic work and burdensome ceremonial events like weddings, and because *fa'afafine* also tended to be better educated and to have higher incomes than straight men, they were well-positioned to help financially. Vasey and VanderLaan argue that this altruistic caring for kin may have been typical of male androphilia among our ancestors, and that kin selection could explain male androphilia. In support of this hypothesis, they found that *fa'afafine*, like gay men in modern cultures, are more likely to be born late in the birth order, when their families already have a number of other children—a finding that suggests that although egalitarian and transgender androphilia look different, they are expressions of the same thing in different cultural environments. If this theory, sometimes called the "Super-Uncle Hypothesis," is correct, then male androphilia, like menopause, is another ingenious trick of nature to ensure a favorable balance of adults (providers) to children (consumers), and perhaps also of males to females in a context in which men and women do different work. More research with groups other than the *fa'afafine* needs to be done, though, before we can say for certain how plausible the Super-Uncle Hypothesis really is.

An alternative hypothesis suggests that the same genes that cause men to be attracted to other men cause women to be more fertile; female relatives of gay men have larger families, and the genes persist for that reason.[48] In this scenario, male androphilia is a pleiotropic byproduct of a gene that makes women more fertile, but it is a fortuitous one, because the extra support from an androphilic sibling or uncle makes the larger family more viable. Perhaps because these advantages have worked together and reinforced each other, male androphilia has persisted at a low but steady rate in our species despite its reproductive disadvantages.

Summing up, what do we learn from these studies of modern foragers? First and most importantly, in every human society, the population of post-reproductive women is a built-in, naturally renewing reserve of alloparental help and productivity that can be used to great advantage in a broad range of circumstances. But other populations can serve similar functions—older sisters and brothers who have not yet married; adult single brothers or uncles in populations like the Ache of the forest period, among whom men outnumbered women; gay or transgender men, according to the Super-Uncle Hypothesis; or any "extra" adults without children of their own. The picture that emerges from studies of foragers is very different from the classical predictions of the Man the Hunter Hypothesis, with its emphasis on the nuclear family, and also somewhat different from those of the Grandmother Hypothesis, with its emphasis on ties between mother and daughter. While I do think that the contributions of grandmothers (and probably also of older men) explain longevity and menopause in humans, the most striking thing about human reproductive strategy is how adaptable it is. In particular, the assumption that families based on the pair bonds between men and women are the organizing principle of human society seems wrong. Foragers clearly form pair bonds, but the residential band, and not the nuclear family, is the basic economic and social unit—nuclear families typically could not be self-supporting in some parts of the life cycle even if they tried. Contractual marriage and the nuclear family are important in the agrarian societies discussed in the next section of this book. But if people in post-agrarian industrialized cultures are returning to a looser form of "serial monogamy," if they rely on grandmothers and communal care to help raise families, and if women do not live with their children's biological father, they are only doing what worked just fine for 95 percent of human history, before the rise of inherited wealth and complex, unequal social systems.

Next, although I think this is obvious, it is best to be explicit here about demolishing any stereotype that "men provide, women consume"—a cultural assumption that probably lay behind the original Man the Hunter Hypothesis. It is much more accurate to say that adults provide and children consume—for a long time. Forager studies show a *division of labor* according to sex (a pattern also found in agricultural societies). Even when women do not acquire as much food as men, they work just as hard (or harder) in other ways. Women of reproductive age probably bear the biggest burden of labor, since they continue to work and forage while also providing most of the care for nursing children.[49] That is, foraging women—and, as we shall see, farming women— contribute to the material support of children, not only to their care, even in cultures in which they do not provide as much food as men. If men's status in foraging societies is higher than women's, this is not be- cause they do more work or contribute more; it is because of the ways in which they use the energy surplus left over from their lower burden of bearing and raising children compared to women—the big game they hunt is highly valued and shared widely, which increases their op- portunity to win prestige and make alliances.

Although there is wide variation among foraging societies, it is fair to say that gender inequality is much less obvious in most of them than it is in agrarian societies, in which several factors discussed in the next section of this book combine to make patriarchy more pronounced and nearly universal. The residence patterns typical of foragers—mixed bands of generally moderate relatedness—suggest societies in which men and women have about equal influence in deciding with whom they will live.[50] Among both the Hadza and the forest era Ache, mar- riages were not arranged; there was no tradition of a wife living with a husband's family after marriage; divorce was common, and either sex could initiate it; there was little serious effort to control women sexually (through draconian penalties for premarital sex or adultery, for exam- ple); lineage was traced through both parents and was in any case not very important, since no resources or status were inherited; neither men nor women controlled property because there was little property to control; there was no property exchange at marriage for the same rea- son; and men did not monopolize leadership roles because leadership was not very important in either society. Among the Ache, though, for ecological reasons, women were more dependent on men for food, and this probably affected their status. Female children were more likely to be killed than males in this population.

Finally, to someone like myself, who has studied agrarian societies all my life, the levels of egalitarianism, cooperation, sexual equality, and (in some ways) peacefulness characteristic of many foraging societies compared to almost all agrarian or pastoral societies is surprising, even shocking. Humans are capable of a much wider range of behavior than a study of only agrarian societies, or only the modern ones that have recently emerged from them, would suggest.

A theory of Christopher Boehm, an anthropologist who has tackled the question of Paleolithic social organization directly, is so interesting on this point that it deserves a paragraph here.[51] Boehm's thesis, which is based on ethnographic reports and on his own observations of modern foragers and chimpanzees—a proxy, in his model, for our common ancestor with other apes—is that forager egalitarianism is really a kind of reverse hierarchy in which the group works together to prevent upstarts (who are almost always male) from gaining power over others. Common ways of doing this are ridicule, chiding, ignoring, and gossiping; Boehm also found examples of more extreme measures like shunning, desertion (in which the band simply abandons an obnoxious upstart), and even execution for especially dangerous and persistent bullies, often repeat murderers. Thus, although foraging bands don't have obvious political structures, they do enforce equality when an individual tries to assert power over others. The ideological basis for this egalitarianism is the value that foragers place on personal freedom. (Many literate historical societies—such as classical Athens—retained a version of this value system, insisting on radical parity for all males within an ethnic group or a privileged group of "citizens.") As Boehm argues, humans have inherited from our ancestors some natural tendency to compete for dominance and can organize into hierarchical societies drawing on that tendency—and we have clearly done this since the end of the foraging period. But for most of our history, this drive was turned upside down—humans developed value systems and formed coalitions to prevent anyone from winning the competition for dominance. This egalitarian strategy of repressing would-be bosses and chiefs was so successful, so widespread, and so long in operation that,

Boehm argues, humans evolved traits favoring cooperation and altruistic behavior that could not have arisen otherwise.

Like most historians, I am a deeply cynical and skeptical person disinclined to romanticize. Because of my research interests I am more aware than most of how the trope of the "noble savage" can infect ethnography, of cultural anthropology's roots in imperialism and colonialism, and of how hard it is to escape the tendency to objectify those we study and project our own values and preoccupations onto them. I would not argue that foraging societies, even very egalitarian ones, are ideal. A huge percentage of children in foraging economies do not grow up; hunger and anxiety about food are a constant theme in research about foragers; eyewitness accounts of child murder among the Ache keep me awake at night. Still, if modern mobile, low-latitude foragers are a guide to our Paleolithic past, we cannot say that human origins lie in patriarchy, male provisioning, the nuclear family, dominance hierarchies, or warfare, no matter how central these things are to most agrarian societies. Rather, the features of human foraging societies that seem most distinctive are food sharing, cooperative childrearing, egalitarianism, and the sexual division of labor.

CONCLUSIONS

Imagine, then, that a hypothetical foraging woman stops giving birth in midlife. She is no longer producing competitors for her dependent children and grandchildren, and is available to provide food or allocare to them instead. If her daughter has many dependent children, she might be more accepted in a group because she has a post-reproductive helper attached. As long as the grandmother gathers at least enough food to feed herself (among the Hadza and !Kung, women reach peak foraging productivity in their 50s and continue to produce a caloric surplus through old age[52]), and is capable of some allocare, the higher ratio of adults to children benefits everyone in the group, including any of her own descendants or relatives living near her who share some of her genes. In this scenario, incorporating the possible effects of transfers, kin competition, and a sharing economy, these older women are helping not only by *producing* but by *not reproducing*.

In this way, nature might select for a longer female lifespan even though grandmothers are not reproducing directly, because longer lives could still improve their inclusive fitness. This thought experiment works best in an economy in which women acquire as much food as men—the kind of "low-latitude" foraging economy that many anthropologists think was typical of our ancestors in Africa. In ecologies where men and hunting supply almost all the food, foragers have resorted to female infanticide to adjust the demographic structure that nature provides. But in any population in which children remain dependent for a long time, and where the ideal ratio of adults to children is therefore high (that is, in all human populations), a mechanism—like menopause—that raises that ratio is likely to be put to good use.

As we have seen, all theories and models of menopause have problems, which does not mean they are all wrong. Some of them could be right, and the objections could be wrong. Or a combination of theories could be right. It is the nature of evolutionary science that we may never know what actually happened. But having read and thought about this problem extensively over the past years, I have some views regarding which of the ideas presented in the preceding chapters are most compelling.

First, I think menopause *must be adaptive* in the sense that its fitness benefits outweigh its costs, and not only because it seems unlikely that humans braved a harsh and violently fluctuating climate to invade and occupy every part of the world while dragging with them a sector of the population—post-reproductive women—that was in any way holding them back. It seems much more likely that the post-reproductive life stage, an unusual human feature, helped in some way. But more than this, if menopause were not adaptive, I do not see how reproductive lifespans would not have lengthened—probably very quickly—to match somatic lifespans. If this were physiologically impossible for some reason, a contingency for which we have no evidence, then women's somatic lifespans would have shrunk to match their reproductive lifespans, again probably very quickly.

Kin selection has compelling power to explain the origin of menopause: women who are not reproducing directly may continue to reproduce indirectly by helping their kin, particularly grandchildren, nieces, and nephews, and even if this strategy cannot explain everything, it is likely that it has been an important factor.

Transfers of resources—not only from grandmothers to children and grandchildren, but among a range of kin and non-kin—are fundamental to our history and probably one reason that menopause developed; indeed, mathematical models have shown that transfers can cause postreproductive life to evolve. Transfer theory allows us to imagine human populations in terms of energy budgets, deficits, and surpluses, and to describe fascinating effects in which the reallocation of energy among kin and non-kin can lead to greater fitness even while sectors of the population—post-reproductive women, "extra" men, gay men—do not reproduce.

Because of the importance of transfers, it is likely that *kin competition* has been important—transfers mean that a relatively high ratio of adults to children is advantageous, and they increase the likelihood that the optimal number of offspring is not the maximum number; in other words, there is some point at which one more child becomes one too many. Also, it seems plausible that our species' strategy of relying on skills that take a long time to master (*embodied capital*) makes postreproductive life even more beneficial—middle-aged people are the most efficient helpers because they are good at acquiring resources.

Finally, I am intrigued by arguments that humans are a *colonizing* species whose normal rate of population growth in the Paleolithic past might have been quite high. Ceasing reproduction in midlife in order to help younger women reproduce faster might be a good way to spread one's genes in a growing population, with the result that humans combine the potential for rapid population growth with the advantages of a long lifespan. Together, these characteristics helped us to survive the volatile climate of the Pleistocene and to rebound from setbacks and population crashes. Our particular combination of life history and reproductive strategy is one of the most unusual things about humans, and menopause is a key part of it.

PART II

·

History

•

CHAPTER 5

Our Long Stone Age Past

HOW GRANDMOTHERS (MAYBE) CONQUERED THE WORLD

BY ONE DEFINITION today, "history" begins with documents—nowhere before about 5,000 years ago, when the earliest writing was invented. History is therefore mostly a story of societies with economies based on agriculture. Societies based on pastoralism—herding domestic animals—have left fewer written traces, although the shift to agriculture and the shift to pastoralism began at about the same time and had some of the same profound effects.

But history is also the chronological story of a people—in this book, all people, the human species. In part I, we discussed *theories* of how menopause evolved. I have framed this part as a history—a narrative story—of reproduction and non-reproduction, and of population more generally, that unfolds through the three great eras described in the prologue—the Paleolithic, agrarian, and modern eras—with special focus on the middle, agrarian period.

I think it is likely that *Homo sapiens'* evolved life history, including menopause, was the main reason that we outcompeted other humans in the Paleolithic and became populous around the world. Our ability to reproduce quickly when necessary, and in cooperation with others, remains with us today. After the invention of agriculture and settled societies, humans organized peasant economies around our evolved life history and reproductive strategies, including menopause. Most peasant economies placed limits on the extent to which families, and human populations generally, could grow. In this era, menopause's function of limiting reproductive competition and kin competition was especially important, and the energy surplus of post-reproductive women was put to use managing family farms. In the modern period, non-reproduction has become critical to our future, and more people than ever are non-reproductive or post-reproductive; the energies liberated in this way have transformed our world.

So, where to begin? When does "human" history start? In particular, we want to know when our species first became long-lived, because the expansion of our lifespan beyond the chimpanzee's 40 years is probably what created menopause. But this change is not easy to pin down.

As recently as about 20 years ago, scholars still debated the idea that longevity might be a modern phenomenon with no biological basis—a product of the industrialized age and the Demographic Transition. In this view, humans live longer than chimpanzees and women outlive their reproductive lifespans for the same reason that some chimpanzees in zoos live longer than those in the wild.[1] The discipline of "paleodemography" seemed to offer support for this thesis: anthropologists who use skeletons to estimate age structures and mortality schedules for ancient populations have found little survival past reproductive age—prehistoric skeletons older than age 50 are quite scarce. Paleolithic humans, the argument went, had lifespans similar to those of chimpanzees.[2]

But paleodemography based on skeletons can be shockingly unreliable for a number of reasons. The statistical methods used to match skeletal signs with chronological age can introduce systematic errors that underestimate the number of old people, and although new methods can correct for this problem, it is too soon to know how they will change our picture of early human population history. Also, many burial sites are poor sources of information about population structure because they do not represent the whole population—they may reflect different customs for the disposal of dead bodies according to age, sex, and class, and some burial sites reflect very specific circumstances, such as an epidemic or a massacre. More importantly for our purposes, the skeletons of old people and young children do not survive as well as those of young adults because of their lower bone mineral density. In a paper published in 1988, a team led by Philip Walker compared the ages of skeletons recovered from the Mission La Purisima cemetery in California, which had been relocated for a building development, with the Mission's records and found that while the cemetery should have contained mostly infants and old people—as expected from normal human demographic patterns—the skeletons recovered were mostly those of young adults. On the one hand, the soil in this cemetery did not favor good preservation, and later studies have shown that infants' bones can be well preserved in some conditions. On the other hand, this cemetery was not very old, dating to the early nineteenth century, and preservation bias may be most significant at the ancient sites most relevant for understanding human evolution.[3]

Finally, when deducing age structure and mortality patterns from remains, most demographers assume a stationary population with zero growth and zero immigration or out-migration. But if Paleolithic populations were usually growing and occasionally crashing, their remains might produce profiles that underestimate average life expectancies and that, in fact, look a lot like some of the assemblages that paleodemographers describe.[4]

In recent years, we have learned more about the demography of both modern foragers and documented agrarian societies before the Demographic Transition. Although there is variation among these groups, traditional societies show a demographic pattern that is not only different from the modern one, but also very different from that of chimpanzees and from the picture painted by paleodemography. While most readers know that average life expectancy at birth was short before the modern era—usually between 20 and 40—these rates were driven by juvenile mortality, which was typically very high in infancy and dropped through childhood. Juvenile mortality also varied much more than adult mortality, so that mortality hazards tended to converge at later ages. Even in traditional populations with exceptionally high death rates, such as the plantation slaves of early nineteenth-century Trinidad, whose life expectancy at birth was only 17 years and whose population did not reproduce itself, or the Hiwi of Colombia and Venezuela, or the Taiwanese farming population that captured Hamilton's attention and inspired the first glimmerings of the Grandmother Hypothesis, many women lived to post-reproductive ages. Among the Hiwi, for example, of girls who survived to age 15, about half lived to be 45, and of those women, about half also survived another 25 years. Mortality among the Trinidadian slaves was even higher, but Post-Reproductive Representation was still almost one-third (.315). Thus we can see that post-reproductive life is a normal feature of human populations, and that the human body is "designed" to last about 70 years. In their comprehensive study of demography in foraging societies, Michael Gurven and Hillard Kaplan write that "for groups living without access to modern health care, public sanitation, immunizations, or adequate and predictable food supply, it seems that still at least one-fourth of the population is likely to live as grandparents for 15–20 years."[5]

That this pattern characterizes foragers all over the world suggests strongly that human life history is based in biology and evolved before the earliest deep divisions between groups of *Homo sapiens* emerged.[6] The answer to our question of when longer lifespans evolved, then, is not, "They didn't." This development occurred at least 130,000 years ago.

An exception to the typical human life history pattern—one that proves the rule, as it were—is the case of the world's many short-stature (Pygmy) peoples (that is, groups in which average adult male height is less than 155 cm, about 5 feet, 1 inch), of which examples can be found in Africa, Thailand, the Andaman Islands, the Philippines, Papua New Guinea, Brazil, and elsewhere. These are some of the highest-mortality populations in the world; in chapter 4, for example, we saw that Efe infants are much less likely than children in most foraging societies to have surviving grandmothers. Short-stature peoples have lower life expectancies at birth (ranging from 15 to 24 years) and at age 15 (ranging from 20 to 32 years) than other foragers. Some researchers have argued that these populations have evolved a different life history—a shorter growth period, an earlier age at maturity, and shorter lifespans—as an adaptation to unusually high rates of extrinsic mortality; that is, their life history is an evolved response to risky and hostile environments in which it is important to start reproducing earlier. This short-stature adaptation seems to have happened independently many times in different locations.[7]

While important in itself, this argument about Pygmy life history also suggests that if *Homo sapiens* populations in the Paleolithic past experienced high mortality rates—greater than those of (for example) the Efe or Agta in modern times—this would be obvious in their skeletons: they would be smaller, since short stature is a common adaptation to high mortality among foragers today. But Paleolithic *Homo sapiens* was larger than humans of the historical era, as average body size has decreased over the last 50,000 years.[8]

Human longevity, then, is not new—it evolved sometime between our divergence from the lineage of chimpanzees and our dispersal from one another. But within those parameters, how old is it? There are many reasons why getting an answer to this question has been extraordinarily difficult. First, and most obviously, few things besides rocks survive well for millions of years. Therefore, our inferences about the early Paleolithic period are based on very little evidence. Second, we cannot study the factors we are interested in—life history and behavior—directly; at best we can make guesses based on indirect evidence, such as brain size or body size, but none of these methods has proved very reliable. Third, genetic methods of analysis have reached a staggering level of sophistication, but they are new. Genetic studies frequently contradict one another, and it can take decades and many studies to arrive at a consensus. In particular, genetic science is still not very good at telling us *when* something happened; this requires knowing the rate at which a genome, or part of one, has accumulated changes, which in turn depends on identifying an event in the genome's past with a known date with which to calibrate the clock. Because calibration is sometimes only a guess, and in many cases there is controversy about which rate of mutation to use, dates based on genetics can have huge error margins of tens or hundreds of thousands of years, or even millions of years for more remote events.

For a long time, researchers have guessed that *Homo erectus*, a species that evolved about 2 million years ago in Africa and spread to several parts of the Old World as far away as Indonesia, was the first long-lived, slow-growing human. More recently, some scholars have argued that *Homo sapiens*, "modern" humans, were the first to live long lives, and I incline toward this view myself. It is helpful, though, to begin the story of human history with *Homo erectus*, our highly successful progenitor, if we wish to understand what is different about the new, subtly modified version of ourselves that is alive today.

Who was *Homo erectus*? The name means "upright person"—*Homo*, the name of our genus, simply means "person" in Latin (in Greek, an unrelated but similar-sounding word means "same"). The boundaries of speciation are fuzzy: some scientists differentiate the species that lived in Africa from what are probably her direct descendants in Indonesia, China, and the Caucasus, where hominins appeared as early as between 1.9 and 1.8 million years ago; however, most use *Homo erectus* to refer to all of these.[9]

Although most scientists believe that *Homo erectus* evolved in Africa and migrated around the Old World from there, there is controversy on this point. In particular, the very ancient site at Dmanisi, Georgia, in the Caucasus region has confused archaeologists with its remains of smaller-brained hominins that may be a very early form of *Homo erectus*, or may belong to another species, *Homo habilis*, that also lived in Africa 2 million years ago (still another possibility is that the bones recovered at Dmanisi belong to two separate species). Was *Homo erectus* really the first species to migrate from Africa, or did an earlier hominin also disperse? Is it possible that *Homo erectus* evolved somewhere in western Asia from a hominin that had left Africa much earlier, and then migrated back?[10]

Although this last thesis is possible, most researchers believe that *Homo erectus* evolved in Africa from a genus of smaller hominins called australopithecines, and that the Dmanisi humans are a group of *Homo erectus* that migrated early to central Asia. It is also widely accepted that *Homo erectus* was our progenitor, either directly or through one or two intermediary species. (There is debate about how to classify the hominin sometimes called *Homo antecessor*, a population known only from fossils in Spain dating to about 800,000 years ago and some footprints in Norfolk, England, as well as the one called *Homo heidelbergensis*, a name sometimes applied to fossils in Europe and Africa that seem intermediate between *Homo erectus* and *Homo sapiens*.) It is possible, and has been proposed, that a species ancestral to humans evolved from *Homo erectus* in Eurasia, and that one branch migrated back to Africa, where it gave rise to modern humans, while other branches continued to evolve into Neanderthals and other archaic human species in Eurasia. That is, while both *Homo erectus* and anatomically modern humans most likely evolved in Africa, an ancestral species in between might have evolved in Asia.[11]

There is a pretty good case that *Homo erectus* and her descendants, including anatomically modern humans (*Homo sapiens*), are really all the same species, and that *Homo erectus* could be called "archaic *Homo sapiens*." Our DNA shows quite clearly, for example, that some of *Homo erectus'* progeny of supposedly different species—ourselves, Neanderthals, and Denisovans, about whom I will write more later—interbred. Still, to avoid confusion I will use conventional scientific names to distinguish *Homo erectus* and her descendants.

While the details may be muddy, one thing is clear: from millions of years ago until quite recently, there were many different kinds of humans. In early Paleolithic Africa, so many species or subspecies of our genus, *Homo*, have been identified that it is difficult to keep track of them from year to year—it is quite possible that by the time this book is published, new species will have been discovered that were unknown when I wrote it. Early *Homo erectus* shared the continent with *Homo habilis*, so named because this species was thought to have made and used the earliest stone tools found at Olduvai Gorge (although anthropologists today aren't sure which species used the tools); *Homo rudolfensis*, a species that may or may not be distinct from *Homo habilis*; several australopithecines, the smaller-bodied, smaller-brained genus from which *Homo* descended and that continued to thrive until about a million years ago; *Homo naledi*, whose bones have only recently been recovered from a treacherous cave in South Africa; and probably a variety of other hominin species not yet known. Similarly, over the period from 2.5 million years ago to a few hundred thousand years ago, new species and subspecies of humans continued to arise in Africa—and a few outside of Africa—and, occasionally, to disperse.

In 2003, *Homo floresiensis*—a tiny creature about one meter tall sometimes nicknamed "The Hobbit"—was discovered on a remote island in Indonesia. This hominin is believed to be a direct descendant of either

Homo erectus, a species that was well-established in Indonesia from an early date, or an even earlier australopithecine ancestor that, unbeknownst to us until now, migrated out of Africa and got as far as Indonesia. Researchers have dated the most recent remains of "The Hobbit" to about 50,000 years ago.[12] We know that until as late as about 40,000 years ago, or perhaps even more recently, two other species of *Homo* also persisted—Neanderthals and Denisovans, some of whom produced children with modern *Homo sapiens* and are thus among our ancestors.

That is, as recently as 40,000 to 50,000 years ago, there were at least four different types of human on earth; now there is only one. Why? After all, our assortment of wailing infants, overburdened wives, babysitting older sisters, single mothers, devoted dads and unreliable cads, showoff adolescent boys, gay uncles, geezers, and grandmothers seems an unlikely candidate for world domination, but that is what eventually happened. It is possible that our oddball life history and reproductive strategy were the decisive differences between anatomically modern humans and other human species, though this is not certain. That is— and this argument can only be speculative with the knowledge we have today—I am going to suggest that the question of "Why us?" is directly related to the question with which we began this chapter, "When did menopause evolve?" The answer is the same—when menopause and the related suite of human life history traits arose, other hominins disappeared and humans colonized the world *because of those traits*.

Homo erectus is usually described as the first hominin with a large body and a relatively large brain.[13] This species may have been the first human to lack fur and, if so, her skin was probably dark, a necessary protection against ultraviolet light in hot, sunny climates. *Homo erectus* had shorter arms than her ancestors and was the first hominin to be terrible at climbing trees, a skill that australopithecines retained for a long time after they became fully two-legged. Behavior is much harder to guess at, however. Many researchers believe that this species could use a form of language, although she lacked the fine motor control of the thorax necessary for many of the sounds humans make today. Analysis of stone artifacts suggests that most *Homo erectus* were right-handed, a development that is linked to language, which depends on the differentiated function of the brain's two hemispheres. (Brain cases of *Homo erectus* and other hominins also show asymmetry, but on its own this bone evidence is much harder to interpret.[14]) Some believe that the control of fire was essential to the evolution of *Homo erectus*, and this species did eventually use fire.[15]

Because there was apparently less sexual dimorphism in *Homo erectus* than among the australopithecines—males were closer to females in size—some scientists have suggested that pair bonding began with this species, but while sexual dimorphism is linked to reproductive behavior, the correlation is loose. Some researchers now dispute that there was a clear difference in sexual dimorphism between *Homo erectus* and our older ancestors. It is of course very difficult to draw conclusions on this point, because researchers usually do not know the sex of the fossil bones they are measuring.[16]

Homo erectus may have been the first of our ancestors to be fat, compared to most other land mammals.[17] Although stronger and more muscular (more "robust") than modern humans, the species was less muscular than modern apes, suggesting that fat stores replaced some muscle mass. Human adiposity, or fatness, is likely an adaptation to changeable environments and uncertain food supplies, and is especially important for infants, who are more vulnerable when food is scarce, and for lactating women, who must supply all their infants' energy. Like other adaptations that increased humans' flexibility in different conditions, fatness may have been one of the advantages that explains *Homo erectus*' widespread distribution.

This was a highly successful species. While *Homo erectus* may not have been the first hominin to migrate out of Africa—if skeletons at Dmanisi and Flores represent traces of earlier migrations—most believe that she was the first to make a major success of it. *Homo erectus* skeletons on Java—which was periodically connected to Asia by a land bridge when sea levels were low—have been dated to 1.8 million years ago, and the species survived there for a very long time, until perhaps about 150,000 years ago, or longer if *Homo floresiensis* belongs to this species.[18] The oldest human skeletons in Dmanisi are about the same age; remains from China date to at least 1 million years ago, and some may be as old as 1.7 million years. Hominins had reached Barranco León in Andalusia, Spain, by 1.4 million years ago (the discovery of a single tooth at that location, published in 2013,[19] pushed back the date for the earliest hominins in Europe; the next-oldest European fossils are at Atapuerca, Spain, and are estimated to be 1.2 million years old). It is likely that over the course of 1 million years and more, many populations of *Homo erectus* dispersed, mixed, and returned to Africa from central Asia and Europe, but the separate technological traditions of East Asia and Indonesia suggest that they remained isolated from the populations farther west.

Around 1.7 million years ago *Homo erectus*, or possibly another hominin species, invented the tool that archaeologists call the hand axe—a stone core flaked on both sides to produce a symmetrical shape, often a teardrop, with a sharp edge. These artifacts, which litter Paleolithic sites all over Africa, western Asia, and Europe, are the chief markers of the "Acheulian" tradition, named after the town in France where they were first discovered. The hand axe was a multipurpose tool that could also be used to produce the sharp flakes that were *Homo erectus'* preferred tool for butchering animals.

Only tools made from stone or bone normally survive long in the archaeological record—the famous 400,000-year-old wooden spears used by Neanderthals near what is now Schöningen, in Germany, and preserved in mud, are a rare exception. One reason that we do not find hand axes in Indonesia may be that hominins there used bamboo tools, which can be as hard and sharp as stone but are perishable in the long run. So hand axes were probably not the only tools that *Homo erectus* produced, but the Acheulian technological tradition is conservative compared to what came later. To humans today who think a few months is too long to wait for the next version of the iPhone, it is hard to imagine a tool that remained the height of technology for more than *a million years*. But *Homo erectus* continued making hand axes, some of which became very beautiful and symmetrical after 500,000 BCE or so, until they were replaced by the Levallois tradition around 250,000 years ago. Although brain sizes grew dramatically in this period, technology did not change very much.

Well supplied with her hand axe, with a large body that could range widely to search for food, with an ability to eat almost anything, and perhaps also with fire, language, and a layer of fat, *Homo erectus* was well adapted to the shifting environment in which she arose. As the Earth's climate swung between the dry, deep freeze of glacial periods and briefer, warmer, wet interglacials, *Homo erectus* survived on three continents, though she did not try to colonize the very cold regions that some of her descendants would later call home.

In order to answer the questions posed in this chapter, we need to know more about the life history and behavior of this species, and particularly whether *Homo erectus* had the characteristics that emerged as important in the first section of this book—allocare, "stacking" offspring, food sharing, long childhoods, and long lives—and which would imply menopause and grandmothering.[20] Unfortunately, the presence or absence of these characteristics is very difficult to assess in the fossil record. For reasons discussed earlier, gauging longevity directly from skeletons is especially challenging, though some researchers continue to try.[21] Others take more oblique approaches. In mammals, two traits that correlate with longevity and long periods of development—although within a broad range, which is important—are brain size and body size. The timing of tooth eruption, if this can be determined, and of the formation of dental roots and crowns can also potentially yield clues about weaning and other life history events, but none of these methods is very accurate or easy to apply to fossil evidence. Schedules of brain growth have also been used to make very subtle arguments about childhood development.[22]

Homo erectus was the first hominin to be about the same size as anatomically modern humans, with a brain that was, at first, about half to two-thirds the size of *Homo sapiens* brains, though with wide variation in both species. This species had a higher "encephalization quotient" (a value based on the relationship between brain mass and body mass) than its ancestors, continuing a trend; at first this difference was small, but over time brain size increased, and the range of cranial capacity—room inside the skull—for late *Homo erectus* overlaps that of modern humans. Because brain tissue uses up a lot of energy and is thus expensive to maintain, researchers have tried to draw inferences about behavior and life history from cranial capacity. *Homo erectus'* larger brain size is most likely linked to changes in diet—more meat, more tubers, the introduction of cooking, a more varied diet generally, or some combination of these (as discussed in chapter 2). One theory holds that *Homo erectus* could not have grown brains so large without allocare.[23]

Because *Homo erectus* was larger and had a bigger brain than earlier hominins, anthropologists long assumed and argued that this must have been the first species to share our longer lifespans and longer childhoods, and to experience menopause. But consensus is now moving in another direction. Our understanding of the early life course of *Homo erectus* depends on close analysis of only two fossils—a skullcap known as the "Mojokerto child," which belonged to an infant perhaps one year old and dates to 1.8 million years ago, and a juvenile skeleton known as the "Nariokotome boy," who died in Kenya 1.2 million years ago, at about age eight. Based on these fossils, and especially on the teeth of the Nariokotome boy, it appears that *Homo erectus* weaned infants earlier than apes did, and that, like modern humans, this species' life history featured an early childhood phase in which adults provided food for offspring past weaning age. But the Nariokotome boy was within a few years of adult size when he died, suggesting that in its later phases, *Homo erectus'* life history had a different course—one with rapid growth, shorter childhoods, and probably shorter lifespans (the latter two items tend to co-vary, for reasons explained in chapter 1)—one perhaps similar to what we see in chimpanzees today, or perhaps different from those of both modern humans and chimpanzees in some ways.[24]

If it is correct, this picture tends to support Blaffer Hrdy's hypothesis that allocare came first in the evolution of the modern human system of life history and behavior.[25] Allocare was probably important if *Homo erectus* was providing food for children older than weaning age. But because childhoods were shorter, periods of dependency were also shorter, and less allocare was necessary than for modern humans; perhaps for this reason, allocare did not immediately lead to other life history changes. Brain expansion followed, as predicted by Carol P. van Schaik and others who have linked allocare to brain development through better nutrition, but this development happened over a long period of time. Long childhoods and long lifespans came last, perhaps as a result of new behaviors—more extensive food sharing, more allocare—that lowered mortality in older children and adults. Once all these elements were in place, they worked together in new ways. For example, grandmothers, once they became numerous, may have provided more allocare, boosting the reproductive rate and reducing child mortality. Exactly when these last changes took place, however, we still don't know. It is possible that a fully modern human life course only appeared with the evolution of our species, *Homo sapiens*.

Homo sapiens most likely descended from *Homo erectus*, but we are not that species' only cladistic offspring. Our best-known cousins are the Neanderthals, who overlapped with us in time and, to a lesser extent, in place—we coexisted in some locations for thousands of years before Neanderthals disappeared.[26] With this species, we enter an era for which more evidence is available. The caves in which earlier hominins sheltered have collapsed or washed out over time, but more recent cave sites are better preserved. Also, it has been possible to recover and sequence DNA from Neanderthals and other hominins of the Middle Paleolithic, with results that have been surprising, and even shocking. In 2010, scientists published a first draft of the Neanderthal genome, and in 2014 they produced a more complete genome, using DNA extracted from the toe bone of a female that died about 130,000 years ago in Denisova Cave, in the Altai Republic in Siberia. As a result, we now know that Neanderthals and anatomically modern humans interbred to some extent—almost all humans outside Africa have a small amount of Neanderthal ancestry, about 1.5 to 2 percent on average. (Because Neanderthals evolved outside of Africa, their genes are less common in people with only African ancestry.) In 2016, researchers determined that gene flow also went the other way: some of the Altai Neanderthal's genes derived from interbreeding with a population of *Homo sapiens*, or with our immediate ancestors. This "introgression" from Africa happened at least 270,000 years ago.[27]

The Neanderthal from Denisova Cave was the child of either half-siblings, uncle and niece, or double first cousins, suggesting a small breeding population, at least in that area. Researchers determined that Neanderthals had low genetic diversity, and that by 130,000 BCE, when the owner of the bone that yielded the DNA for the 2014 study died, this population had been declining for some time.[28]

Neanderthals are not our progenitors; we diverged from a common ancestor some time between 550,000 and 765,000 years ago, evolving along different, but in some ways parallel, paths. The earliest skeletons that anthropologists classify as Neanderthals are about 230,000 years old, suggesting that this species took shape around the same time as our own. Most or all Neanderthals were extinct by about 30,000 years ago.

Ranging from western Europe to central Asia and surviving long periods of icebox conditions in the last two glacial eras, Neanderthals were well adapted to cold, with bodies slightly shorter and more robust than the tropical anatomy of *Homo sapiens*. But Neanderthals also thrived in warmer areas around the Mediterranean, in southern Spain, Italy, France, and the Levant. Speaking generally, Neanderthals were prodigious hunters of large game, especially deer and horses, but also woolly mammoth and woolly rhinoceros. Chemical isotope studies of their bones suggest that Neanderthals in all climates ate mostly meat, and some researchers have suggested that the entire population hunted, including women and children who were old enough.[29] The skeleton of one Neanderthal child who died in France around age 11 shows a healed broken jaw, wear on the front teeth from using them as tools, and wounds to the head that might have killed him.[30] Hunting as most Neanderthals probably practiced it, at close quarters with spears, was dangerous, and some researchers read the record of this hard lifestyle in Neanderthal skeletons, comparing them to those of rodeo riders in their number of injuries and fractures.

This portrait of Neanderthals as burly hunters reliant on a single food source has been used to explain their eventual disappearance: they failed to adapt when modern humans arrived in Europe about 45,000 years ago and began competing with them for large game. Some researchers, however, dispute this image of Neanderthals, arguing that they do not show more injuries than anatomically modern humans, that they adapted their diet to their environment, and that they hunted a range of small game as well as the large herbivores that survive so well in the archaeological record. Evidence of plant foods can be hard to recover but does survive from several Neanderthal sites, and dental plaque from one Neanderthal in northern Spain preserves traces of many plant foods but no meat at all. There is evidence that some Neanderthals hunted small, fast animals, including rabbits, birds, and even (though this is more dubious) fish. Some birds seem to have been captured mainly for their feathers. There is other evidence that Neanderthals used ornamentation or jewelry—ochre, which has other uses but is also a decorative bright orange pigment, is common at Neanderthal sites, and two sites in Spain have produced pierced and pigmented shells some 50,000 years old. The Neanderthal "encephalization quotient" was slightly higher than that of anatomically modern humans.[31]

Neanderthals are our best-known cousins, but they are not the only other descendants of *Homo erectus* who were alive as late as 40,000 years ago. The discovery of another recent human species, the Denisovans, is one of the most surprising developments of the last decade.[32] In 2008, archaeologists recovered the finger bone of a female child who had died in Denisova Cave—the same cave that produced the Neanderthal toe bone from which scientists extracted a complete genome. The finger bone came from a layer in the cave dating to about 50,000 to 30,000 years ago, and scientists expected to find that it belonged either to an anatomically modern human or to a Neanderthal, as both species were present in the region around that time. But the genome, once sequenced in 2010 and more completely in 2012, showed that the bone came from a hominin that was neither human nor Neanderthal but belonged to a different, unknown hominin more closely related to Neanderthals than to modern humans. Researchers named this population, known only from their genome, the "Denisovans" and guessed that they diverged from *Homo sapiens* about 800,000 years ago. They also found that modern people indigenous to Oceania (Australia and the South Pacific) have inherited some of their DNA—about 5 percent on average—from Denisovans. Some researchers believe that the Denisovans crossed Wallace's Line, a barrier of open water in the South Pacific that has been impenetrable to most mammal species other than rodents. When the first *Homo sapiens* arrived in Australia, perhaps they found the Denisovans already there.[33]

The Denisovan girl's DNA was so well preserved that researchers were able to sequence the genomes of both her parents. Based on the differences between them, scientists concluded that genetic diversity in this species was low, and that its population, like the Neanderthal population of the same era, was relatively small and in decline. In contrast to the Neanderthals, some of whom were probably pale-skinned and red-haired, Denisovans had dark skin, hair, and eyes.

Recently, researchers were shocked to discover that when they tested mitochondrial DNA from skeletons found in a cave at Sima de los Huesos, in northern Spain—very far, that is, from Denisova Cave—they found that the skeletons there were more closely related to Denisovans than to Neanderthals or to anatomically modern humans. The Sima de los Huesos skeletons are more than 300,000 years old, and researchers had previously classified them as *Homo heidelbergensis*, a species supposedly intermediate between *Homo erectus* and the more modern lineages of Neanderthals and *Homo sapiens*.[34] To make things even more complicated, the *nuclear* DNA of the Sima de los Huesos humans is closer to that of Neanderthals than to that of Denisovans. Some geneticists have explained this as the result of a complete population-wide replacement of Neanderthal female lineages, with their mitochondrial DNA, through interbreeding with another group, ancestral to modern humans, that migrated from Africa perhaps 300,000 years ago. That is, a group ancestral to modern humans helped shape the Neanderthal lineage as it diverged from the Denisovan line.[35]

Finally, it seems that Denisovans also interbred with a third, unknown hominin species different from both Neanderthals and *Homo sapiens*—perhaps late-surviving *Homo erectus* or another hominin we have not discovered yet.[36]

Our species, *Homo sapiens*, appeared last of all the hominins. For a long time, there was debate between scientists who believed that *Homo sapiens* evolved in Africa and spread around the world, displacing all other human populations, and those who argued that modern humans evolved separately in different regions from ancestral populations of

Homo erectus dispersed throughout Africa, Asia, and Europe. The first group, those who supported the "Out of Africa" Hypothesis, gained a decisive victory in 1987 when *Nature* published an article by Rebecca Cann and colleagues that compared mitochondrial DNA from 147 humans all over the world to trace their most recent common female ancestor ("mitochondrial Eve"). Mitochondria are organisms living in our cells that we inherit from our mothers, and their DNA is often used to reconstruct ancestral relationships because, unlike the DNA in our cell nuclei, it does not "recombine"—mothers pass identical copies to their children, and any changes are due to mutation only. Mitochondrial DNA mutates faster than nuclear DNA, and because there is no obvious way for natural selection to act on it, researchers assume that it changes at a steady rate and thus can be used as a "clock" to trace population history through the female line. It is also much more abundant, and thus easier to find and isolate, than nuclear DNA. Cann and her colleagues argued that mitochondrial Eve probably lived in sub-Saharan Africa about 200,000 years ago. (Other women were alive at that time, but their lineages died out if they produced only male descendants or no descendants.)

Genetic studies of more recent decades have confirmed Cann's picture, dating mitochondrial Eve to about 160,000 to 200,000 years ago. (Most recently, however, some genetic scientists have argued for earlier dates for the origins of *Homo sapiens*—perhaps 300,000 years ago.)[37] Studies of the Y chromosome, which also does not recombine, have traced male lineage with more variable results. Most show a common ancestor more recent than mitochondrial Eve, but with the rise of commercially available testing for genetic ancestry, researchers have discovered a rare male lineage deriving from western Cameroon that diverged from other lineages more than 300,000 years ago—before the emergence of modern humans! Fossils discovered recently at Iwo Eleru, Nigeria, show evidence of some "archaic" features (similar to those of *Homo erectus* and Neanderthals) as recently as 13,000 years ago, and may have been a late-surviving remnant of a very ancient people; it is possible that this population mixed with modern humans to produce the mysterious male lineage from Cameroon.[38] It is exciting to imagine what new discoveries await us as more remains are uncovered and techniques for genetic testing improve.

In any case, Cann's 1987 research clinched the view already held by many anthropologists that anatomically modern humans evolved recently in Africa and migrated from there to replace all previous human populations—the Out of Africa Hypothesis. We know now that older hominins were not displaced completely; instead, some of them interbred with anatomically modern humans (so that the theory is now sometimes called the "Mostly Out of Africa" Hypothesis). Nevertheless, our ancestry is overwhelmingly that of a population that evolved recently in Africa, the species called *Homo sapiens*.

Scientists agree that the oldest anatomically modern human skeleton found so far dates to about 195,000 years ago and comes from Omo Kibish in Ethiopia.[39] While life history traits are, as always, very difficult to reconstruct from ancient remains, most researchers accept the conclusion—based on a study of the teeth of a seven-year-old child who died about 160,000 years ago in Morocco—that *Homo sapiens* of this early era had long, modern childhoods and probably modern lifespans.[40]

The people who separated earliest from other living populations are speakers of Khoi and San languages—the Khoisan peoples—of southern Africa, many of whom are still foragers today (the !Kung are in this category), and who appear to have diverged from other groups as early as 130,000 years ago. Human genetic diversity is greatest in sub-Saharan Africa, probably because our history is longer and deeper there than anywhere else.[41] All non-African human populations share a common ancestry in the recent past, about 80,000 to 50,000 years ago. Some time within that period, a group of modern humans migrated from Africa, traversed southern Asia, and crossed the sea to the ancient continent of Sahul (now Australia, Tasmania, and Papua New Guinea), picking up Neanderthal and Denisovan genes along the way; branches reached Europe somewhat later, and the Americas last. This was not the first migration of anatomically modern humans out of Africa. Archaeological evidence shows the presence of *Homo sapiens* in the Levant, India, and China as early as 100,000 to 120,000 years ago, and Papuans today trace about 2 percent of their DNA to an older population of modern humans that left Africa long before the main dispersal 80,000 to 50,000 years ago. But for the most part, earlier populations of humans around Eurasia sputtered out, returned to Africa, or were replaced by humans from the last migration out of Africa.[42]

Around the time of this last migration—around 50,000 years ago— technology and culture became much more complex and also more varied among groups. By comparison with previous human history, technology changed very fast, though by modern standards the full development of the "Upper Paleolithic" (as it is called in Europe) or "Late Stone Age" (in Africa) took a long time, with a transition period of tens of thousands of years. New "microlith" technology—very fine points, blades, and other stone tools—became common, as did tools made of other materials and figurines, jewelry, and musical instruments— evidence of what is sometimes called symbolic thought.

It is possible that the emergence of new technologies reflected a new level of specialization in labor—perhaps the division of labor by sex that is universal among foragers today began in this period. Many of the new tools invented in this era were for food preparation, sewing, and catching fish and small game, tasks often performed by women, reminding us that while "technology" is a male domain in some modern societies, the same was not necessarily true of the Paleolithic. (Also, among chimpanzees, females are more likely than males to use tools to forage for termites and crack nuts.)[43] Weapons for hunting advanced too, especially with the invention of the bow and the spear-thrower, a tool that greatly improved power compared to throwing a spear by hand.

Neanderthals disappear from the archaeological record within a few thousand years of the last dispersal of *Homo sapiens* from Africa, around the same time that technology became more complex. Of the several types of humans alive in 50,000 BCE, only one was still around a short time later. The last wave of anatomically modern humans migrating out of Africa replaced them all.

The most common explanation for this striking change in the fossil record has been that *Homo sapiens* had some cognitive advantage over Neanderthals and other human species contemporary with them that explains their more sophisticated technological development and, through that, their dominance. This idea that Neanderthals were just not very smart is intuitively appealing, but it has been hard to prove. Neanderthals did not share in the full-scale technological revolution of the Upper Paleolithic, but why not? For a long time, anatomically modern humans and Neanderthals used very similar technology—stone points, scrapers, and notched "denticulate" saws, often made by the Levallois technique of knapping flakes from a stone core, and sometimes hafted to (probably) wooden handles. In the Levant, where Neanderthals and humans coexisted for some time about 100,000 years ago, the technologies used by the two different populations are indistinguishable. Similarly, the Châtelperronian culture of the Neanderthals of southwestern France is difficult to distinguish from that of the anatomically modern humans who also lived in the area at the same time, about 45,000 to 40,000 years ago. Some Neanderthals used late Paleolithic artifacts like jewelry, painted their bodies or other objects, and buried the dead—that is, they showed some of the same evidence of symbolic thought as anatomically modern humans. They may even have produced cave paintings found at several sites in Spain, some of which are much older than any evidence for anatomically modern humans in the area.[44] So the case that modern humans replaced Neanderthals because they were naturally smarter is not so strong.

The search for a natural advantage of *Homo sapiens* over Neanderthals in intelligence has also taken other forms. Some researchers argue that the different shape of Neanderthal brains, which were longer and flatter than the globular brains of anatomically modern humans,[45] gave *Homo sapiens* an advantage, but no one disputes that Neanderthals' brains were at least as large as those of modern humans. Efforts to find genetic evidence of cognitive differences have not panned out. Theories inspired by the discovery of the FOXP2 gene, which is necessary for complex language and seems to have evolved recently, were deflated when Neanderthals proved to have the same gene. A study of 51 genes related to cognition that differ among humans and other primates found no obvious differences between *Homo sapiens*, Neanderthals, and Denisovans.[46]

There is also the problem that anatomically modern humans, and their brains, evolved more than 100,000 years before the technological revolution. Some researchers have speculated that a new intelligence gene must have been introduced around 50,000 years ago to explain the changes that occurred shortly afterward. In this model, such a mutation would have had to sweep through populations of *Homo sapiens* that had diverged up to 80,000 years previously and had remained mostly isolated from one another. This is theoretically possible, but genetic analysis of the human population alive today finds no evidence of such a "selective sweep."[47]

Although the "better brains" model still has advocates, it seems a little bit silly to me. Imagine for a moment how human history since the Stone Age will appear to future archaeologists if they are unable to date objects more precisely than within a window of a few thousand years. Humans were a thinly scattered species herding goats and making coarse clay pots, and then suddenly—overnight!—a sprawling population of 7 billion, wielding smartphones, flying drones, cruising around the planet in airplanes, filling the seas with garbage, and transforming the chemical composition of the atmosphere. Archaeologists living thousands or millions of years in the future might suspect that aliens invaded from another planet, or that new biological changes to the human brain caused the sudden appearance of very advanced technologies. We know that the answer is different—humans have been biologically *capable* of producing drones and smartphones for a long time, and it is social, cultural, economic, and demographic factors that explain why technology has changed so rapidly in the last few centuries. But these processes will be hard to trace unless future archaeologists can reconstruct human history on a fine scale.

Our ability to do this for the Paleolithic is limited, but the more evidence accumulates, the more complicated the story becomes. Technologies characteristic of the late Paleolithic revolution appear much earlier in South Africa and parts of northern Africa; they either sputtered out or were the seeds of later change. Similarly, new technologies like beads and bone tools arose and sometimes disappeared in Europe and elsewhere, both before and after the arrival of anatomically modern humans. Some scientists avoid talking about a technological "revolu-tion" in the Upper Paleolithic at all.[48]

More persuasive explanations link the cultural changes of the late Paleolithic to other factors, such as population density and migration among groups. Perhaps *Homo sapiens'* decisive advantage was not intel-lectual, but demographic—that is, there were more of them.[49] Nean-derthal and Denisovan populations seem never to have been very large.[50] A larger population could have produced more new ideas, more opportunities to exchange ideas, and more potential experts to master and teach technologies. Also, a dense population can generate pressure to innovate if it overtaxes its food supply—an important theory often applied to the agrarian period, discussed further in the next chapter.

Neanderthals and Denisovans did not die out completely—they in-terbred with modern humans, and many researchers believe that we acquired valuable genes in this way, including some that strengthened immunity to new diseases not encountered by our ancestors in Africa.[51] If modern humans partly displaced other populations by outcompeting them for habitat and partly "swamped" them genetically with their larger numbers, this would explain the result that we see today.

Several scientists have argued that population size explains the rise to dominance of anatomically modern humans, and they make a good case, but as far as I know, none of them has linked demographic advan-tage to life history. This, however, seems an obvious point to make—if modern human life history is unique to us, at least in its full package, it would explain a lot. Modern humans are a fast-reproducing, colonizing sort of animal in a way that Neanderthals perhaps were not. Lower mor-tality and higher fertility meant that *Homo sapiens* could rebound faster after crises than other humans, and colonize territory quickly in favor-able circumstances. Longer lives for both men and women, and longer post-reproductive lives for women, meant that some people had more time to master, advance, and teach culture and technology, creating a higher level of knowledge and experience in the population. That is, life history, and not better brains—or rather, life history (unique to us) in combination with big brains (shared with other humans)—might explain our success.

That our life history pattern is unique to anatomically modern humans and was not shared by our closest relatives is, however, a difficult thesis to prove. Even if *Homo erectus* did not fully share the modern human life course, it might have evolved in an intermediary species or independently in Neanderthals (or in Denisovans, though we have no way of investigating this). Did it? Many researchers have tried to reconstruct Neanderthal life history based on their bones, skulls, and teeth, but there is no consensus on the question of whether they shared our long childhoods, longer lives, lower extrinsic mortality, low interbirth intervals, and post-reproductive life stage, or whether they shared the behaviors that probably accompanied or caused these changes—modern patterns of food sharing, allocare, and stacking offspring.[52] One innovative study that compared ratios of older to younger adults in the fossil record based on the amount of wear on their teeth concluded that longer lifespans evolved only with anatomically modern humans, but the results were anomalous in ways that suggest a problem with the methodology.[53] Researchers have also tried to describe early phases of the Neanderthal life cycle based on studies of skulls and teeth, but again with only equivocal success, as the methods used are very uncertain. Tentatively, one could conclude based on current research that Neanderthals, like modern humans, weaned children earlier than apes, and that adults likely provisioned children after weaning, as their brains continued to grow for several years after weaning. But at least some studies of Neanderthal bones have suggested that this species grew faster than *Homo sapiens* in adolescence, and reached maturity earlier (around 13 or 14). This evidence is not conclusive, but it does allow for the possibility that only *Homo sapiens* possessed a fully modern life history.

Why such a long gap, then, between the evolution of anatomically modern humans—with their modern life histories—and the boom of the Late Stone Age? Perhaps rapid change had to wait for the right combination of factors—not only more people, but also the stimuli for new ideas and behaviors. There are several reasons why major technological revolutions might occur, or appear to occur, both rapidly and rarely. When a process is logarithmic (increasing by multiples of itself) rather than linear (increasing by a fixed amount over time)—both population growth and information transfer can have this quality—changes begin slowly and then build speed; to archaeologists examining the record, these changes will be invisible until a strong signal appears, sometimes seemingly out of nowhere. (In Roman history, the spread of Christianity, whose signature in inscriptions expands explosively in the later fourth century, is such an example.) When two factors are working together—here, population density and technological innovation— changes can arise and then sputter out if the conditions necessary to sustain them are not met, as when population size or density falls below a certain threshold. (The most famous case of this effect occurred on the island of Tasmania, which was settled by migrants from Australia and then cut off from the mainland by the rising sea at the end of the last glaciation. At the time of contact with Europeans 10,000 years later, the population of Tasmania was only a few thousand, and its people made only 24 kinds of tools, compared to the hundreds of tools used by foragers on the mainland.[54]) Conversely, positive feedback loops, such as might occur if technological advances boost population growth, can escalate the pace of change very rapidly. As modern human populations probably fluctuated in boom-or-bust fashion, it makes sense that cultural change might suffer frequent setbacks for long periods of time. The potential for rapid change that eventually sweeps the entire world population is always there, but it comes to fruition only rarely.

The role of life history in the eventual domination of anatomically modern humans is not certain—I emphasize it here because it is the explanation I find most persuasive and because it fits with the thesis of this book, but not all (and probably not most) anthropologists would agree with it. Some continue to argue that "better brains" explain *Homo sapiens'* rise, although this position is starting to sound old-fashioned. Some argue that neither brains nor life histories were very different among anatomically modern humans and other humans, and that factors like climate change, local ecology, and migration to new environments caused the changes of the Upper Paleolithic. The burden of explaining why these external factors applied differently to *Homo sapiens* than to other humans is difficult to overcome, however, and I do not

find the explanations offered very convincing—in a sample of three recent articles, they boil down to an old and problematic assertion that
Neanderthals relied too much on big-game hunting, a circular argument
that their population was too small, and a confession that we just don't
know.[55] This debate is likely to continue for some time.

The most momentous change of the late Paleolithic, besides the extinction or near-extinction of other human species, was that technology
and culture became new, fast, and effective ways of adapting to different
environments and climate changes. By the end of the Paleolithic period,
Homo sapiens lived on every continent but Antarctica. They could survive everywhere, even in eastern Europe and Siberia in the depths of the
last glacial maximum about 26,000 to 16,000 years ago. They colonized
the Americas last, and probably very quickly, within a few thousand or
even a few hundred years.[56]

Everywhere outside Africa, and particularly in Australia and the
Americas, humans brought with them a wave of destruction. About
two-thirds of large mammals in the Americas died out shortly after the
migration south from Beringia (although whether humans were the sole
cause of their disappearance is debated, and current consensus blames
a combination of humans and climate change).[57] It was at this point—
when humans had colonized all six habitable continents, developed
advanced technologies, and extinguished much of their preferred
prey—that the last major sudden warming event of the Earth's climate
history occurred, at the end of the period called the Younger Dryas,
about 11,000 years ago. The strategies that had worked so well in the
past—the great migrations across steppe, tundra, and forest, the evidence of which is written in our genetic code and revealed in scattered
remains buried deep in the ground—gave way to new methods of
adaptation.

•

CHAPTER 6

The Age of Farmers

PATRIARCHY, PROPERTY, AND
FERTILITY CONTROL

OUR CURRENT WARM, interglacial period, the Holocene, began about
11,000 to 12,000 years ago, or 10,000–9000 BCE in the system I will be
using from now on. All of human recorded history, plus all of the story
of our species' development after we mostly abandoned the foraging
way of life, has taken place in this period—that is, we remember only a
very small fraction of our history as a species, all of which has taken
place in a period of warm and *relatively* stable climatic conditions, although some climate fluctuations have occurred in this era (figure 4).

In the first four chapters of this book, I presented the evidence that
menopause evolved in our Paleolithic past as part of a flexible reproductive strategy that allowed humans to multiply quickly in good conditions while still investing a great deal in their offspring. In chapter 5,
I suggested that this combination of qualities may explain our dominance
over other hominin species, some of whom were likely as smart as we
are. But menopause also evolved as a way of increasing the ratio of
adults to children and the number of producers to consumers, and of
reducing competition among one's offspring and other kin—factors
discussed in the parts of chapter 2 that address kin competition. That is,
menopause is part of a highly *flexible* reproductive strategy that works
both in good times, when fast reproduction is an advantage, and in hard
times, when resources are tight and non-reproduction is important. I
think that both of these advantages of menopause—fast reproduction
and reduced competition—were important in the agrarian era of our
history that followed the Paleolithic, *but especially the second.*

Menopause was part of a larger picture in which fertility control was crucial to survival and prosperity; menopause and middle-aged women were foundational to the peasant economies that developed in this era. In order to see how this worked, we first need to understand social structures in the agrarian period and how they were different from those of foragers. Peasant societies were more patriarchal and more unequal than foraging societies, and they centered on the family rather than on the co-residential band.

Humans began to raise domesticated plants and animals in several different parts of the world soon after the rapid warming with which the Holocene began. We are not really sure why this happened. A simple explanation is that populations boomed when temperatures warmed, while the cultural changes of the late Paleolithic gave humans more advantages. As first-choice foods were used up and there were fewer empty places to move into, "intensification"—increasing the productivity of the land with extra labor or new technologies, in this case domestication—became a better strategy than moving. This kind of strategy was first described in the context of agricultural societies by the Danish economist Ester Boserup, in her 1965 book *The Conditions of Agricultural Growth*, as a necessary complication and correction to Malthus's theory that human populations must either limit their own growth or expand until checked by malnutrition, disease, and war. Intensification is a third alternative. That the domestication of plants and animals was an early example of intensification is a classic and plausible theory of the Neolithic revolution, best known from Mark Nathan Cohen's 1977 book *The Food Crisis in Prehistory*. I incline to something like this view myself, although many archaeologists and anthropologists today favor a different scenario that will be discussed later. Cohen himself has modified his ideas to concede that climate change in the Holocene—which in his view had the immediate effect of reducing resources for late Paleolithic humans—may have been more important than population growth in forcing humans to adopt agriculture.[1]

At first, most people who farmed practiced horticulture—"slash-and-burn" cultivation with digging sticks or hoes, and without plows or animal traction—and some societies in Africa, South America, and Southeast Asia still do so today. This form of cultivation is more mobile than intensive, technology-dependent methods. It depletes the soil quickly and requires long fallow periods of 20 years or more, before the forest can be cleared again by fire for cultivation. In places where the population grew too dense to be supported by slash-and-burn horticulture, where forest had turned to grassland because it could not be left long enough to regrow, people began to use plows and animal power to cultivate permanent plots. These plots had to be cleared of tree stumps and other obstacles at great cost in labor. In some places, through further intensifications, people began planting the same land every year with no fallow period, or even growing two or three irrigated crops per year.[2]

As land became a limiting factor in production, property became important; families controlled and inherited specific plots of land. People began making pottery vessels to store goods, and families with more land could accumulate surpluses. In Mesopotamia, a system of clay tokens was invented to track produce for storage and trade, which eventually evolved into the cuneiform writing system.[3] Over the very long term—starting around 3000 BCE at the earliest—states arose, collecting surplus produce as taxes and redistributing it to rulers, administrators, temples, workers on building projects, craftsmen, and armies.[4] This story is hugely oversimplified, of course. But the changes that followed the turn to agriculture had profound effects on human social structures, with some patterns emerging across cultures. These patterns were so widespread and lasted so long—originating before writing was invented to record them—that many researchers consider them the normal or natural human condition. It is better, however, to see them as typical of agrarian societies but not necessarily of foragers, of modern societies, or of humans in general.

The agrarian era was the era of the family. We have seen that among foraging peoples, monogamous marriage plays much less of a role than it typically does among agrarian peoples—the fluid, cooperative, residential band was more important than the nuclear family, which rose to central significance in the agrarian period. The agrarian era was also the era of inequality, and the two institutions went together.[5] Once wealth could be accumulated, humans, who have always invested intensely in their offspring, wanted their children to inherit it. Some people then started out with an advantage—not only the material advantage of inherited resources, but also social advantage in a world increasingly stratified into classes, particularly free and unfree classes. Societies thus became much more unequal.

Inequality arose where there were concentrated resources that could be defended, inherited, and stored. Some foraging societies met this condition and had more inequality than those without them—for example, those foragers whose economies were based on the salmon that spawn in rivers of the Pacific Northwest—but the potential for inequality was much higher in farming economies.

In horticultural societies, inequality was typically not very great compared to what developed with the advent of intensive farming. While cultivating abundant forestland in long rotation, ethnic groups claimed rights to certain territories but still shared a great deal among themselves. Private property was not so important; land was generally not bought and sold, as each family could cultivate only a small area; and families did not accumulate much wealth. Unlike most foragers, however, sedentary or semisedentary horticulturalists and small-scale agricultural societies can be very violent and warlike, and conflicts with neighboring peoples over resources or women are common.

The huge, mysterious Neolithic town of Çatalhöyük, in Turkey, 9,500 years old and housing a population of up to 8,000 at its height, is an unusually large example of an egalitarian horticultural settlement. Houses agglomerated together, sharing party walls (people entered their homes through the roof), and changed little over time—they were rebuilt in the same way every 60 years or so. There is some evidence that moveable goods like cooking supplies belonged to individual houses—that is, they were private property—but the town's inhabitants shared millstones, which were large and expensive to produce. Most of the millstones that have been discovered were deliberately broken, suggesting that the people of Çatalhöyük took steps to prevent them from being inherited and increasing inequality. They buried the bones of the dead under the floors of the houses, and analysis shows that the remains were not especially related to each other, strengthening the case against the inheritance of property in this village and suggesting that the family was not so central to social organization as it became later.[6]

But as wealth, especially land and animals, began to be inherited, the family became the main unit of production, a nearly universal pattern in farming societies. Thus it was not the domestication of plants in itself that led to the most important social changes, but the kind of intensive cultivation of permanent plots that is achieved with tools and animal labor.

MARRIAGE AND FAMILY IN AGRARIAN SOCIETIES

More than 100 years ago, in his book *The Origin of the Family, Private Property, and the State*, Friedrich Engels, the famous collaborator of Karl Marx, argued that monogamous marriage and the male-headed family arose with the invention of property and displaced systems in which women had more power.[7] It has been fashionable for many decades to reject this theory, although we know little about what social organization might have been like before recorded history began. (In countries dominated by Marxist ideology, however, Engels's theory has remained standard.) As anthropologists are learning more about foraging and horticultural societies, Engels's idea that the rise of the nuclear family and of strict monogamy for women is linked to the institution of property has begun to seem more plausible, and I for one think that he was right. If we substitute the words "foraging," "agricultural," and "modern" for his Victorian terminology (he identified "savage," "barbarous," and "civilized" stages of development), and if we see in his concept of group marriage the residential band that is the basic social unit of foraging societies (a move that I acknowledge does violence to the history of sexuality central to his ideas), much of Engels's theory holds up quite well. I am not sure we can demonstrate, as he tried to do, following the anthropologist Lewis Morgan, that matriliny and matriarchy were typical of foraging societies, but the link between patriarchy and agriculture is convincing.

Historically, in most agrarian societies, men have owned and controlled all or most property and passed it preferentially to their male children. This generalization holds true even for most societies in which inheritance is reckoned through the female line (that is, in "matrilineal" societies).[8] There are several reasons why this preference for male heirs might have emerged. Once property existed, it had to be defended. The agrarian economy created opportunities for big payoffs from raiding, rustling, kidnapping people for slavery or sex, and other types of aggression. As the tasks of aggression and defense fell to men—who used weapons anyway for hunting and who now found a new use, in banditry, feuding, and warfare, for the energy surplus discussed in part I of this book—male alliances became more important. Men needed to be near their brothers and other kin, explaining why, in most agricultural societies, women move away at marriage and become part of their husbands' families, and not vice versa.[9]

Under these circumstances, resources invested in daughters eventually leave the family, a constant complaint in those societies that bestow dowries on daughters. The priest of the medieval village of Montaillou once made a passionate defense of incestuous marriage for this reason— if only his brothers could have married his sisters, his family would have avoided ruin! Almost all societies have shunned brother–sister marriage, most likely because of humans' natural aversion to marrying their siblings—the so-called "Westermarck effect"—but uncle–niece marriage and marriage between half-siblings are not uncommonly allowed, and cousin marriage is preferred in many cultures. In Roman Egypt, brother–sister marriage was common, probably because land was scarce and inheritance practices divided property among sons and daughters almost equally.[10] That daughters leave the family and contribute labor and reproduction only to their husbands' families is the main reason for female infanticide in some societies, discussed later. Thus the constraints of property and inheritance in the agrarian economy have driven humans to some extreme behavior. In any case, because women in agrarian societies leave the family, property settled on them is typically moveable property, such as cash or goods, and not productive property (land), which has often limited women's ability to live independently.

Another reason why agriculturalists might prefer sons over daughters as heirs is that, among humans, males compete for mates more than females do—there is normally more differentiation in male reproductive success than in female reproductive success.[11] Males therefore benefit more from the advantages of inherited resources. Finally, when societies become organized around inherited wealth and men are heavily invested in their own children, they have a compelling interest in making sure that they are the biological fathers of their wives' offspring. It becomes more important for men to control women, and rendering them dependent by controlling most property is one way to do that. For these reasons, it is not surprising that marriage is a much more formal, rigid, and legally complex concept in cultures in which property is accumulated and inherited, or that men in these cultures have typically controlled most property.

The married couple and their children is the fundamental economic and social unit in agrarian societies. Strict monogamy is important for women and less so for men; most agriculturalists have allowed some men to have more than one wife or condoned sex outside marriage for men. Nevertheless, in agrarian systems most men have a strong motivation to limit the number of their legitimate heirs, so that even where polygyny (marrying more than one wife) is allowed, only the rich usually practice it.[12] Other patterns typical of the agrarian period are virilocal marriage (in which brides move in with their husbands' families), patriliny (the practice of reckoning kinship and inheritance through the male line), male inheritance, and male control of property, especially the productive property on which subsistence depends. It is this system that I am, for convenience, calling "patriarchy."

Because the scope of this book is large, I will mostly talk about dominant patterns like patriarchy. I note here a few of the many exceptions and limitations to these patterns. There are agricultural societies in which most property is inherited by women and marriage is uxorilocal (men move in with their wives' families; "uxor" means "wife" in Latin), such as the Khasi of northeast India and Bangladesh.[13] Class, and not only gender, is a source of inequality in agrarian societies, and the inequalities of class cut across those of gender and complicate them. Where rich classes exploit the poor, or in imperialist societies, where one ethnic population oppresses another, elite women participate in those systems even as they are oppressed within their own group.

Finally, some societies are more patriarchal than others. In general, women's well-being and status relative to men are higher if they can inherit or control at least some property (if, for example, a woman's dowry or bride-price functions more as her property than as the property of her kin, husband, or husband's kin), if they have opportunities to support themselves and their children independently of men (though usually at a substantial disadvantage compared to men), if they retain close ties with their natal families after marriage or continue to live with them, and if the chastity of women is less of an obsession and they are subjected to fewer strict controls. Other factors, such as polygyny compared to monogamy or bride-price compared to dowry, can be relevant to women's well-being but do not always have an effect or do not have the same effects everywhere.

While in this chapter I have adopted the view of many scholars that the "conjugal unit" of husband and wife was the basis of the family everywhere in the agrarian period, this generalization works less well for much of sub-Saharan Africa. Compared to Eurasian societies, polygyny has been more frequent in this region, and wives have been more independent from their husbands. In this system the basic economic unit is the mother and her children, and wives provide for themselves and their children by trading their farm's produce or engaging in other types of production. Agriculture is less intensive, and women do a larger share of the agricultural labor and produce more than they consume, making the pattern of polygyny and "matrifocal" families centered on the mother more advantageous. Thus European colonial and "development" projects have often foundered on assumptions about the family that do not hold true in Africa. However, in sub-Saharan Africa, as in Asia or Europe, most real property has been held and inherited by men, and women have had access to land mainly through their relationships to men.[14]

Despite these limitations, most scholars of the family take the conjugal unit to be the foundation of the family and classify families according to how many conjugal units they contain—"nuclear" families contain one couple and their children; "joint" families contain more than one married couple; and "extended" families contain some members related to the married couple, but not another couple. They also distinguish households according to the rules by which they form: in a virilocal marriage, as described earlier, a bride moves in with her husband's parents, and married brothers often share a household with their parents and with each other; in an uxorilocal marriage, the husband moves in with the wife's family; and in a "neolocal" pattern, couples form their own households when they marry, and the nuclear family is typical.

Since the early 1970s, research on the family has emphasized that households in agrarian societies were usually small, around five individuals on average, no matter what rules of household formation they followed; because mortality was high, most families were "nuclear" even in virilocal societies, with the difference that most virilocal families passed through phases as joint families, while neolocal families did not.[15] Although it is important to correct widespread assumptions of a primeval era of vast extended families, these generalizations about small family size may oversimplify: in some regions of traditional China, for example, most people lived in joint families, and even ordinary peasant households could become large.[16]

Wherever a family based on married couples is the main unit of production and inherited wealth is important, this raises the stakes of paternity—men, and their parents and families, want to be sure that they are the fathers of the offspring who consume their resources and will inherit their wealth. The more strongly patriarchal a society is, the more a family has to gain by guaranteeing its daughters' chastity before and after marriage. Families that can afford to keep daughters inside, doing indoor domestic labor only, do this as a way of securing an advantageous match; similarly, men who can keep their wives at home do so in order to control them. Values exalting purity, virginity, modesty, and chastity for women, as well as their signs or symbols, such as pale skin, veiling, bound feet, and extreme youth, are typical of agrarian societies. An image of purity is more obtainable for families with more resources, and values about chastity are usually linked to class—elites perceive poor people as less virtuous. On the other hand, the more repressive practices of elites can expand throughout the population when families aspiring to upward mobility imitate the higher classes, as happened with the practice of foot-binding in China.

Historically, one function of marriage, possibly the main function,[17] was to identify legitimate heirs to property—it served both to guarantee the chastity of the wife by fencing her around with restrictions on contact with other men, and to distinguish those offspring who would inherit the father's property from those who would have no claim. Children within marriage are a major investment for men and represent a reproductive strategy of "quality over quantity"—fewer descendants, but higher investment in each child. However, many agrarian societies also offer men opportunities to pursue an opposite "quantity over quality" strategy at the same time—that is, they offer ways of producing children in which men may invest little or nothing. Thus they make distinctions, obscure to modern observers, between wives, concubines, and slaves, with additional shades of difference within those categories. Rich men of the Roman aristocracy, for example, could have hundreds of slaves in their households and many natural children among them; it was common for these elite men to free slaves in their wills and leave them property, at which time the former slaves would become Roman citizens and take their deceased master's name. The Roman system, with its large enslaved population, was especially well adapted to a double reproductive strategy for men who could afford it, but many large agrarian states have had similar institutions that allowed elite men to reproduce much more successfully than others.[18]

Marriage is a tool for channeling the transmission of property, and when a new couple marries, property must usually be transferred or allocated to them. In most agrarian and pastoralist societies, some kind of property exchange between families takes place at the time of marriage.[19] This transfer can go from the groom's family to the bride's family (in which case anthropologists call it bride-wealth or bride-price), or from the family of the bride to that of the groom (in which case it is called dowry), or both kinds of exchange can take place. Bride-wealth is attested in all regions of the world; dowry has been most common in Eurasia, especially among the middling and higher classes; and many societies have practiced both. A bride's dowry often constitutes her share of inheritance from her family. It serves to secure a good match so that her descendants can benefit from the extra resources, and as protection for the bride if it buys the goodwill of her in-laws, or if it must be returned to her or her family upon divorce or upon her death. In some societies brides have had little control over their dowries or the inheritance that will eventually, through them, pass to their heirs; in others, the dowry has been a source of autonomy and economic power.

As an example of the former pattern, we know a great deal about property and inheritance in classical Athens of the fourth century BCE, because these matters were often the subject of lawsuits, and a large number of speeches survive.[20] As in many societies, Athenian women brought their share of the patrimony to their marriages as dowry. This passed from their male kin to the control (*kyrieia*) of their husbands, who had authority over it, with the important restriction that the husband was obliged to return it to his wife's family in the case of divorce, or else pay a high rate of interest on it. On the wife's death, her children inherited the dowry.

Law and custom in classical Athens strictly limited women's ability to dispose of property on their own authority—their husband or closest male relative, usually a father or brother, managed their dowry. Women inherited property in their own right only in restricted circumstances. Elaborate rules governed the situation in which a father died leaving only daughters to inherit—the father's nearest male relative, usually a brother, was entitled to claim the daughter in marriage, and uncle–niece marriages are well attested in sources from classical Athens. The husband then became *kyrios* or master of the property until the couple's sons reached adulthood and took over. The other situation in which women inherited was when their only brother died without heirs: if they had no remaining brothers, their claim would precede the claim of their paternal uncles and cousins. In all cases in which women inherited, there was a strong expectation that one of their sons would be adopted back into their father's family so that the property would continue to be held by that family.

Athens' rules about heiresses are only one of many possible creative solutions to the dilemma a family faced when it had no male heirs. Adopting adult sons was another common strategy, in Athens and elsewhere. Other solutions have included the female-to-male gender-switching practices attested in Albania, among the Igbo in Nigeria, and no doubt in other cultures (see chapter 4) and the varieties of uxorilocal marriage that emerged in traditional China, which I shall discuss in chapter 7.

By custom, Athenian women had no right of inheritance from their husband or his family, though husbands sometimes left them small bequests. Husbands also sometimes left their wives dowries for remarriage and even designated new husbands for them—as in the case of Cleobule, the mother of the famous orator Demosthenes, who was supposed to have married her husband's nephew by the terms of his will, though she managed to remain single by forfeiting control of the dowry to the nephew. This path was open to her because she happened to be an heiress from a wealthy family with no living male relatives and thus had some leverage.

While scholars have rightly emphasized that Athenian women made many economic transactions in everyday life despite the restrictions imposed on them, and that custom was flexible and restrictions not always enforced, it is nevertheless fair to say that classical Athens was more patriarchal than most other cultures; and the Athenians acknowledged the contrast between themselves and other societies, especially the Spartans, on this point. Athenian women had little economic independence from their husbands or close male relatives. In contrast, Roman law recognized similar property rights for women and men. Although in legal theory only the ascendant male, the *paterfamilias*, could own any property, the law was almost the same for adults of both sexes who did not have living fathers or grandfathers, and it strictly separated the property of spouses in marriage.

Customs about dowry and bride-wealth have varied in important ways—this kind of property has sometimes functioned as a gift to the bride herself, or to one of the families; it has sometimes been returnable in the case of divorce or (less often) not. These variations have important implications for women's well-being. A large gift of bride-wealth made to the bride's family and returnable on divorce can severely limit a daughter's options. Daughters may be perceived as assets to be traded and sold, and the rights of the husband who buys a wife for a great price may be more absolute. A woman wishing to escape a marriage must buy her own freedom, because the property a family receives for its daughters is often used to secure brides for sons and cannot be paid back without great sacrifice. There is thus some overlap between marriage with bride-wealth and slavery, although societies with both institutions make careful distinctions of status between them. Also to the detriment of women's well-being, it is advantageous for a family in this system to cash in early by marrying daughters at young ages. But if the customary bride-price is endowed irrevocably, if it is not too large, and especially if it is under the control of the bride herself, she has more autonomy.

While patriarchy is typical of agricultural societies, this is not the same as saying that men as a group oppress women as a group. Societies have been structured—usually very rigidly—to support male dominance, but women have participated fully in reproducing that structure; some have resisted and some have escaped, but most have done their best to help their families prosper within those structures and have taught their principles to the children they have raised. In strong forms of the patrilineal, virilocal system outlined earlier, women can only achieve security by producing sons, who will become the husbands and fathers of the families they will eventually, as mothers-in-law, command and direct; daughters will move away, their labor and offspring lost to the family. It is hardly surprising, then, if women prefer sons over daughters in this system and if mothers-in-law exploit daughters-in-law to get some return on their investment of long sacrifice. Whether women in agrarian societies have *felt oppressed* is a vast topic on which I do not wish to comment. But it is safe to say that most have not thought themselves in a position to do much about it, and have instead done the best they could within the constraints of the only world known to them, the societies into which they were born.[21]

MEN'S WORK AND WOMEN'S WORK

As foragers divided labor by sex, farmers and pastoralists have universally done the same thing, defining "men's work" and "women's work."[22] It is a reasonable guess that women were the original domesticators of most plants, since women in foraging societies specialize in collecting plant foods, and in early horticultural societies it may have been common for women to farm while men continued to hunt, which is the practice among some peoples today. In historical farming societies, the plant–animal divide is not very important, and there is a lot of variation in the farming tasks assigned to men and women. One problem that has long impeded the study of women's work in farming societies is that, because men are socially dominant, women's work is often ignored or devalued, both in ancient sources and by the modern historians who study them; sometimes it is not even considered work. This perspective pervades much of the scholarly discussion about women in agricultural societies, even Ester Boserup's famous study, discussed next, though in recent years researchers have begun to correct for it.

In 1970, in her book *Woman's Role in Economic Development*, Boserup argued that as societies moved from horticulture to agriculture, sex roles changed and women's social standing deteriorated. Because intensive agriculture is much more work than horticulture, men took over most of the tasks of cultivation, especially plowing. Also, because inherited property was important in agriculturalist societies, they were more stratified; they included landless people who could be hired as wage laborers, which allowed some better-off families to exclude women from outdoor labor. Boserup argued that, as a result, women lost status and autonomy and were valued mainly for reproduction. While much of this argument has held up well, some modifications are important.

Boserup did not consider work other than cultivation, but changing agricultural methods created other kinds of work that were almost universally performed by women. One of these tasks was textile production: once humans began raising plants and animals for fiber, and indeed even before this, women took up the labor- and skill-intensive jobs of spinning, dyeing, weaving, and sewing; in imperial China, raising silkworms was also women's work.[23] Across cultures, food processing remained mainly women's work and also became more time-consuming, as grains had to be threshed, hulled, pounded, ground, milled, boiled, baked, and so forth; animals had to be milked, and the milk turned into cheese and yogurt; harvested food had to be preserved by drying, pickling, or smoking; and plants were made to yield oil or fermented into alcoholic drinks. Intensive agriculture with plow and animal traction increased workloads for both men and women as compared to horticulture; for women, the amount of domestic labor increased while the amount of outdoor labor stayed about the same.[24]

Besides the production of food and cloth, certain types of medicine and healthcare were also the province of women in agricultural societies, which, in contrast to foraging cultures, had complex plant-based medical traditions. Older, post-reproductive women were often well represented in the "herbalist" profession, as well as in the profession of midwifery that is nearly universal in agrarian cultures. An example is the stereotype in Ming-era China of the "six grannies"—healers, herbalists, and midwives who mostly treated women and children in a society with rigid gender divisions. The grannies are ubiquitous in the literature of that era, although the male elite classes who wrote it viewed them with suspicion and contempt. Although some grannies were young (and wet nurses were a kind of "granny"), the stereotype was of an older woman.[25]

Divisions of labor by sex were not always rigidly observed in agrarian cultures, however. Wherever the family is the unit of production and residence is fixed, each sex must be able to function without the other. Foraging groups with too many men, women, or children could rebalance themselves by exchanging members, but although agrarian families could resort to methods like fostering or exposing children, forming joint households with brothers, hiring help, acquiring slaves, and delaying marriages, these strategies were not as easy or flexible, and some were available only to the well-off. Most peasant farms operated close to the margins of subsistence without much room for strict specialization of labor in the household; partners often died; men in particular might be absent for long periods at war or at wage labor jobs, while women from poor families might also hire themselves out as wage labor.

Divisions of labor by gender also have an ideological dimension, and are closely linked to class and privilege. Ideals espoused in surviving sources of the agrarian period reflect the aspirations of elites, but not necessarily the lived experience of all women. Enslaved women often performed the same tasks as men, because cultural ideals about the division of labor were less rigidly applied to them, or not applied at all.

While the Athenians of the classical period prided themselves on keeping women of citizen status at home and indoors, the scant evidence about rural women that survives from the ancient Mediterranean world shows them reaping and threshing grain, picking olives and grapes, harrowing, hoeing, digging, tending animals, hauling water, and generally performing all kinds of backbreaking labor. Even among the foot-bound women of late imperial mainland China, many survivors interviewed in the 1990s remembered doing outdoor farm work, picking cotton, and carrying water and wood. A striking example of how arbitrary and flexible distinctions between men's and women's work can be comes from Huayangqiao, near Shanghai: there, rice planting was considered heavy skilled labor for which men were paid three times what women got for their work pulling up paddy shoots. After the Chinese Revolution, however, both women and men planted, and women proved better at it, with the result that it became low-paid "women's work" in some villages.[26]

Most agrarian societies are deeply invested in the subordination and control of women, for reasons explained earlier. Distinguishing male and female spheres is important to the patriarchal agenda, and these values—rather than economic efficiency—may explain why the sexual division of labor is so pervasive in agrarian societies. For example, cross-culturally, plowing is usually done by men, but women can plow and often do, as many observers have noted. One intriguing guess about why plowing is men's work is that it is a task performed upright and not, as so many other agricultural tasks, bent over—an argument that might also explain why, in Europe, harvesting with a sickle was women's work, but once the scythe was introduced, it was used mainly by men.[27] Also, a very commonly articulated reason for giving women less outdoor work and protecting them from wage labor is that domestic work keeps them out of the sight of unrelated men. In societies that value female seclusion, families that can afford to keep their own women indoors doing domestic work and use wage laborers or slaves to help with outdoor work have done this as much as possible.

As men in agricultural societies collaborated for wars and raids and controlled women's movements to ensure their children's paternity, they came to dominate the public sphere of leadership and community organization. Again, there have been exceptions; even where they are most rigidly excluded from men's political activities, women may maintain separate networks of relationships and institutions less visible to outside observers and in the historical record, and in some societies, especially in sub-Saharan Africa, women's political institutions have been very important. But almost everywhere that farming or herding has been the basis of the economy, men have held most of the economic and political power.

As a result, it can be hard to recover information about women generally, and especially about older women, who have been less interesting to the men who produced most of the literary and documentary record (older men, however, who have often held high status, are well represented). They have also been less interesting to anthropologists, making comparative methods hard to apply. We catch glimpses of a few individuals, and I note some of them here for the benefit of those who might be interested in further reading. Genghis Khan's mother, Hoelun, is an important character in *The Secret History of the Mongols*, and her story is told in the prologue to this book.[28] The records of the Inquisition in France from the early 1300s preserve the testimony of the noblewoman Béatrice de Planissoles, widow of the feudal *châtelain* of Montaillou, a tiny hillside village in Languedoc in southern France and hotbed of the Cathar heresy. Under interrogation, she divulged many details of her private life; these included a long love affair with the priest of the village of Montaillou and an impulsive adventure with a younger man, also a priest, that began, as she herself stated, when she was past menopause.[29] The earliest autobiography in English is that of Margery Kempe, who was over 60 years old when she dictated the story of her life to a scribe in 1438. Margery was 40 and had borne 14 children at the time of her spiritual conversion, after which she convinced her husband to join her in a vow of chastity and became, as she saw it, the bride of God. Known and often reviled for her spectacular displays of mystical weeping, she spent much of the rest of her life making pilgrimages to European cities and caring for the sick, especially her husband, who suffered from a traumatic brain injury and incontinence after a bad fall on the stairs.[30] Particularly interesting is the story of Ahebi Ugbabe, an Igbo woman of colonial Nigeria. She fled from her homeland at the age of only 15, when her father planned to dedicate her as a bride to the goddess Ohe in compensation for some unknown transgression for which the family, as they saw it, was being punished. To survive, she worked as a prostitute, became a successful trader in horses, networked with British officials, and eventually returned home with the support of the British government. She ruled as warrant-chief, and eventually as transgender king, from 1918, when she was 38, until her death 30 years later.[31] This example brings us into the modern era, when the lives of post-reproductive women are much better attested.

While stories of a few famous women like these survive in literary sources, documents like wills or legal records often preserve details about the lives of women of more humble status. While I do not focus too much on elite women in this book, I offer several examples of the lives of ordinary women in the next chapter.

FERTILITY AND MORTALITY IN AGRARIAN SOCIETIES

Social structures were very different, then, in agrarian societies compared to foraging societies. What about reproductive strategies and the closely related subject of population dynamics? How did the evolved reproductive strategies of the Paleolithic play out in the agrarian period? We have seen that menopause is part of a complex cooperative breeding strategy in which humans delegate reproduction to one part of the population. It is an extraordinarily adaptable and flexible strategy that can help populations "boom" in good conditions, but also, in other conditions or even at the same time, limit kin competition and maintain a high ratio of adult producers to child consumers. People made use of these same reproductive strategies in the agrarian period, but within the patriarchal family system just described, and with somewhat different emphases, resulting in different demographic patterns.

The contrast between the demography of the agrarian world and that of the modern one has always been obvious: both fertility and mortality were much higher during the agrarian period. Total Fertility Rates (the number of children borne by an average woman surviving to age 50) in agrarian systems could be as low as three and as high as eight or so, with figures from five to seven being most typical. Mortality before age 15 could vary between lows of about 20 percent and highs of over 50 percent, sometimes with huge differences between areas within a few miles of one another if, for example, one area's microclimate was hospitable to the mosquitoes that carry malaria.[32] Cities, with their concentrations of people and of filth, were usually more deadly than the countryside. Life expectancy at birth, driven by infant and child mortality rates, was typically in the mid-20s to mid-30s, though we know of some populations with average life expectancy rates below 20 or above 40, or even 50. These are very wide variations, but they still describe a pattern that differs sharply from that of the modern industrialized world. Most "excess" mortality was at ages younger than 15, and especially younger than 5. The innovation of the modern era has not been to lengthen the natural human lifespan at its end, but to reduce child mortality very dramatically. Thus, while the mortality hazard—the risk of dying—for a 60-year-old forager is, to be sure, four times that of a person of the same age in the modern industrialized world, the mortality hazard for a toddler is *100 times higher* among foragers.[33]

What, then, if we compare agrarian societies to foraging ones—do we find the same stark differences in mortality and fertility? Many researchers believe in something like this thesis, contending that both fertility and mortality were higher in agricultural societies. An influential theory today, the chief exponent of which has been Jean-Pierre Bocquet-Appel of the National Center for Scientific Research (CNRS) in Paris, argues that fertility increased sharply with the Neolithic revolution and caused a leap in population before higher mortality rates, largely due to infectious disease, balanced higher fertility rates. This phenomenon, as Bocquet-Appel and his colleagues argue, happened worldwide, and they have called it the Neolithic Demographic Transition, seeing it as a sort of parallel and mirror opposite of the modern Demographic Transition. Some identify a "signature" of increased fertility in the larger percentage of children among the skeletons recovered from Neolithic sites compared to Paleolithic sites.[34] Others have used a technique called Summed Calibrated Date Probability Distribution—based on aggregated radiocarbon dates from multiple archaeological sites—to reach similar conclusions about a population boom in early Neolithic societies. Many of these researchers reject the theory of Cohen and others that population pressure caused humans to turn to agriculture in the Holocene; instead, they argue, human populations were low at the end of the Pleistocene, and people adopted agriculture because farming was a great new idea that they liked better than foraging, or because farming populations grew so fast that foraging populations were crowded out and had no choice but to convert.[35]

Although the idea of a Neolithic Demographic Transition is intuitively appealing and not incompatible with the themes of this book, and although it is fair to say that scientific consensus accepts it, I am not convinced that it happened, at least not in the way described by the people who use the term. Inferences from the age distribution of skeletons are problematic for the reasons discussed earlier. In particular, it would be unsurprising if burial patterns, or patterns in the survival of skeletons, changed in ways that were similar across cultures when humans began living in settled villages. Methods that rely on the distribution of radiocarbon dates also have many problems, which intrepid readers can learn more about in the note.[36]

More importantly, there is not a lot of evidence that foragers have either lower fertility or lower mortality than agriculturalists. Fertility is high among today's foragers, like the Hadza and Ache, and there is a broad range of variation in fertility among both foragers and agriculturalists. Some analyses have shown no significant difference between the groups, while other researchers have argued that only people practicing intensive agriculture—a development that might occur thousands of years after a society first domesticates crops and starts living in villages—have somewhat higher fertility than foragers.[37] Regarding mortality, an analysis of all the best demographic data on foragers shows no obvious differences from the pattern observed among agriculturalists (and also finds very high mortality among some short-stature foragers, discussed in chapter 5).[38] While most scientists agree that infectious diseases, parasitic diseases, and probably deficiency diseases became more common and more virulent with dense settlements and dependency on agriculture, it does not necessarily follow that mortality was higher. Also, genetic studies of bacteria have sometimes surprised us by failing to confirm that the most widespread diseases of the agricultural period arose with the Neolithic transition—it is probably true that the most deadly type of malaria, falciparian malaria, emerged around this time, but researchers have found that some diseases (like lice, tuberculosis, syphilis, and "benign tertian" malaria, *P. vivax*) arose earlier, and others (like measles and dysentery) appeared much later. Furthermore, while we once thought that humans acquired tapeworms and tuberculosis from living close to domesticated animals, it now seems that the reverse is true—animals acquired them from us![39] To put it another way, while there are extremely obvious differences in fertility, mortality, and causes of death between the agrarian and the modern periods, we cannot say the same about the differences between the foraging and agrarian periods.

On the other hand, even a subtle shift in rates of fertility or mortality—though not reliably detectable, as I see it, by the methods we currently use—could have caused large changes in the pattern of our demographic history. In particular, a pattern of slow but steady population growth with narrower oscillations seems to have been typical of the agrarian period, rather than the boom-and-crash model that may have prevailed in the Paleolithic. Population did increase in the agrarian period, worldwide and in all regions, and this is not surprising, because agriculture, once adopted, can support more people per square kilometer than foraging. Also, agrarian populations could grow rapidly and opportunistically in some circumstances, although these frontier-type situations were not the norm. But, as in the Paleolithic, the net rate of increase over time was low. The long-term average growth rate of the world population in the agrarian period was between 0.01 and 0.1 percent per year, varying over time within those parameters.

How should we understand this balance? I have suggested that, in the agrarian era, we do not see boom-and-bust cycles of rapid growth punctuated by frequent episodes of catastrophic decline—that in this era, average fertility rates matched average mortality rates closely, so that population grew (more or less) steadily but slowly. Most scholars accept this view, but is it true? The theme of environmental crisis and "crisis mortality" has recently become popular among historians of the agrarian era, so that we should perhaps look at the possibility of boom-and-bust cycles more carefully.[40]

A few major large-scale population setbacks are obvious in the historical record and well known, including, in the thirteenth century, the Mongol conquests; in the fourteenth century, the Black Death; in the sixteenth and seventeenth centuries, the depopulation of the Americas; and in the nineteenth century, the Taiping Rebellion, which decimated whole regions of China. There was also some sort of collapse in the eastern Mediterranean at the end of the Bronze Age, around 1100 BCE, but we know much less about this.[41] In most of these cases, it took 200 to 300 years for population to recover. In the case of the Americas, indigenous populations did not recover; they were replaced by European populations over the course of 300 to 400 years. But we know of only a few catastrophes on this scale in the historical period, and there is no obvious evidence of the more frequent events that would be necessary to offset a substantial persistent imbalance between fertility and mortality.

Infectious disease was the most common cause of death in the agrarian period, and the suppression of infectious diseases is the main reason that average life expectancy is so much longer in the modern era. Infectious diseases were not just a steady contributor to baseline mortality; they could also strike in epidemics that killed more people than usual. Epidemic "crises" could, in theory, cause boom-and-bust population cycles, if populations grew quickly until (more or less) randomly occurring epidemics struck them down. In Abdel Omran's classic 1971 theory of "Epidemiologic Transition," the first stage is the "age of receding pandemics," in which mortality spikes that previously kept the population in check become infrequent and disappear.

We know of only a few great pandemics like the Black Death, but how important were occasional epidemics on a smaller scale? To answer this question, we must look more closely for evidence of crisis mortality in individual regions. For some populations—notably England after 1500 or so—it is possible to do this. Demographers have long concluded that crisis mortality was important there and elsewhere in Europe during the same era, but they have used the term "crisis mortality" in a somewhat different sense than the one I have suggested. Years in which mortality was higher than normal by 10 percent or more were frequent—every few years on average, and sometimes consecutive years—but only one crisis in which mortality rates doubled occurred between 1500 and 1750 (in the years 1557–1560, a poorly attested epidemic of what was probably influenza). Extra deaths from crises were only a small percentage of all deaths over the long term. England's population was not seesawing in a boom-and-bust pattern, although frequent mortality spikes were a fact of life.[42] We would like to know more about the effects of crisis mortality on individual parishes or towns, and a few such studies have been done, but small sample sizes make analysis harder.[43]

There are theoretical reasons why growth should have been more smooth in the agrarian period than it presumably was in prehistory.[44] By domesticating plants and animals, humans mitigated the cycles of depletion and recovery thought to have occurred when they were dependent on wild foods—agriculturalists regenerated their own resources to maintain a constant supply. Agriculturalists could also store food as security against hard times, although storing food is more difficult than is commonly assumed. Because these societies were less egalitarian, wealthy people could hoard resources and survive at the expense of poor people, which may have resulted in fewer deaths during crises than occur in food-sharing cultures, in which people all starve or prosper together. Finally, compared to most of our prehistory, climate has been unusually stable for the last 10,000 years. Thus, while it is a good guess that human population fluctuated between boom and bust in the Paleolithic—growing rapidly, then being decimated by catastrophe—in the agrarian period crisis mortality *in this sense* was a less important factor. Despite cyclical fluctuations and occasional sudden and extreme events, births and deaths were in close balance over the long term for most of the agrarian era, even as both rates varied greatly among different populations. How was this balance achieved?

In 1798, the Reverend Thomas Robert Malthus first published his famous argument about how agricultural populations might grow and then stagnate or decline as they exploit and then outstrip their resources. Population, he argued, grows geometrically, or exponentially, but growth in food production is much more restricted. Therefore, human populations grow until malnutrition, famine, poor working conditions, and disease cause mortality to reach high rates among the poor—a mechanism that he called nature's "positive check." As population grows, labor becomes cheap; the price of food, rent, and land rises; and people either succumb to misery and mortality or find ways to produce fewer children.

Malthus's theories apply better to the agrarian period than to the modern one—at least at this point, early as we are in the history of the modern period. As early as 1844, Friedrich Engels argued in his "Outlines of a Critique of Political Economy" that science could increase the soil's productivity potentially without limit, and called Malthusianism "the crudest, most barbarous theory that ever existed."[45] The class dimension of Malthus's theory, which argued that support for the poorest classes would only encourage them to reproduce more, made him unpopular with Marxists. In the modern period, factors like mechanization; an improved understanding of soil chemistry; crop hybridization; revolutions in transportation, including the invention of the shipping container and long-haul refrigerated trucking; and genetic engineering have changed the relationship between population and food supply, at least temporarily.

For the agrarian period, however, the most important modification to Malthus's theory has been the thesis of Ester Boserup in her book *The Conditions of Agricultural Growth*, published in 1965. She emphasized that "carrying capacity," a concept often cited by demographers, is flexible: land can be more or less productive depending on whether a population is practicing, at one extreme, slash-and-burn agriculture with long, 25-year fallow periods or, at the other extreme, irrigated agriculture growing crops twice or even three times per year on each plot. If population rises, farmers respond by doing more work and using more intensive methods. Boserup challenged Malthus by describing population growth as an engine for progress rather than a cause of misery. Her theory explains why the trend of population over time is upward—intensification acts as a ratchet. Once a higher population is being sustained with more intense agricultural technologies, people have to continue using them or starve.[46]

Today, scholars studying the demography of agrarian societies mostly agree that both Malthusian and Boserupian forces have shaped our history and that food supply is central to population dynamics.[47] In both Malthus's and Boserup's theories, populations cannot grow beyond their food supply. When the food supply falls short, people can work harder, bring more land under cultivation, or use new, more intensive techniques, all within limits dictated by circumstances. As an alternative, people can control fertility, an option subject to fewer external restrictions, although costly in other ways. Or, if they do none of these things, mortality rises to the point at which population and food supply match—modern scholars identify the main mechanism of this increase as higher infant and child mortality from disease.[48] Migration over short and long distances has also been important throughout agrarian history, but in general, agrarian populations have been more sedentary and less mobile than the foragers of our Paleolithic past.

A recent demographic model of agrarian societies developed by researchers at Stanford University explores how populations might grow, decline, and suffer in two conditions: frontier-type conditions in which new land is abundantly available, and conditions in which land is limited.[49] When a population is expanding into new territory, extra work can increase both population size and well-being. Fertility control can increase well-being within parameters, but frontier populations, which are early in their history and usually small in number overall, must maintain high rates of growth or be vulnerable to extinction when catastrophe strikes. In some ways, the frontier condition is similar to the normal situation of our Paleolithic ancestors. Things change, however, when a farming population runs out of space for expansion. In this phase, the model's grim prediction is that the population's size stabilizes only at a level at which many people are hungry. Working harder or pressing children into service at younger ages can increase population size and improve well-being temporarily, but does not alleviate hunger at its equilibrium level; that is, at the point where population restabilizes, people are just as hungry as before. Technological innovations that increase yields have the same effect. Eventually these populations reach a point at which extra work does little to increase yields, and more people take specialized, nonfarming jobs. Fertility control becomes a central and critical strategy for increasing well-being, allowing populations to reduce hunger and extend average life expectancy. Crises causing high mortality increase the well-being of those who survive, but at terrible cost, and in general, a population must trade off well-being and size at equilibrium; larger populations suffer more. Needless to say, farming populations usually spend most of their history in this second, land-limited phase of the cycle, although they may revert to the frontier condition temporarily after a catastrophe.[50]

I have outlined the Stanford model of agrarian population dynamics in some detail not only to illustrate how researchers reconcile Malthusian and Boserupian ideas, but also because it explains so much of what we see in the history of agricultural populations—why children work from young ages, why people learn specialized jobs, why fertility is so much higher in frontier societies, and especially the role of fertility control in most peasant societies. As we saw in part I of this book, humans are capable of fast reproduction and population growth when conditions are favorable—the "frontier" conditions of the first Stanford model—but our evolved strategy that relegates reproduction to specific groups can also work well in situations with limited resources, like the scenario explored in the second Stanford model. Just as the Ache of the forest period modified the reproductive strategy that all humans have inherited to adjust the ratio of men to women in their society, creating more male providers and non-reproductive male helpers, peasants during the agrarian era also modified the same inherited strategy to include more types of fertility control. This thesis is controversial, but I think by the time my discussion is over, the case will be convincing.

Fortunately, we have much better sources of information for the agrarian period than we do for the foraging era. While our best data about fertility and mortality among foragers come from long-term observations of the few populations that still survive, many agricultural peoples have left good records, enough for statistical demographic studies. The University of California at Berkeley maintains a Human Mortality Database with historical information on 38 countries. Scholars have published books and articles on the demography of Roman Egypt, early modern England, nineteenth-century Sweden, Tokugawa Japan, the enslaved population of Trinidad, the Mormon settlers of nineteenth-century Utah, the peasant population of Liaoning in nineteenth-century China, and many other nonindustrialized societies.[51]

While it is normal to describe all societies in the range of the traditional high-mortality, high-fertility pattern as "natural fertility" populations, most historical populations, even very stable and well-nourished ones, fall short of what should, in theory, be "natural" fertility. A woman with a 30-year reproductive lifespan giving birth every two years would produce 15 children; to produce as many as 24 children over a lifetime is not beyond human biological capacity. But the highest Total Marital Fertility Rates (TMFRs) of which we know—this term refers to the average number of children borne by a hypothetical married woman who survives to the end of her reproductive life with her marriage intact—are around 11 or 12. The Hutterites of the mid-twentieth century are often cited as a benchmark of real-world maximum fertility, and a few other historical populations with rates this high are known. But marital fertility is much lower in most traditional societies and can be less than half the Hutterite rate.[52] Total Fertility Rates (TFRs), which include unmarried adult women and are always lower than Total Marital Fertility Rates, have been close to replacement level for most agrarian populations.

High fertility appears to be common in colonizing frontier peoples with access to unlimited land, like the Mormon population of Utah, whose rate of reproduction for a few decades in the mid-nineteenth century was as high as that of the Hutterites, or the English colonists of North America.[53] Similarly, we saw that the Ache of Paraguay had very high fertility in a period when they were expanding into new, depopulated territory. But in all populations that have been observed, including these, at least some of our reproductive potential, and often most of it, is held in reserve.

Some reproductive control is biological: human fertility responds to signals about the environment, especially nutrition and physical labor.[54] Where there is plenty of food, as in modern Western societies, girls reach puberty early, whereas in agrarian populations, average age at menarche (when menstruation begins) might be 16 or 17. Where mortality risk is unusually high, as among the Efe and other short-stature populations, girls mature earlier—the body is responding to the greater odds of an early death by reproducing early. Many agrarian societies can be described as low-nutrition and medium-risk, with a late age at menarche. Whether chronic malnutrition and hard physical labor of the type common in the agrarian period have important effects on adult women's ability to conceive and bear children is less well understood; populations with similar fertility rates can vary widely on these points, and populations with similar lifestyles can have different fertility rates. However, researchers have found an effect of poor nutrition and hard work on ovarian function in some populations, and these conditions may lengthen the period during which women are infertile between births. During famines, when food shortages are extreme, women can stop ovulating and fertility may drop dramatically (though usually a rebound follows). Like menopause, these "natural" adjustments to fecundity—a woman's ability to conceive and bear children—probably evolved in our prehistoric past, and in peasant economies, they may have functioned as a "preventative check" that helped keep populations close to the limits of their resources.

It is harder to identify environmental factors associated with variations in age at menopause, and we do not really know whether nutritionally stressed women, for example, might reach menopause earlier or later than well-nourished women—as we will see in chapter 10, there is some evidence that women in modernized countries reach menopause a few years later than those in traditional societies, but methodological problems make comparisons hard, and the reasons for any differences that might exist are unclear. Average age at last birth in traditional societies appears to vary less than age at menopause and is around the same—between 38 and 42—even in cultures that are very different from one another.[55] Thus there is not a lot of evidence that the age at which reproduction ends is sensitive to environmental factors. But compared to those mechanisms that adjust age at menarche and fecundity to environmental factors, menopause is overall a much more important natural limitation on human fertility.

Many scholars, following criteria first established by French historian Louis Henry in 1961, have tried to draw distinctions between "natural fertility" and controlled fertility, and some strongly deny that most or any traditional populations practiced fertility control.[56] However, it can be very difficult to distinguish the effects of natural limits on fertility from the more deliberate methods humans have used to control reproduction. These include delayed marriage, long periods of breastfeeding, customs or laws that discourage the remarriage of widows, celibacy for part of the population (that is, some percentage of men or women do not marry), and infanticide. Also, while most scholars since Malthus have dismissed the idea of reproductive restraint within marriage, assuming that sexual passion is an insuperable obstacle to this, others have argued that abstinence within marriage or *coitus interruptus*, the "withdrawal method" for which societies the world over have coined so many colorful metaphors, have been common practice. Abstinence after childbirth, to avoid births too close together, has been a custom especially common in sub-Saharan Africa.[57]

A few scholars have argued that some methods of birth control or abortion used in traditional medicine were effective in controlling fertility, and there I am more skeptical. Recipes for contraception and, more commonly, abortion are part of most folk medical traditions, but the methods attested are extremely various, and it is hard to demonstrate that any of them, as actually used, would have worked, or how available the necessary substances were, or how many people used them.[58] Still, it is possible that some of these methods, if used widely, had an effect on fertility at the population level. By the mid-nineteenth century in Europe, birth control was becoming more effective: rubber condoms were widely available, and the diaphragm had already been invented.[59]

In traditional societies, behavior has probably been a more important cause than external factors for the wide variations of fertility that historians observe, because behavioral interventions have much stronger effects. For example, if age at menarche declines by about two years in a society with abundant food, the corresponding increase in fertility is about 4 percent, but delaying marriage for women until their mid-20s reduces fertility by *30 percent*.[60]

A study conducted in The Gambia by Caroline Bledsoe in the 1990s illuminates some ways in which fertility may be controlled through behavior, in a society that values large families, has a Total Marital Fertility Rate of between seven and eight, and meets most criteria and tests for "natural fertility." By means of abstinence, traditional remedies, and some Western contraception, Gambian women (and usually men too, as most decisions were made by women and their husbands together) sought to avoid bearing children too close together or becoming pregnant before the end of the long nursing period thought important for each child. Most women retired from childbearing before menopause, with the support of their husbands or sons (or, more specifically, with the support of new young co-wives or daughters-in-law who took over their reproductive functions as well as the household's more physically demanding work). While Gambian women saw themselves as conserving fertility rather than limiting it—the idea of choosing to limit families to a specific number of children was not meaningful to them—without birth spacing and early retirement from childbearing, TMFR would have been higher.

An important point here is that in agrarian societies, women have not always had control over decisions about when to marry and whether to raise the children they give birth to—these decisions affecting their fertility have often been made by their parents, husbands, or in-laws. Women might have more control over decisions about how long to breastfeed their children and about whether to use any available traditional methods of contraception and abortion.[61] Whether and how often to have sex with their husbands might also be a subject of negotiation, not necessarily under the husband's sole control.[62]

But why control fertility? Theoretical models tell us that fertility control is important to well-being in land-limited agrarian societies, and demographic evidence tells us that most agrarian societies probably controlled fertility in one way or another. But to explain individual decisions and behavior, it is important to answer this question at the level of the peasant household. The foundational work on the economy of the peasant family farm is a book by the Russian scholar Alexander Chayanov, which he based on an exhaustive body of research on pre-revolutionary Russian farms. This work was published in Russian in 1925 and later, in 1966, in English as *The Theory of Peasant Economy*. Chayanov's theory is limited by its restriction to the Russian system, which used almost no hired labor and in which families could easily rent more or less land as their structure changed. His arrest in 1930 and execution in 1937 prevented him from extending his theory to other systems. But many scholars have expanded on his work, and I draw on that body of literature for what follows.[63]

Small children are a burden in peasant societies, just as in foraging societies, and families with many small children struggle to survive. To be sure, as children become older, they contribute more; even children as young as five or six can help with farming tasks, and unmarried teenagers undertake wage labor, herding, or clearing lands, investing in their future patrimony or dowries in the hope of one day having a family of their own. They also help care for their siblings, which makes it easier for their parents to have more children. Thus the thesis first proposed by John C. Caldwell and others in the 1970s that a large family is critical to provide the labor force necessary for farming is both plausible and important.

However, most studies have found that children in agrarian societies do not produce more than they consume while they are at home—as among foragers, resources usually flow downward through the generations, and children do not make their parents richer. At the same time, because land must be divided among heirs, even a prosperous family can be impoverished in a single generation if the number of heirs is too great. Setting up a child for adult reproductive life with a dowry or patrimony may require enormous sacrifices from parents and siblings. In families with many children who reach adulthood, some may not marry or reproduce because they lack inheritance or dowry.

Rather than having many children to become prosperous, it may be closer to the truth to say that families in agrarian societies have as many children as they can afford. One reason this conclusion is not more obvious and straightforward is because of the role that social factors play in individual decisions about what is affordable. For example, although wealthy families have more resources, they also have more power and status to lose by dividing those resources among children. For this reason, rich families do not necessarily have more children than poor families, although analyses that control enough variables may show that, within groups, there is a correlation between resources and family size.[64] It is also important to remember that where wealthy men practiced a double reproductive strategy, there is usually no documentary record of their illegitimate children for historians or demographers to find; analyses are based on legitimate children only, which can obscure differences among economic groups.

The easiest and most prosperous part of the life cycle for a peasant family comes after menopause and as a result of it—when the parents stop reproducing and have only older children and teenagers at home.[65] This is true of single-generation families; in multigenerational families, the burdens of reproduction are spread more equally through the life cycle, as couples with small children live with post-reproductive parents. Couples can marry younger in this system, because they share work, land, and resources with the previous generation. For the first few years after marriage, before children are born, the daughter-in-law's work helps support a family still burdened with many underage consumers; later, as the younger family's own burden grows, it benefits from the previous generation's more favorable worker-to-consumer ratio. That is, the multigenerational strategy tends to smooth out fluctuations in the ratio of producers to consumers over the family life cycle.

Although the effects of fertility control were weak during the agrarian period compared to the modern one, they were still very important; Total Fertility Rates could vary by a factor of two or more in these societies, and no population ever sustained the maximum fertility of which humans are capable. Menopause, then, is part of a larger picture of non-reproduction in the agrarian period. Even if its benefits are hard to prove directly (see the discussion of tests of the Grandmother Hypothesis in chapter 2), it is reasonable to interpret the history of the agrarian period as one in which the evolutionary benefits of menopause were adapted to a new economic system. People reproduced rapidly and gobbled up vast territories and resources when the opportunity was there, but they also made good use of their ability to limit fertility when it was advantageous to do so, and of the skills, experience, and labor of older, non-reproducing women. Age at last reproduction did not, in fact, increase in the agrarian period; menopause did not disappear despite profound changes in the way humans lived.

•

CHAPTER 7

Reproduction and Non-Reproduction in Some Agrarian Societies

MALTHUS, THE FOUNDING father of the science of demography and among the first to address the question of how populations might grow and decline in relation to resources, imagined the thoughts of a typical young man considering marrying and raising a family, in his famous *Essay on the Principle of Population*:

> Impelled to the increase of his species by a . . . powerful instinct, reason interrupts his career, and asks him whether he may not bring beings into the world, for whom he cannot provide the means of subsistence. . . . Will he not lower his rank in life? Will he not subject himself to greater difficulties than he at present feels? Will he not be obliged to labour harder? and if he has a large family, will his utmost exertions enable him to support them? May he not see his offspring in rags and misery, and clamouring for bread that he cannot give them? And may he not be reduced to the grating necessity of forfeiting his independence, and of being obliged to the sparing hand of charity for support?[1]

Malthus looked with horror on the squalor, disease, malnutrition, and poverty that he attributed to what demographers of the mid-twentieth century would later call overpopulation. He argued that humans either followed their natural inclination to breed until mortality reached high rates among the poor—nature's "positive check" on population growth—or exercised the rational "preventative check" of self-control by marrying late or not at all. A third group of checks on population Malthus categorized under the heading of "vice"—these included war, prostitution (which he assumed to be common in his society), luxury, and infanticide. The social system that Malthus recommended was an austere one in which men, who were the only audience he addressed, would practice abstinence before marriage and delay marriage until they had saved enough money to raise a family. He acknowledged that when economic times were hard, poor men might have to wait a long time to marry, and some might never marry.

Malthus was right to observe that, in his time, delayed marriage probably served the function of limiting fertility in England and in northwestern Europe more generally.[2] More important, however, than the age of marriage for men was that for women, because women are the limiting factor on fertility—a population can reproduce at maximum fertility even if all men do not have children or if some delay marriage, as long as almost all young women are marrying. This might happen if polygyny is common, if women marry outside the group in patterns that make it harder for men to find wives, if the population is shrinking so that younger cohorts are smaller than older cohorts, or if sex ratios are unbalanced so that there are more males than females. In any case, because women are most fertile in their early 20s, some demographers have calculated that delaying female marriage to age 26 reduces marital fertility by about 30 percent.[3]

The world into which Malthus was born, England on the verge of industrialization (scholars often apply the term "early modern" to the era between about 1600 and 1800 in Europe, an appellation confusing for our purposes but that I have sometimes been unable to avoid), is an exceptionally well-documented agrarian society and has been the subject of a great deal of scholarship on family and population history. For these reasons, I have chosen it as an example to illustrate the themes of the previous chapter in more depth. My other main example, selected for the same reasons, is late imperial China, ruled by the Manchu Qing dynasty from 1644 through 1912, at the opposite end of Eurasia. Together, these two societies give us a sense of the variation that is possible in agrarian family structures and the ingenuity that agriculturalists have brought to reproductive strategies, while also showing how the constraints of the peasant economy create patterns across widely different cultures. That Malthus himself saw China of his day as the ultimate "other" lends a poetic element to this choice of examples.

In England in this era, monogamous marriage was normal, there is no evidence of a serious imbalance in sex ratios, and women married relatively late—in their mid-20s, on average.[4] Until then, it was common for young people to work as servants in other households, and most households were "nuclear family" units of a married couple, their children, and servants if the couple could afford them—English families of this period were founded on the principle that a husband and wife should be an independent, self-supporting economic unit. Parents and their adult offspring preferred to live apart if they could (although it is easy to underestimate the number of widowed parents and other relatives in English households, as they are often missing from records or listed as "lodgers"). Because there was no social expectation that women would marry as soon as they were sexually mature, the age gap between husbands and wives was lower than in most other agrarian populations, and wives were not infrequently older than their husbands (widows who married bachelors might be much older). The Total Fertility Rate was at the low end of the range for traditional societies, hovering between four and five births for the average woman surviving to menopause, and it sometimes dropped below four. In some periods, one-quarter of the English population never married.[5] By the time Malthus composed his famous *Essay* at the end of the eighteenth century, however, marriage was changing in England—more women were marrying at younger ages, and fertility was increasing as the modern population surge began.

While many English men and women who did not marry were probably self-motivated by the kinds of considerations that Malthus described, social coercion was also a factor. Banns had to be published for weeks before the church would sanctify a marriage, during which time anyone in the community could object to it, and they often did, on the grounds that the parties were too poor to marry and that the parish was in danger of having to support their families (per England's laws mandating poor relief). Although the church did not officially recognize poverty as grounds for an objection (incest—a family relationship within the very broad range forbidden by the church—was the only official impediment to marriage and the reason for the institution of the banns), these objections were often effective. Thus, while many scholars have emphasized English individualism and its role in fertility control—the idea that each married couple needed to be self-supporting—community pressure was another reason that people delayed marriage or did not marry.[6]

What about other methods of fertility control? We have seen that some foragers practiced infanticide as a reproductive strategy. This practice was much more obvious and pervasive in the agrarian period than among foragers, as the demographic theories discussed earlier would predict. Infanticide became a capital crime in England in 1624. Formal legal accusations of infanticide were rare, but it was widely believed to be common practice. Those accused of infanticide—that is, those for whom we have records—were usually unmarried women. Young servants who became pregnant (employers were frequently the culprits) lost their jobs, became unemployable, and had little recourse, and it would be unsurprising if they resorted to this crime more often than legal records reflect.

Infanticide was a difficult charge to prove—one had to be able to show that the infant had been born alive and had not died of illness or accident. A common cover story throughout Europe in the medieval and Renaissance eras was that a parent had rolled over on the baby in the night. (Similarly, some readers will remember the conviction of Waneta Hoyt in 1996 for killing five of her children and claiming that they had died of Sudden Infant Death Syndrome.) More common than killing infants outright in European and Mediterranean cultures, however, was the practice of exposing or abandoning them, which is widely attested in all eras. For example, in Egypt of the Greek and Roman period, the traditional place to leave unwanted infants was the village dung-heap, from which some were rescued and raised as slaves or, occasionally, as adopted children. Even though Christian law and custom forbade infanticide from an early stage and Christian societies imposed sometimes brutal penalties for this offense, abandonment was usually considered a different and lesser crime. Not all abandoned children died, and in the medieval period, besides supplying the ranks of servants, slaves, and prostitutes, some abandoned babies were raised in convents and monasteries.[7]

Infant abandonment is certainly attested in Malthus's England. At the time he wrote the first edition of his *Essay*, the Foundling Hospital of London, which had opened its doors in 1741, was receiving about 100 infants per year, operating under strict quotas. Earlier in the century, when for a few years the hospital had accepted all infants with the support of a grant from Parliament, admissions had surged to over 4,000 per year, and the grant had been withdrawn as the amount of money needed to meet demand seemed to spiral out of control. While it was once thought that most English foundlings were illegitimate, scholars now argue that many were children of the married poor, who often abandoned them under pressure from their fellow parishioners.[8]

Foundling hospitals became popular in northern Europe in the eighteenth century but were common much earlier, starting in the fifteenth century, in southern Europe, where some accepted thousands of infants per year. Before the advent of modern nutritionally fortified formulas, treated water, and sterilization, the vast majority of infants in these institutions died of disease or starvation within a year of their arrival. Survival depended on being placed quickly in the home of a paid wet nurse; still, it was a lucky infant who survived the low-investment care typical of these substitute mothers. Infants forced to remain in the hospitals nearly all died. The London Foundling Hospital, of whose charges slightly more than one-third survived to apprenticeship age of 10 or 11, stands out as a relatively successful example of such an institution, but still most foundlings died. This hospital's favorable outcomes were tied to its small number of new admissions; during the period of general reception in the late 1750s, mortality rates for new infants topped 80 percent, similar to those of other European foundling hospitals. So high was mortality in these institutions that, as Malthus wrote, "if a person wished to check population, and were not solicitous about the means, he could not propose a more effectual measure, than the establishment of a sufficient number of foundling hospitals, unlimited as to their reception of children."[9]

Most infants left at the Foundling Hospital came from London or nearby parishes, and England had no other such hospitals. Outside of London, abandoned children were supposed to be cared for by the parish, which, under the Poor Laws, was charged with finding wet nurses and paying for their support. Recorded numbers of these cases are low, which could mean either that infant abandonment was not widespread in rural England, or that most abandoned infants were not turned over to the parish, an act that required mothers to acknowledge the birth publicly. This last guess is not implausible; it was the sight of the decaying bodies of exposed infants in the countryside around London that inspired the founder of the Foundling Hospital to take that step.[10]

Malthus, with a chauvinism typical of his age (and of most ages), considered the preventative check of what he called "moral restraint"— that is, delayed marriage—particular to European culture; the rest of the world, he believed, was subject only to the checks of "misery and vice." Like other Englishmen of the time, Malthus was fascinated by reports about China, especially those of Sir George Staunton, who lived many years there as a diplomat and employee of the British East India Company, and whose publications Malthus incorporated into later editions of his *Essay*, which he continued to revise until the sixth edition of 1826. Especially impressive was China's enormous population, thought to be above one-third of a billion (the real figure was even higher: more than a third of the people alive in the world in 1800 lived in China). Malthus, following his sources, attributed China's large population to favorable climate, fertile soil, the industriousness of its people, the cultural importance of agriculture, a generally egalitarian social structure, and, finally, the value placed on descendants and the practice of early and universal marriage. In Malthus's description of China, wages were rock-bottom low, families lived in extreme poverty and misery, and—surely a sign of their desperate condition—the Chinese ate mostly plants and little meat. Infants, he wrote, were frequently abandoned or killed by their desperate parents, and famines causing widespread mortality were common.[11]

For a surprisingly long time, scholars accepted Malthus's assumption that, among traditional populations, only western Europeans controlled their fertility while everyone else bred like rabbits to the point of total immiseration; some have even cited this as a crucial factor explaining Europe's lead in the Industrial Revolution. But researchers in recent decades have become more critical and in particular have questioned Malthus's contrast between Europe and China, or at least sought to add some nuance to it.[12]

One team of scholars has analyzed records from a nineteenth-century community of military state servants in Liaoning, on China's northeast frontier. This peasant population owed labor, grain, and military service to the state, which kept registration records for that reason. Demographers James Lee and Cameron Campbell argue that this population, like that of early modern England, adjusted fertility to economic circumstances: fertility was higher when grain prices were low, and vice versa.[13] The most important method of fertility control, however, was not delayed marriage or infant abandonment, but female infanticide (femicide)—about 20 to 25 percent of girls born were likely killed on their first day of life. The practice not only affected reproduction directly by eliminating some children, but also lowered the reproductive potential of the next generation by reducing the number of future mothers.

Average age at marriage for women in Liaoning was about 18, and virtually all women married at least once; remarriage after widowhood was rare for women but more common for men. Even though most women married young, a wife who survived to the end of her reproductive life with her marriage intact (that is, not necessarily a typical woman) had an average of about 6.3 children (a figure that does not include infants killed at birth, who do not appear in the record). Thus, while China's Total Fertility Rate was higher than that of England in Malthus's era, its Total Marital Fertility Rate was lower, even though people married earlier. In Liaoning, the average time delay between marriage and the birth of a first child was 4 years, and delays of up to 10 years were not unusual. The average age at which women had their last child was about 33.5, compared to about 40 in most traditional populations.

Liaoning does not necessarily represent all of traditional China, which was economically and ecologically very diverse. Resources were more constrained in Liaoning, state land was limited, industry other than agriculture was minimal, and the population was forbidden to migrate. Most scholars do, however, agree that some of the patterns described for Liaoning—early marriage, low to moderate fertility within marriage, near-universal marriage for women, and female infanticide—are typical of much of China during this period. Traditions of arranged marriage were strong in China, and if, as some have argued, the fact that husband and wife were often strangers at their wedding partly explains the delay between marriage and first birth, this is another example of a fertility-limiting behavior.[14] "Minor" marriages, common in some times and places in China (discussed later), also had lower fertility than major marriages.

Infanticide was illegal under Chinese law as far back as sources allow us to trace, and was piously condemned by moralists and philosophers over centuries, but it was nevertheless practiced widely, just as infant abandonment was practiced widely in Europe. Patterns of infanticide differed from those in Europe in that the child's sex was important: parents killed daughters much more often than sons, typically by drowning.[15] In scenes imagined by the authors of moralizing stories and essays, the decision to kill an infant is made immediately after birth, in a tense atmosphere with only women present—the midwife, mother, mother-in law, and female relatives (midwives and mothers-in-law receive the greatest share of blame in these stories).

While we were able to explain why girls in foraging societies were killed more often than boys among some peoples, it is more complicated to explain the same phenomenon in agrarian societies—why female infanticide was important in some, like imperial China, and not in others, where infants were killed or abandoned without regard to sex. Typically, scholars point to cultural explanations: for example, while values in early modern England were more individualistic, family identity and connection to ancestors were (and still are) much more important in Chinese culture. In China, lineage, inheritance, and ritual connection with ancestors were passed on through sons. Marriage customs were strongly virilocal—because brides moved into their husbands' families and ceased to contribute their help and labor to their natal families, they were perceived as a drain on family resources. This was especially true in places where parents provided dowries, sometimes at great sacrifice. Even in places where it was more expensive to marry off sons, who not only had to pay a bride-price but also needed enough patrimony to support a family, the bias against raising daughters could be strong; because sons remained with the family, expenditures on them were perceived as an investment, not a waste. In this family pattern only sons, and not daughters, could support a parent who was too old to work, and old-age support is still a common reason for son preference today. Mothers who bore only daughters could face rejection and abuse from the husbands and in-laws on whom they were dependent; their own well-being in the present and future depended on sons.

Cultural patterns favoring sex-selective infanticide are strong in some agrarian societies and negligible in others, even if they also practice infanticide on a large scale. Sex-biased infanticide is most common in societies with especially strong patrilineal, virilocal customs, because raising daughters is so disadvantageous in these conditions—by the time they are able to work enough to produce more than they consume, they marry out of the family and their contribution is lost. The larger reasons why some societies are more strictly virilocal than others are not, I think, well understood; internal warfare, rigid sexual divisions of labor, lack of opportunities to be economically productive outside marriage (that is, sources of income besides subsistence labor on a family farm), large-scale social organization (for example, kingdoms or states as opposed to clans or villages), and stable environments have all been proposed as contributing factors. Whatever the explanation may be, these reasons have been especially strong in China, northwestern India, South Korea, and a few other regions of central, southern, and eastern Asia, where femicide has been an important phenomenon.[16]

A complication is that, although in China both commoners and elites practiced femicide, elsewhere its prevalence could vary across social classes. Because of the "hypergamy" typical of agrarian societies—that is, parents tried to place their daughters in families of higher status, trading their reproductive capacity for social advancement and better chances of success for their grandchildren—higher-class families had to sacrifice more than lower-class families to secure desirable matches for their daughters. Among elite families, it might be almost impossible to find such a match, and they might prefer not to raise daughters at all. This effect of social status on rates of femicide has been most obvious to European observers in parts of northern India. The historical evidence for it comes from reports of the East India Company dating to the late eighteenth and nineteenth centuries, when the British, shocked at the evidence for femicide in this part of their colony (a response somewhat hypocritical, as the practice of infant abandonment was reaching grue-some heights in Europe in the same period), set out to eradicate it.[17] British sources and their informants described femicide as a practice specific to certain peoples, notably the Rajputs, a group of elite warrior clans. This group partly defined itself by its restrictive rules for women, including seclusion and veiling, widow self-immolation, strategic hyper-gamy for daughters, and a disinclination to raise daughters.

The effects of hypergamy may overlap with those of evolution. A fa-mous theory, the Trivers-Willard Hypothesis, predicts that families in relatively poor "condition" should invest in daughters rather than sons because their sons will have few chances to reproduce, while families in good condition should invest in sons because they will out-compete other males and produce more offspring than daughters. If high-status families are in better condition than low-status ones, their preference for raising sons makes evolutionary sense according to this theory. Studies looking for effects of the Trivers-Willard Hypothesis in humans, and indeed in other animals, have produced mixed results, and in general people assume that this hypothesis applies only to biological processes from conception through birth and not to parenting decisions after birth, but it is possible that the effect it describes has been a factor in sex-selective infanticide.[18]

Because of femicide, and because widows were discouraged from remarrying, there were usually more men than women eligible to marry in China, and many men could not marry. Among the nobility, some men had more than one wife—Zhang Yimou's 1991 film *Raise the Red Lantern* depicts the stereotype with high melodrama—and polygyny probably exacerbated the problem. China's unbalanced sex ratio persists today, despite modernization and the Communist government's aggressive assaults on patriarchal customs and values in the twentieth century. Once technology made it easier for parents to abort unwanted daughters before they were born, this method, although illegal, mostly replaced infanticide, with the result that the sex ratio at birth in China in the early twenty-first century was about 120 boys per 100 girls, compared to the range of 102–107 that most scientists consider normal; in recent years this rate has declined to 115 or so.[19]

I have discussed infanticide in England and China in this chapter, and among foragers in chapter 2, to show how critical non-reproduction and family limitation have been to human survival in many circumstances. Infanticide and infant abandonment (which for most of human history amounted to the same thing) were common and effective methods of fertility control in both societies and in many others. It is fair to assume that people killed their children not because humans before modernization were immoral or unfeeling, but because they were forced by circumstances; in many situations, they had better chances of survival and future reproduction without a new baby that was likely to die anyway. It is rare to find primary sources that describe a parent's feelings or motivations for this choice, but the memories of one woman, Ye, whose long life spanned the latter half of the sixteenth and first half of the seventeenth centuries in China, were recorded later by her son in his memorial biography of his father, Ye's husband. Even in old age, her anguish at recalling how she had drowned her newborn daughter with her own hands—an act that clearly traumatized her—is obvious and wrenching to the reader. At the time, as she remembered it, she despaired over her life of poverty and her family's inability or unwillingness to help.[20]

I have deliberately avoided the suggestion, easy to read between the lines of many demographic histories, that Chinese society was somehow more barbarous, more primitive, or more amoral than European society because it practiced different methods of infanticide, and I do not believe that this is true. Abandoning an infant to die of exposure, starvation, animal predation, or disease is no less cruel than drowning, and rates of infant abandonment could be very high in Christian European countries. In Florence in the first half of the nineteenth century, a

staggering 30 to 40 percent of infants baptized in the city were foundlings, most of them doomed to a rapid death.[21]

Modern advances in contraception and abortion have spared most
people today from these choices, although infanticide is still practiced
throughout the world in both industrialized and "developing" nations.[22]
The incidence of infanticide in the United States is estimated at about 8
per 100,000 newborns, or about 2,000 total per year (and an additional
31,000 newborns are abandoned in hospitals each year). In the United
States and elsewhere, infanticide is underreported and often not prosecuted, and courts are reluctant to convict mothers of the charge or to
impose harsh sentences on them for it. In the United Kingdom, the Infanticide Acts of 1922 and 1938 abolished the death penalty for women who
killed their children under the age of one year while in a disturbed mental
state, reducing the punishment to the level of that handed down for man-
slaughter, and today very few women convicted under this law, or under
similar provisions in several other Western countries, are sent to jail.

Infanticide thus seems to be understood and accepted more than one
might expect in cultures in which public rhetoric exalts and romanti-
cizes children, infants, and even fetuses. This acceptance, in turn, reflects the long, deep history of infanticide as a reproductive strategy
among humans (and not only, I think, a romanticism about motherhood that prevents us from seeing mothers as murderers, which is the
usual interpretation). Repeated studies have shown that women who
kill their infants within a day of giving birth—the most common
pattern—are typically young (in their teens or early 20s), unpartnered,
and unemployed or dependent on their parents; have no mental illness;
and have concealed an unwanted pregnancy. These characteristics,
along with other evidence, have led some researchers to conclude that
infanticide is an evolved strategy—perhaps one that arose with allocare,
as parents "stacking" offspring faced complex decisions about what resources to invest and the willingness of others to help.[23] (This kind of
reckoning up of the commitment of others is exactly what we see in the
decision of Ye, whose parents and in-laws both failed her at the birth of
her daughter.) While infanticide is common among mammals, most
other animals kill infants that are not their own—only a few primates

attack their own children, and these are the ones that also practice co-operative breeding. Whatever its prehistory, infanticide was practiced in a wide range of circumstances in the agrarian period.

Chinese family planning was not limited to reproductive restraint and infanticide.[24] Families that lacked sons, for example, had to make up the deficit somehow. A common strategy was to arrange for various types of uxorilocal marriage, in which a son-in-law surrendered some of his parental rights and rights in his own lineage, and adopted respon-sibilities toward his wife's lineage, often living with his in-laws for a time or forever. Some or all of his children would inherit from his wife's fa-ther; in return, his family paid little or no bride-price or received a "groom-price" instead. These marriages were considered less honorable than traditional "major" marriages and were hard on the groom, who had to accept a subservient position and often give up any hope of an inheritance or head-of-household status, but for families with too many sons, they could be an alternative to worse outcomes. Conversely, fami-lies eager to secure a daughter-in-law from the limited supply, especially if they were too poor to afford a bride-price, sometimes adopted girls as children or babies; if they were not yet weaned, the future mother-in-law often nursed them herself. Selling an infant daughter to a future son-in-law's family was an alternative to femicide for families who could not afford to raise a daughter. (This option was especially popular in the Hai-shan region of Taiwan, part of the Taipei Basin in the north of the island, whose well-documented, mostly mainland Chinese peasantry is described in the classic study *Marriage and Adoption in China, 1895–1945*, by Arthur Wolf and Chieh-shan Huang.) Called "minor marriage," this practice was perceived as a recourse for the poor, though it is well-attested among all classes. Minor marriage reduced tensions between daughter-in-law and mother-in-law, which could otherwise be formidable, but was often distasteful to the bride and groom, who had to marry each other after being raised together as siblings. Minor marriages had lower fertility than major marriages for this reason.

Finally, another type of marriage called *zhaofu yangfu*, "getting a hus-band to support a husband," is attested in some parts of China. In this form of marriage, a wife took a second husband with the permission, or at the insistence, of her first husband. In return, the second husband helped support the family. As with uxorilocal marriage, parental rights were negotiated at the time of marriage and often involved some sharing of rights between the two male lineages. Typically, couples who resorted to this type of polyandry (multiple husbands) had children too young to be productive, and the husband was ill or disabled. Second husbands were usually extra, unmarried migrant laborers with few prospects of finding a wife for major marriage. Though illegal and, like uxorilocal marriage and minor marriage, not considered ideal, this type of marriage was recognized and accepted by custom. Couples marrying this way often used matchmakers and witnesses and sometimes drew up written contracts, and the term *zhaofu yangfu* was widely used and understood.[25]

Chinese families did not necessarily, as Malthus thought, reproduce until they suffered the results of poverty and disease; instead, they made strategic, often heartbreaking decisions and sacrifices. While Malthus worried about the consequences of people following their hearts and instincts to marry early, Chinese marriages were arranged by families and were not impulsive, individualistic decisions. Chinese couples married early but controlled reproduction within marriage. One team of scholars argues, though their case is controversial, that these traditions explain why China made the transition to modern contraception and lower fertility faster than the Western world did.[26]

Families in imperial China formed on a different pattern than those in England—they had few servants or lodgers, and were more likely to include relatives outside the nuclear family. Both men and women married young instead of working as servants for a decade or more before marrying, as they did in England. By tradition, a bride moved to her husband's household at marriage, and the couple continued to live with his parents for as long as the latter were alive, as well as with the husband's brothers, sisters-in-law, and unmarried siblings. This could theoretically result in very large family groups and sometimes did, but because mortality was high and large families had a tendency to divide, average family sizes were about the same as in England, and most individual families were "nuclear" families at any given time. However, most people lived for at least some part of their lives in joint households with their married parents, children, or siblings.

Early in marriage, then, it was common for young brides to live with their mothers-in-law, and for most women in midlife, the role of mother-in-law dominated their identity. Before modernization, this transition to the status of mother-in-law was, along with the end of childbearing, by far the most important event of a woman's midlife years—that is, menopause was most significant as a social, rather than biological, phenomenon. The status of mother-in-law was a leadership position, while the role of bride and wife in a multigenerational household was one of subservience: mothers-in-law assigned the work they disliked most to their daughters-in-law and controlled household decision-making. This virilocal, patrilineal family structure so typical of peasant societies was hard on young women. As already described, mothers-in-law and daughters-in-law have some competing reproductive interests, and conflict between them is very common; in Chinese culture of the imperial period, it was both normal and expected.[27] Chinese literature, legend, song, and myth were sympathetic to the plight of the young daughter-in-law and usually portrayed the mother-in-law as ruthless, selfish, exploitative, and domineering (modern historians and social scientists also share these biases). While there is likely some truth to this portrait, it is much harder to recover the mother-in-law's perspective from historical sources. I also note here that when the typical economic unit is a family farm, household economies are extremely complex, and it is therefore not unreasonable to place a high value on decades of experience.

The stereotype of the malignant mother-in-law is balanced by evidence that paternal grandmothers can improve the chances that their grandchildren will survive. Although the evidence for this correlation is not as straightforward as that in favor of maternal grandmothers, in many cases paternal grandmothers do seem to help child survival. A few qualitative studies of the role of mothers-in-law in childrearing can add nuance to our image of the mother-in-law in traditional societies.

In Nepal, Tamang grandmothers interviewed by researchers saw themselves as sources of advice and care for their daughters-in-law during pregnancy, birth, and breastfeeding and were enthusiastically committed to this role. They attended births, encouraged breastfeeding, and made sure the mother-to-be ate the right kinds of traditional foods and that the right traditional healing rituals were performed when needed. Most had positive views of modern medical practices that improved outcomes for mothers and babies, even if these contradicted tradition and thus undermined their own position as sources of advice and care. While daughters-in-law had little power in this society, mothers-in-law perceived their role as benign and helpful, and their openness to modern medicine was probably important to the health of their families, given that they made most decisions about childbirth and new baby care.[28]

A picture that is similar in some ways (but different in others) emerges from a study conducted in rural Malawi, in southeastern Africa. There too, mothers-in-law attended at childbirth, helped establish breastfeeding, made most decisions about when to administer traditional medicines or supplementary foods, cared for children when mothers were busy, grew food for them, brought newly weaned children to live with them, provided advice about pregnancy and childcare, and fostered children whose mothers had died. Like mothers-in-law in Nepal and elsewhere, they dominated their daughters-in-law in authority. But unlike their Nepalese counterparts, the mothers-in-law of Malawi resisted advice from modern medical practitioners, who recommended exclusive breastfeeding of new babies for six months, among other measures. As they saw it, modern practices increased their own workload, made their daughters-in-law lazy, and encouraged daughters-in-law to ignore traditional periods of sexual abstinence after childbirth, which they believed caused dangerous health problems in their grandchildren and increased the risk of HIV in a society devastated by that disease. The result of these conflicts was that young mothers usually pretended to comply with hospital advice without actually doing so. In this case study, mothers-in-law bossed their daughters-in-law and sometimes gave bad advice about infant feeding, but also worked hard for their families and contributed a great deal to childcare.[29]

Most "development" programs aimed at improving health for mothers and infants are based on Western social models; they look down on the kind of traditional medicine practiced by grandmothers (and, indeed, on the older women themselves), focus on women of childbearing age, and take a "transmission-persuasion" approach, essentially telling people what to do. But a few programs have reached out to grandmothers, recognizing their authority in the household and the importance of their advice to the family. The best known of these is an experimental program of the late 1990s that involved 13 culturally conservative Serer villages in Senegal. There, as elsewhere, mothers-in-law had authority in decisions about health and childrearing, were repositories of knowledge about traditional medicine and health practices, and worked hard at household and childcare chores. According to this program's researchers, the stereotype of the grandmother/mother-in-law was a positive one of wisdom, patience, benevolence, and commitment to the well-being of grandchildren. To the researchers' surprise, grandmothers in this study were eager to be included in health classes and to update their traditions. Traditional wisdom advised hard work and not too much food for pregnant women, so that their bodies would be strong and the child would be small and easy to deliver; also, as is common in traditional societies, most people believed that infants could not survive on breastmilk alone and needed supplementary water. Researchers focused on changing the advice given by mothers-in-law on these points and a few others. They used a dialogue-based, problem-solving teaching method more appropriate to the senior women's status and authority, and they used familiar communication methods, especially songs. This approach was so successful that almost all village grandmothers participated in the program; by its end almost all were giving good nutritional advice to their daughters-in-law, and almost all women of childbearing age were following it, compared to a much smaller fraction in the study's control villages that offered health education only to younger women. One of the male community leaders included in the interviews commented, "Grandmothers are very respected and everyone seeks their advice, men as well as women. To succeed in promoting changes in health habits it is essential that you work with grandmothers who are the guardians of tradition."[30]

Another source of "qualitative" insight on the traditional mother-in-law and many other subjects discussed in these pages is the story of

Ning Lao T'ai T'ai ("Venerable Mrs. Ning"), whose autobiography was compiled from the stories she told in the 1930s to her friend, a teacher named Ida Pruitt, who had been raised in China by missionary parents from the United States. Ning Lao was born in 1867 to a family that had lost its land. Her father supported them by selling baked goods and doing other jobs in the town of Penglai, in Shandong province. At 13 she was married to a man in his 30s, who was considered a catch because his mother was dead; Ning Lao would not have to endure a mother-in-law. Her older sister had so many problems with her own mother-in-law that she suffered a nervous breakdown and had to be brought home for a while, although in other respects her marriage was good—she liked her husband, and her father-in-law provided for the family. Ning Lao's husband turned out to be addicted to opium and stole everything of hers that wasn't nailed down. She and her daughter survived on the food sent by her own mother and her brother, the family's only son, who never married. (Much as Ning Lao might have hated a strong mother-in-law, perhaps such a person might have kept her husband in line.)

At first, Ning Lao tried to conform to social norms that restricted respectable women mainly to the house, but when her mother died and her brother joined the army and was killed, she was reduced to begging in the streets. She allowed her second daughter to be sold into adoption, although she resisted that decision for a long time. She worked several different jobs as a domestic servant and a peddler to support herself and her older daughter, and eventually had two more daughters, both of whom died, and a son. The son brought her great satisfaction because he assured the continuation of her family.

Ning Lao most definitely controlled the size of her family. She left her husband frequently and reconciled with him when she thought she should have more children to carry the family forward. (She even did the same for her daughter, calling her son-in-law home when she thought she could support grandchildren.) In the end, three of her five children survived to adulthood, though the daughter sold into adoption died young of cholera.

Ning Lao's daughter never forgave her for the marriage her mother arranged for her, to a young cobbler who seemed promising, but turned out to be another addict who stole and did not support his family. His parents decamped for Manchuria soon after the marriage, and Ning Lao struggled to help her daughter and her grandchildren, all girls, to survive. In her world, men were supposed to earn income and support their families, wives were supposed to be supported by their husband's parents, and children were supposed to support parents in their old age. But in the world she actually described, husbands were not much help; parents continued to support adult children, nephews, nieces, younger siblings, cousins, and grandchildren throughout their lives; and few received substantial support from their descendants. The main exception was Ning Lao herself. Her oldest granddaughter, Su Teh, went to school, studied in the United States, and became an educator. With her income from this job she supported her own mother, her grandmother Ning Lao, her uncle's wife, and her young cousins, but never married herself; she rejected her grandmother's one effort to arrange a match for her, and Ning Lao, understanding that times were changing, did not try again. Eventually Su Teh became involved with the resistance movement against the Japanese and the revolution that created the Republic of China.

In Ning Lao's family, the transition to modernity temporarily reversed the normal flow of resources as a younger generation benefited from changes such as mass education. ("Surely it is a great thing to be able to read," Ning Lao said.) Also, the satisfaction that Su Teh took in her job and in helping to build a new China were among her reasons for rejecting marriage and children.

When her son married, Ning Lao quarreled with the bride's family, and it is clear that her daughter-in-law, Mei Yun, who was stuck in the middle of the conflict, was unhappy for a time. But Ning Lao also took her duty to help and support her daughter-in-law very seriously, once turning down a good job that would have taken her out of town because it was unthinkable to leave Mei Yun alone while pregnant. As time went on, her relationship with Mei Yun seems to have improved, and in any case they remained loyal to each other. When Pruitt knew Ning Lao, she was in her late 60s, enfeebled by a stroke but still doing everything she could to help her children and grandchildren, though her son was by then the main breadwinner for the family. "The old must think for the young," Ning Lao told Pruitt. "A family cannot be held together unless the older ones sacrifice for the young."[31]

Older women might not only be mothers-in-law, but also widows (though widows can also be young in environments in which adult mortality is high). Widows are often more visible in the historical record than other women, for reasons that will soon become clear, and so they are promising subjects of study as we seek to understand more about the lives and roles of post-reproductive women in peasant societies. Values about widowhood were especially interesting and complex in imperial China.[32] In some societies, an older woman's power in the family depended on her husband's survival, but Chinese widows had rights not available to married women or unmarried daughters. Under Qing law, a widow had the right to administer her dead husband's property until it passed to her sons or to an heir appointed by the husband's lineage. This special status depended on remaining faithful to the dead husband; if a widow remarried, she left behind her children and her former husband's patrimony to join her new husband's lineage. If her in-laws could prove that she had sex with any man, they had the right to sell her into marriage. (Some widows fought for their rights by insisting that they had become pregnant without having sex—and won!)

Widows who were chaste could not legally be forced to remarry, but their in-laws often pressured them to do so, because they wanted control of the widow's property or the bride-price they would receive for her. Those widows whose husbands did not leave enough property to support them had no choice but to remarry. Widows with some land, and whose husbands had no surviving brothers, were in the strongest position. These lucky widows might marry uxorilocally, live with an unmarried partner, or do whatever suited their advantage, as there was no one to benefit from making them forfeit their property rights.

An example from a case heard by Qing magistrates in 1876 in the province of Jiangxi illustrates some of the forces at work.[33] Wu Luo Shi was a widow whose only family was a grown son of 20. She had hired a landless laborer to help with the farm and had then begun a sexual affair with him (a common pattern in legal cases involving widows). The son found out but said nothing, not wanting to bring scandal on the family. The laborer wanted to become the widow's uxorilocal husband; she would have preferred to remain unmarried, but he threatened to expose their affair and so she consented. Her son did not try to intervene. Years later, however, the former laborer tried to sell some of his wife's property, and the son, whose inheritance was now in danger, finally took action: together with a gang of his friends, he murdered his stepfather, which is how the case came to court.

Another widow, Chen Shi, in a 1762 case from Henan, was even less fortunate. She had three small sons and two brothers-in-law, who had divided their inheritance with her husband. When she fell in love with her hired laborer and wanted to marry him uxorilocally, one of her dead husband's brothers refused to allow it. She began an affair with the laborer anyway and gave birth to a baby girl, who died. When her brother-in-law found out, she confessed the affair, hoping he would accept the marriage rather than endure a scandal, but he became enraged and threatened to have her prosecuted under the laws against illicit sex. The story's tragic climax is the reason the case ended up in court: Chen Shi drowned herself and her three children.

In Wolf and Huang's anthropological study of the Hai-shan region under Japanese rule, more than half of widows under age 30 remarried.[34] Among those, a large percentage married uxorilocally—about a third of all young widows under 24, with the percentage increasing as the women aged. Another large percentage of widows in this population lived with men without marrying them, and almost half of unmarried widows aged 25 to 29 bore additional children. Widows past 40, who were likely to be mothers-in-law with adult sons and high status in their households, rarely remarried, but we cannot tell how many may have lived with unmarried partners in the years past childbearing. On the surface it seems hard to reconcile this picture with the harsher one painted by Qing legal sources. It may be that customs were more permissive in Hai-shan, or that most widows there lacked interfering in-laws and had the power to do what they wanted; because no one brought them to court, we don't have legal records about them.

Studies of marriage and family patterns can give the impression that these institutions are rigid, timeless, and rooted in mysterious forces rather than flexible and adaptable to circumstances. Although for the sake of brevity I must simplify when discussing reproductive strategies

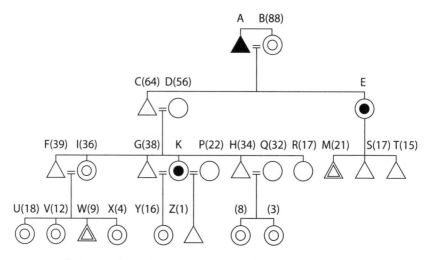

FIGURE 5. The Tan Family on March 28, 1933. From Wolf and Huang 1980, 361. In parentheses is the age of each living family member on that date.

in early modern England and imperial China, I have also tried to mention some of the ingenious alternatives and loopholes to which families have resorted in order to ensure that there are heirs, to keep a patrimony intact, to optimize sex ratios for the sake of both of those goals, to achieve the right balance between "men's work" and "women's work" on the farm, or just to survive. Rules about chastity and propriety have often been selectively enforced. While there is not space in this book for a lot of what historians call nuance, the story of one rural family, as recorded in the exhaustive registers of Taiwan's Japanese rulers in the early twentieth century and reconstructed by Wolf and Huang, will serve as an example. Even here I will have to oversimplify, as providing even a bare-bones narrative of this family requires several pages and multiple diagrams in Wolf and Huang's book (figure 5).[35]

In 1906, the Tan family consisted of a grandfather and grandmother in their 60s, an adopted son (a common but expensive strategy for continuing the lineage), his wife, and an adopted daughter. They lived in a U-shaped compound with another family headed by the grandfather's brother. The adopted son and his wife had already produced four children and had adopted a daughter-in-law for one of their sons. This couple would eventually have 12 natural children, the last born when the mother was 42 years old, and would adopt another daughter-in-law. Of these 14 natural and adopted children, 4 girls and a boy died in early childhood, and 2 daughters were given out for adoption. Four of the children—2 natural sons and 2 adopted daughters-in-law—married each other. When the grandfather died, his adopted son became head of the family.

The grandfather's adopted daughter never married but bore three sons by an unmarried partner. These sons all stayed in their grandfather's lineage and were still children when their mother died. They were raised by their maternal grandmother, the patriarch's widow, who lived into her 90s and saw her great-great-grandchildren. Eventually, when the first of the sons was old enough to head his own family, this branch of the lineage separated, becoming its own economic and social unit with its own stove and ancestral altar, though its members continued to live in the U-shaped compound.

Meanwhile, between them, the patriarch's seven remaining grandchildren through his adopted son produced a total of six natural children, four of whom died in infancy or early childhood, and adopted eight girls and one boy. Only one natural son in this generation survived; he married the adopted daughter of his father's mistress. The only natural daughter was sent out for adoption. The family lacked enough sons to do the "men's work" required to support them, so at least three of the adopted daughters were hired out as prostitutes (the economic exploitation of adopted children is a theme that deserves more attention than I can give it here). Before they married, these three daughters bore a total of six children and adopted one daughter. Two married out of the family, and one married uxorilocally; among the six children they bore as prostitutes, two girls were sent out for adoption, and two sons remained in their mother's lineage. This family had a total of 22 members when it was divided in 1936, most of them adopted.

Just as in foraging societies, not everyone in agrarian societies reproduces—the system only works if there are groups of people who do not reproduce. Besides post-reproductive women, foraging societies have other non-reproducing members—gay uncles, adolescent men, and "extra" men in groups with unbalanced sex ratios or high rates of polygyny. In the system that prevailed in early modern England, servants were the largest group of adult non-reproducing people. Some were domestic servants, precursors of the class romanticized in *Downton Abbey* or *Remains of the Day* (though more of these individuals worked in cities, especially London—as in *Upstairs, Downstairs*—than on rural estates). The majority were agricultural workers providing supplemental labor on farms, more familiar to us today from the novels of Thomas Hardy. Many or most people worked in service for part of their adult lives, and little social stigma attached to this practice. It was traditional to leave service at the time of marriage, with the result that few servants were married. Service was a non-reproductive phase of the adult life cycle— not a "natural" one, like the phase that menopause provides, but one that functioned to control fertility and increase productivity in this particular society. The English pattern was probably unique—we don't know of other societies outside of early modern northwestern Europe with the same system. Malthus was probably right to argue that this system was flexible and potentially sensitive to economic conditions— when these were unfavorable to the establishment of new households and families, people could stay in service longer. The institution of service meant that women were more independent than in most other systems, because they could accumulate their own savings (though female servants were paid less than men) and enter or leave contracts on their own authority, and also because they married later and thus escaped subjection to the legal authority of a husband for some length of time. Marriage in this population usually followed courtship and was not arranged by parents.

It was a weakness of the English system that, once children reached an age at which they could make a net contribution to the household economy, they left to serve other households for wages, which they mostly kept for themselves. While their older offspring were serving other households, some parents hired young adults from other households as servants for their own homes. While this practice might seem paradoxical, it probably helped distribute the labor of young adults more efficiently, as extra children from large households moved to smaller households where they were needed.

In preindustrial England, many people never married, and the interesting class of never-married, non-reproductive women has attracted scholarly study.[36] Daughters in this society did not normally inherit land if they had brothers, although there were many exceptions and regional variations, and women who did not marry rarely headed their own households until middle age, as law and custom strongly discouraged this. Single women were expected to, and by necessity usually did, live lives of pious, frugal dependency, caring for elderly parents, raising orphaned nieces and nephews, keeping house for bachelor brothers and overburdened married sisters, and tutoring young relatives. Those not living on farms worked as housekeepers, companions, and seamstresses and, when education for women became more popular in the eighteenth century, as teachers, tutors, and governesses. Those with property, discouraged by law and custom from forming their own households, shared it generously with their kin and left it to needy relatives when they died, preserving resources in the family that would otherwise have been dissipated. In a darker version of this pattern, brothers charged in their parents' wills with paying inheritances and marriage portions to their sisters sometimes resisted or refused to do so—it was in their interest to exploit the services of a single sister rather than give up her share of the family property.

Despite the fact that it was common not to marry in early modern England, this society narrowly restricted single women's economic options to discourage them from competing with householders with dependent children. Women's wages for all jobs were much lower than men's. Women who had never married rarely received poor relief. In England's prisons women greatly outnumbered men, and most of these female inmates were single women—servants and prostitutes arrested for crimes like shoplifting or petty theft. Many single women who were fired from a service position, orphaned without resources, or unable to survive on a seamstress's wage worked as prostitutes, at least for a time. Despite the tirades against prostitutes common in the public discourse of the era, this trade did not carry ruinous social stigma, and many who worked for a period as prostitutes eventually returned to more respectable (if lower-paid) jobs, or married.

Although they were often mocked and despised in literature, like the widows discussed next, and although society offered them scant means to prosper economically or even to live self-sufficiently, single women were nevertheless bulwarks of the family and community, like the non-reproductive *fa'afafine* of Samoa, as well as a ruthlessly exploited source of labor in the sectors of service and education. Indeed, reading scholarly studies of single women in preindustrial England, it is hard to imagine how families got by without at least one of them. An example is the case of Elin Stout, the spinster sister of a grocer, William Stout, whose autobiography is the source of many details about her life. Born in 1629 and of frail health from an early age, Elin never married but instead helped her widowed mother raise her younger siblings. She lent money to her bachelor brother William to buy a shop, helped him run the shop, nursed him when he was sick, and finally moved in with him to keep his house. She also raised two children through school age in William's house for their married brother, Leonard, and his overburdened wife, and nursed her mother in her declining years, before she herself died at age 65.[37]

Offspring like Elin Stout might care for elderly parents even though this role was not institutionalized as it was in China, and it was common for women too old and frail to work to move in with a grown child. English society offered a few other options for people too old to work. Sometimes farmers retired by handing over their property to one of their children on the condition that the offspring continue to provide food, lodging, and support for the rest of the parent's life. Also, beginning in the late sixteenth century, parishes were required by law to provide for the poor. However, as I have tried to emphasize in this book, the study of women past menopause is not the same as the study of the very old and infirm—the latter is an important subject in itself, but for most of human history infirmity and the inability to work have been typical of a short, final stage of life. In England, as in other societies, most middle-aged and older women (and men) worked, often until they died.

Older women in preindustrial England are, as in imperial China, most visible in the documentary record as widows. Because English law, which was draconian on this point, declared that almost all of a woman's property would become her husband's upon marriage, the lives of ordinary married women are obscure in records. Widows, however, owned property in their own right, which was recorded in wills and probate documents.[38] Though widows faced many economic disadvantages compared to men, their range of options was nevertheless much less narrow than that of women who never married. A recent study of widows' work based on documents dating from 1534 to 1699 shows that widows typically continued with the same kinds of work they had done while their husbands were alive—livestock farming, arable farming, textile production, food processing (making cheese, butter, malt, and beer), keeping inns and taverns, renting rooms or (if they were wealthy) land, and lending money.[39] Widows' households were, unsurprisingly, not as wealthy as married households—on average, they had about 70 percent of the wealth held by married households. About two-thirds of the 75 widows studied were still doing hard physical work when they died—agriculture, textile production, food production, or hospitality work—and only 7, or 16 percent, were "retired" and doing no obvious work for money. Five had not been left enough property to live independently, and all of these women were working when they died, raising livestock and producing malt.

The documents, though jejune, sometimes tell stories of remarkable resilience and enterprise. Elizabeth Grimes of Burmarsh inherited and administered the estate of her husband James, a farmer, when he died in 1685. He left £151 in inventoried assets and £266 in debts. When his wife met all the obligations of his will, she faced a deficit of £166. But when she herself died five years later, she still had the farm, plus inventoried property worth £77—she had been able to support herself and, presumably, pay off the debt accumulated during her marriage in the meantime. Another widow, Margaret Greaves of Chesterfield, inherited all of the marital property when her husband died in 1620; at that time their six children were still under the age of 21, too young to inherit. The couple had made a good living in tanning while married, but Margaret gave up the tanning business—which may have been her husband's work rather than hers—and supported her large family by farming, brewing, making cloth, and running a restaurant and bakery. When she died in 1635, her inventoried estate was still worth two-thirds of what her husband had left, and she was owed £23 in money she had lent at interest.

Many widows were poor, partly because of laws and customs that disadvantaged them economically. Because married women did not, for the most part, legally own property, when a wife died, the marital estate remained intact, but when a husband died, the estate had to be settled, and debts and legacies paid. A man could remarry and acquire any property belonging to his new wife, whereas women surrendered their property on marriage. Although it was normal for women to inherit part or most of their husband's estate—in parts of the country, by ecclesiastical law, the minimum bequest to a widow was one-third of the estate, and most husbands left the bulk of their property to their wives—many men left more debts than assets. Women who worked for wages were paid much less than men, which further limited their ability to support their families; in cities, widows and spinsters, both related and unrelated, sometimes lived together in "clusters," the better to survive on the starvation wages paid to women who made clothes or lace. In preindustrial England, as today, most people on poor relief were women with dependent children. Statistics are hard to reconstruct, but in one parish, one-third of all widows were on poor relief (on the other hand, only a small percentage lived in institutions like workhouses).[40] Still, poor as they might be, provided they had inherited any sort of property, widows were the most liberated class of women in England in the centuries before industrialization (and they were often lampooned in literary stereotypes of the brash, sex-crazed "merry widow"). The great majority of unmarried widows, about three-quarters, headed their own households, and about 13 percent of all households were headed by widows.[41]

Because women married later in England than in most agrarian societies, they spent less time on average as grandmothers. Also, since parents did not typically live with their adult children, there may have been less opportunity for grandmothers to help their grandchildren. However, a small but significant number of households consisted of grandparents and their grandchildren, most likely reflecting a tradition in which grandparents picked up the slack when their children died or pursued opportunities far away. Despite the very large amount of data available and the exhaustive demographic studies of family and population structure that have been produced for early modern England, scholars have not attempted the types of studies that have tested the Grandmother Hypothesis in some other societies, and in general have published little on the role of the grandmother in this population. We do know that despite many institutional disadvantages imposed on unmarried women, widows worked hard to support themselves and their dependents; unmarried sisters and daughters played highly useful, sometimes critical roles in many families; and the average number of children per family was relatively low.

Here I would like to digress to make a point about the "female-headed household" (by convention, most modern sociology assumes that all families in which an adult male is present are "male-headed"). In any society in which men control most of the resources and power, female-headed households will struggle compared to their male-headed counterparts; this is obvious in the historical examples of households headed by widows discussed in this chapter. In the late twentieth and early twenty-first centuries a great deal of research was devoted to proving this tautology, and in particular to showing that female-headed households were inferior environments for raising children. Stable, monogamous marriage was often proposed as the solution to social problems perceived to arise from the growing number of female-headed households. In the United States of the late twentieth century, this argument had a racial dimension, as female-headed households were more common among African Americans, partly as a result of high rates of unemployment and incarceration among men in that group; this was the era of the "Moynihan Report" and its intellectual fallout.[42] As a result of centuries of oppression, black families also had (and still have) much less money than white families, which had the same predictable effects on children's well-being as institutional sexism had on female-headed households. It is very unfortunate that the results of poverty and oppression were for so long attributed not to inequality in an unfair socioeconomic system, but to "cultural" factors such as values and single parenthood.

A better solution than reinforcing an institution (marriage) that took shape along with the rest of the patriarchal apparatus of agrarian societies would be to address the problems that create hardship and inequality for women and oppressed groups. An example of this approach can be found in Iceland, where gender equality is mandated by government policy, and which since 2009 has ranked in first place, with the lowest degree of inequality, on the World Economic Forum's Gender Gap Index. Two-thirds of Icelandic children are born to unmarried mothers, but the nation's rate of child poverty is the lowest in the world. Icelandic family policy promotes the equal participation of men and women in parenting as part of its commitment to gender equality; its policies, which include long paid leave for both parents and subsidized childcare, support families and children without necessarily being pro-marriage. In the United States, on the other hand, despite a strongly pro-marriage culture and laws and policies that make single parenthood difficult (for example, the federal government requires no paid parental leave, allows health benefits to be tied to marital status, and taxes married couples in which partners make equal incomes at a higher rate than those in which one partner, usually the wife, is economically dependent), more than 40 percent of children are nevertheless born to unmarried mothers. Meanwhile, by most measures, the United States ranks highest or nearly highest among modernized nations in child poverty (Romania is higher by some measures). The strategy of promoting marriage to fix social problems has proven ineffective, and thankfully, the sociology of the family is now moving in new directions.[43]

VIRGINS AND EXTRA MEN: NON-REPRODUCERS IN THE AGRARIAN PERIOD

Non-reproduction was as important as reproduction in sustaining English society in the centuries before industrialization. In this system, *reproductivity* was relatively low and *productivity* for women both early and late in adult life was relatively high. In China, a higher percentage of women married, age at first marriage was lower, and almost all women married at least once. Even so, there were some non-reproductive women. Married women stopped having children earlier than in other societies. Not all widows remarried, even if they were young. (Many unmarried widows continued to have children with unmarried partners, but their rate of fertility was lower at all ages.) Some female children were purchased as slaves. While their status otherwise resembled that of daughters-in-law bought for a bride-price or, indeed, that of brides, slaves were not expected to marry or reproduce. In China, as in many agrarian societies, professional entertainers and prostitutes reproduced at a lower rate than wives. These women, of whom Sappho of archaic Lesbos is probably the example most familiar to readers, are fascinating historical subjects to whom I wish I could devote more space. In imperial China, the number of women in most of these categories, who were of reproductive age but not actually reproductive, was small and probably did not affect fertility that much, but they did exist. The category of married women of reproductive age who had either stopped childbearing or not yet borne children, however, seems to have been substantial.

While a full catalog of non-reproducing or low-reproducing groups in peasant societies is not practical here, I should mention the most visible of these: the celibate monks, nuns, and saints of traditional religions. These are attested widely and from a very early date. The *naditu*, unmarried women dedicated to temples by their families in Old Babylonian Mesopotamia, lived together in cloistered groups separated from men; records of many of their contracts survive, and they appear in the Code of Hammurabi. *Naditu* women typically received a share of land from their families and appear to have had some control over it, including the rights to sell it or bequeath it, though they remained dependent on their brothers for their maintenance and the management of their property.[44]

In numbers, these groups might be small or large. There were only six Vestal Virgins in Rome, but about 10 percent of all contracts that survive among the documents of Old Babylonian Nippur involve *naditu* women. In societies that severely restricted women's economic independence, religious celibacy was often the only appealing alternative to marriage and was important for that reason, even if the overall numbers of women who adopted that life were small. Sources often explicitly contrast the typical female life cycle and destiny of marriage and childbearing with the alternative of celibacy, which they frame as a liberation from drudgery, a spiritual marriage to a supernatural being, an indefinite extension of the state of virginal purity, a renunciation of the evils and weaknesses of femininity, or all of these.

To illustrate the subject of religious celibacy for women, I offer a brief discussion of the female Hindu saints of medieval India, many of whom were part of the *bhakti* movement that spread through Hinduism beginning in the twelfth century CE.[45] The term *bhakti* signifies a range of diverse traditions that have in common the theme of personal devotion to a god, and a tendency to be egalitarian and critical of traditional Hindu social hierarchies. Its saints produced poetry that was mostly performed orally, as songs; some of these circulated widely, became very popular, and have been preserved and handed down along with stories and legends about the saints who produced them. In some cases it is possible or likely that songs attributed to a particular saint were actually composed by followers, so that the body of their work contains more voices than just those of the named authors. In these ways, although the medieval female saints were few in number, they spoke, and still speak, for a broader population. I choose this example because it is from a region I have so far neglected, and because one theme in the poems of the Hindu saints is their resistance to or escape from the traditional life cycle of marriage and childbearing. Marriage was perceived as incompatible with spiritual liberation, and the stories of women saints emphasize their struggle against subjection to their husbands, parents, and in-laws.

Female Hindu saints were usually either widowed or unmarried because of their station in life (among the best known were a courtesan and a household servant), or else they managed to renounce marriage in some way. One of the earliest and most famous, the Tamil saint Avvaiyār, is said to have been miraculously transformed into an old woman so that she would not have to marry any of her wealthy suitors. The fourteenth-century yogini Lalla of Kashmir married at 12 but abandoned her husband and cruel mother-in-law and wandered naked for the rest of her life, in defiance of the ridicule she suffered for this violation of female modesty. Some saints described themselves in their poetry as adulteresses or prostitutes who had abandoned their husbands for love of another (that is, a supernatural man or god), a metaphor so potent that male saints used it too.

One of the best attested of the female *bhaktas* is Bahiṇābāī, who lived in the seventeenth century in the region of Maharashtra and left hundreds of poems, including a long series of autobiographical poems, in the regional language, Marathi. She was married at age 3 to a 30-year-old widower and began married life with her husband at 11. But early in life she became a devotee of the *bhakti* saint Tukārām, who appeared to her in dreams, and of the god Viṭhobā. Her husband was violently abusive and beat her unconscious on her spiritual conversion at age 11. When he tried to abandon her, however, he fell sick, and interpreted this as divine punishment for his rejection of his wife's faith. In the meantime Bahiṇā, who had become pregnant, decided to affirm her subjection to her husband:

> I'll serve my husband—he's my god
> My husband is the supreme Brahman itself.[46]

Unlike most female Hindu saints, then, Bahiṇābāī remained married apparently until her husband's death, the time of which is not recorded; she herself died at age 72. In her poems she makes a special point of the notion that it is possible to reconcile the frequently incompatible principles of religious devotion and married life. Her life shows that despite the tradition that women, like men, could be saints devoted to a god, her world provided no acknowledged escape from the normal female life course. "A woman's body is a body controlled by someone else," she writes. "Therefore, the path of renunciation is not open to her."[47]

The story of Mīrābāī, the daughter of an elite family of Rajasthan, who lived in the first half of the sixteenth century, is particularly interesting because of her origin among the highly patriarchal Rajputs, discussed briefly earlier.[48] Legendary accounts of her life tell us that she was married against her will to a prince but was a terrible wife, refusing to be subservient to her new family, following a low-caste guru, and devoting herself to worship of Krishna. She was widowed while still young. In some versions of her story, her in-laws pressured her to become a *sati*—to commit suicide on her husband's funeral pyre—but she refused, determined to dedicate the rest of her life to her true love, Krishna. Her in-laws plotted against her and tried to poison her, until finally she left and became a homeless wanderer, performing pilgrimages and eventually disappearing in a temple, having finally merged with the god.

The tradition about Mīrābāī holds that she rejected not only marriage, but all the conventional roles of a Rajput wife, daughter-in-law, and widow. In her poetry, which survives in the Brajbasha dialect of Hindi and in several other regional dialects, she explicitly renounces the social customs among which she was raised. For example, rather than romanticizing her natal home and family, a common theme in Indian women's poetry, she dissociates herself from all kinship ties as well as ties of marriage:

> Like casting off a veil
> Honor, shame, family pride are disavowed
> respect, disrespect, marital, natal home
> renounced in the search for wisdom . . .
> of father, mother, brother, kinsfolk have I none.[49]

In modern times Mīrā is reviled among the Rajputs, though some Rajput women interviewed privately have said they admired her courage but could not condone her rejection of her duties as a wife. Her songs are sung and celebrated among untouchable castes, peasants, widows, beggars, and other oppressed groups in Rajasthan, and Gandhi invoked her as a hero of social justice.[50]

Although a few communities of female disciples of Mīrā exist today,[51] medieval Hindu society offered no well-recognized institutions similar to the Buddhist nunneries of China or the Catholic convents of medieval France, where women could reject marriage and also live independently from their families, if not from the discipline of a religious order—Mīrābāī, Bahiṇābāī, and other female Hindu saints were exceptional individuals. Their poems show that Hindu women have been well aware of the strict social and economic bonds that constrain their lives; telling the saints' stories or singing their poems is just one way in which they have commented on those constraints.

Classes of non-reproductive "extra" men were also important in agrarian societies and could be quite large, especially where rates of polygyny or female infanticide were high, or where sons other than the oldest could not count on inheriting enough land to support a family. Historians have most often seen extra men as a problem, and have tended to assume that high sex ratios (that is, populations in which men outnumber women) cause men to compete more aggressively for mates and increase the level of violence in a society. But it is obvious that many extra men helped their families by working on the family farm or by sending money home from wage-labor jobs. Like single Ache uncles who helped support their nephews and nieces, extra, unmarried brothers and uncles in many agrarian societies contributed labor without increasing a family's burden of dependents. At Liaoning, for example, brothers who were not the heads of households waited longer to marry and had fewer children, playing the role of low-reproducing helpers. A similar pattern has been described for the farming families of nineteenth-century Sweden.[52]

The position of these unmarried brothers was generally, however, unenviable. The situation of China's *guanggun*, or "bare branches"—a term invented by Qing dynasty officials in the late seventeenth century to denote those men, usually poor, who were shut out of the marriage market by the sex imbalance and therefore still unmarried at an age when most men were raising families—is still difficult in the twenty-first century. Mostly poor and living in rural areas, the *guanggun* are marginalized and despised in a world in which one must have a spouse and children to be considered a fully adult member of society. But even though, as a result of the state's family planning policies, they usually have no brothers to help, and despite their reputation as lazy and shiftless, these bachelors perform a crucial function: they support their parents. Thus most village officials, in spite of their prejudices, describe bachelors as helpful and filial. So far, this group has not fulfilled the dire predictions of sociologists who see them as a destabilizing, potentially violent threat to society.[53]

Opportunities to try one's fortune at something other than farming—as a miner, sailor, bandit, scholar, artist, or bureaucrat, for example—were more available to men than to women in peasant societies. Many of the professions open to extra, non-reproductive men were

materially productive and nonviolent, and some were culturally important. Among the higher classes in imperial China, it was common for
extra sons in families with too many heirs to study for the clerical exams
and join the ruling bureaucracy. In many societies, extra sons joined
monastic orders, which, while neither as idyllically peaceful nor as
chaste and non-reproductive as one might think, did create, preserve,
and transmit a great deal of high culture.[54]

Extra men, then, are not necessarily the problem that they are sometimes made out to be. A recent review of a large number of studies on
the relationship between sex ratios and violence finds mixed results and
no evidence that high sex ratios and increased competition for women
lead to more violent crime. These authors also make the point that not
all sexual competition is violent, and some can be productive.[55] Anyone
with a gym membership or job in a male-dominated field like technology or academia can think of countless examples of men competing
nonviolently.

This is not to say, however, that extra men have no historical relationship to violence. States in all eras have enthusiastically put surplus male
energy to work in war and conquest, activities that can also help correct
the sex ratio by killing off some of the men and capturing and enslaving
new women.[56] Organized violence has been an attractive option for
extra sons with few other means of acquiring women and property. In
his famous article "Youth in Aristocratic Society," Georges Duby painted
a striking picture of the younger, noninheriting sons of aristocratic families in twelfth-century France, who wandered the countryside in bands,
competed in staged tournaments, caused the endemic raiding and feuding that plagued the region, and eventually sought their fortunes in the
Crusades.[57] Similar situations in which younger sons of high-status
families banded together to prey on the countryside or to fight one another, or were co-opted by a higher authority to fight external wars, have
been described for fourteenth-century Portugal and for Uttar Pradesh,
in northern India, before its annexation by the British. In a different
pattern of violence, the young apprentices of sixteenth- and seventeenth-
century London rioted predictably on annual holidays despite special
curfews and the deployment of hundreds of guards and militia; these

were great, noisy, armed disturbances, in which they attacked immigrant tradesmen, broke windows, burned crops, and destroyed theaters and brothels.[58]

Banditry, a type of predatory rural violence common in agricultural societies, was an obvious option for unmarried, extra men. To be sure, banditry was a complex phenomenon, especially because states usually labeled any organized opposition or groups that flouted their control as "bandits" or an equivalent in the appropriate language (there are similarities with the modern category of "terrorist"). A few bandits (the "social bandits" of E. J. Hobsbawm's famous work[59]) had ties to local peasant communities that might perceive them as champions of the downtrodden, like Robin Hood; others collaborated with rich landlords who used them to rob, bully, and extort their neighbors; still others occupied inaccessible terrain that insulated them from the reach of the state or local authorities. To the extent that bandits were groups that sustained themselves by rustling, looting, extortion, kidnapping for ransom or to sell their victims into slavery, highway robbery, and murder, they were mostly male and, it is reasonable to assume despite a lack of much hard demographic evidence, largely unmarried. (There were female bandits too, and some famous female bandit chiefs, but they were a small minority everywhere). Hobsbawm argued that bandits hailed from "the rural proletariat," by which he meant the class of mobile laborers dominated by unmarried young men, the sons of poor or landless families. In the Roman Empire, many bandits were enslaved shepherds operating in rural highlands on the margins of civilization, cashiered soldiers, or deserters from the army—that is, mobile, unattached, and more or less desperate populations of men.[60]

In southern China during the Qing dynasty, most people convicted of banditry were landless laborers who formed temporary gangs to commit one crime or a few. Here most bandits were older than the stereotype popularized by Hobsbawm—over 30, with a median age of about 32. Both contemporary sources and modern historians associate banditry in imperial China with the *guanggun*. A majority of convicted bandits in nineteenth-century southern China—about 55 percent, according to court records—were bachelors.[61]

Among history's extra men, one of the most interesting is the heretic Pierre Maury of the village of Montaillou, who kept sheep in the high pastures of Languedoc and, across the Pyrenees, in Catalonia. Pierre was not the eldest son in his family, and was destined never to succeed his father as head of household. He spent his life moving his herds over vast distances with the seasons, traveling from village to village, staying with peasant families, or living with other shepherds in remote huts where they cooked together, made cheese, and entertained friends. Shepherds were a law unto themselves, and Pierre operated as an independent agent, contracting for work with different employers; he also owned his own sheep, sometimes as many as 100. A fierce defender of his friends and siblings, he once kidnapped his sister Guillaumette to save her from an abusive husband.

Though most transhumant shepherds were bachelors, marriage was not an impossibility for Pierre: as a young man of 20, he was tempted by an offer of engagement to the 6-year-old daughter of one of his employers, pending a long wait until she reached puberty. This offer included the hope of being adopted by his father-in-law, who had no sons, and inheriting the patrimony of the household. Later, and despite his protests that he could not afford to support a wife, he was briefly tricked into marrying Raymonde Piquier, the mistress of one of his friends. This marriage was dissolved within a few days at the friend's insistence, and its purpose was no doubt to legitimate the baby Raymonde was already carrying and that was born a short time later. In the meantime, we hear hints of flirtations, hookups, and mistresses; close friendships with other shepherds; and associations with a long string of employers, including several female landowners.

Shepherding was a humble profession and often a last resort for the poor and dispossessed, and Pierre's life was hard, exposed to weather and accidents, but it was not especially lonely or unsatisfying, and he was free. With the advantage of mobility he evaded the Inquisition for longer than most of his family and friends. But its tentacles were long, and eventually, in 1324, he was arrested and put in prison. We have the testimony he gave under interrogation, but we do not know his fate.[62]

•

CHAPTER 8

The Modern World

IN MANY WORKS of what is tastelessly called "Big History," humanity's massive primeval past is hurried over quickly, with progressively more attention given to progressively smaller units of time as one approaches the modern period. The past is just background for the Industrial Revolution and everything that followed. But in this book, the modern era is only a coda to a much deeper, darker, more obscure, and more portentous past, the submerged part of an iceberg of which modernity is only the tip. Our review of this past, oversimplified to be sure, suggests some broad approaches to understanding the present. We find that the agrarian economy gave rise to many institutions that might be unsuited to an economy based not on family farms and inherited property but on wage labor, capitalism, factory-model production, mechanization, and the more recent communication and information technologies that have "globalized" that economy. In particular, production of the most basic necessity of life—food—can now be done by a small number of specialists, and in modernized economies, few people depend for survival on inherited rights to farm a plot of land. In the world's fully modernized nations, only about 4 percent of people are employed in farming, although in the rest of the world, the figure is still close to half. Globally, about 40 percent of the population still farms for a living.

The most obvious of the institutions rendered obsolete by modernization is the territorial nation-state, which arose when land was the basis of the economy and had to be competed for and defended. Other institutions arising in the agrarian period and not necessarily relevant in the modern one include, as we have seen, warfare, economic inequality, patriarchy, and formal marriage. The division of labor by sex dates to the Paleolithic period and persisted in the agrarian era, but with wide variation, and probably for social rather than economic reasons. None of these things is a necessary feature of human society. I don't think we yet know what is possible in the modern economy, because it is so new and because the attitude of the most dominant nations has been to avoid imposing intentional outcomes on the modern economy.

If the basic economic and social unit of the foraging period was the residential band, and that of the agrarian period was the family, in the modern period this unit is the individual, and the concept of individual rights is part of the intellectual armature of modernization. This last development brings obvious advantages but also raises many challenges. It is difficult to organize a social system around the individual while providing for the collective good, and this tension has not been completely resolved.

Modernization is a new phenomenon in our history, but it has been enabled by some of our oldest evolved traits as humans. The investment in experience, technology, and culture that has flourished with modernization is an extension of our evolved life history. Our fast reproduction rate compared to other apes has caused a massive population boom as resources have expanded and mortality has dropped; on the other hand, our tendency to invest heavily in our children and grandchildren, and in some circumstances to favor a quality-over-quantity reproductive strategy, has caused fertility to plummet in the wake of the boom, though not, as yet, far enough to stabilize population in most regions. While we are used to thinking of modernization as a liberation of energy from the work of producing food—and it certainly is that—it is equally, and perhaps even more importantly for women, a liberation from the work of reproduction, as families can be reproduced with few children. The modern age is the first age in which mothers and grandmothers have an energetic surplus comparable to that of men, and in which most people can draw on relatively large reserves of energy to invest in activities besides bare subsistence and reproduction. These changes have not all been good and have caused problems—problems that we should be able to solve with the energetic surplus available to us, but on which we have not made enough progress.

I want to make sure that I do not appear to be naive about modernization, though I believe in its potential. Some of its benefits (or potential benefits) are so important that extending them to everyone and preserving them should probably be our highest priority—these include low mortality; widespread or universal education; the gradual recognition of women (and even children) as individuals rather than subordinate members of a family; relief from hunger, backbreaking labor, and the ravages of infectious disease; lightning-fast communication, reaching all over the world and increasing the efficiency with which we can pursue and share knowledge; and a vastly expanded role for nonsubsistence activities (art, science, entertainment, and "culture" generally). But the costs and dangers of modernization are also very great, including the exploitation of workers; the rapid degradation of the environment; the expanded opportunities for surveillance, coercion, and corruption that modern technology affords; and the dispossession of large numbers of people in the transition from agrarian to industrial economies.

From its origins, modernization has been linked to Western exploitation of the rest of the world. The Industrial Revolution arose first in Europe, and early-industrializing nations sought new raw materials, markets, and outlets for their booming populations by acquiring colonies all over the world. In the wake of decolonization, the process of "globalization" has been dominated by the United States, and its rules favor rich nations.[1] For most of its history since World War II, globalization has been closely linked to "neoliberal" economic theories, based on the idea that businesses and markets, if left alone to pursue profits (or, more accurately, if supported by governments in their pursuit of profits), will automatically benefit everyone—principles that have been imposed worldwide by institutions like the World Bank, and have been linked to a strategy of U.S. political and military domination. In its practical application, neoliberal ideology has tended to make a few people very rich, to encourage the formation of giant transnational corporate conglomerates, and to increase inequality between rich and poor, within and between nations. (The latter trend has started to reverse by some measures as the populous "emerging" economies of India and China begin to catch up, but inequality within nations is getting worse.)

Globalization has often been a "race to the bottom" as corporations relocate factories wherever labor is cheapest and regulations are few.

The modern economy, in contrast to the peasant economy, has the potential to eradicate poverty and guarantee a high minimum standard of living for everyone, and it is good news that "extreme" poverty (now defined by the World Bank as income less than $1.90 per day or its equivalent purchasing power) is falling worldwide. The rate of extreme poverty has been reduced from very high levels in the agrarian and early industrial periods, when almost everyone lived at subsistence level—this rate was still as high as 44 percent in 1981—to about 13 percent in 2012.[2] (Although there are problems comparing poverty in industrialized and peasant economies, I think the point is a fair one.) On the other hand, 13 percent is still an enormous number of people, about the same absolute number as before industrialization and the modern population boom, and much of the global workforce still labors in misery. Some regions have fared better than others. About half of all people in extreme poverty live in sub-Saharan Africa, and more than a quarter of a billion live in India. It is fair to guess that hunger and extreme poverty might be rare by now if their abolition had been a higher priority in the process of modernization.

Besides the stratospheric inequality that it has fostered, its subversion of democracy to the interests of business, its links to a new sort of colonialism under the leadership of the United States, and the enormous amount of waste and pollution that it generates, another problem with modernization in its current form is its dependence on exponential economic growth over time. Even for rich nations, an annual economic growth rate of 1 percent is almost always described as "anemic." These expectations are obviously unsustainable for the same reasons that exponential population growth is unsustainable, and especially because economic growth is the more direct cause of environmental problems like carbon emissions. Also, while demographers can at least foresee a way in which population might stabilize or decline once the Demographic Transition is complete, economists offer no comparable vision.

Finally, there is the question of how the modernized world will cope with a potential reversion to the more variable climate of our ancient past. To return to the metaphor I used at the opening of this chapter, the underside of the iceberg tells us that for the first 200,000 years of our existence as a species, external catastrophes played a large role in keeping our population in check. Using our ability to multiply quickly and to colonize vast distances in a few generations, we kept two steps ahead of extinction. In the agrarian period, climate was much more stable, and constraints on resources played a greater role than before, along with high background mortality from infectious disease. Malthus's "vice and misery"—poverty, malnutrition, and war—limited population growth, and our problems were of our own making. The long view, however, tells us that we are unlikely to escape the forces of natural catastrophe indefinitely. While we can see a reversion to older conditions coming in the form of climate change, and while it is fair to blame ourselves, in Malthusian fashion, for an overconsumption of resources and overproduction of waste that is also causing many other problems, Earth's history tells us that it was always only a matter of time. Agrarian and modern institutions arose in a period of unusual climate stability and may not have the flexibility to adapt. Our species, which once traversed central Asia like it was the parking lot at Waffle House, now lives in dense settlements dependent on a great deal of infrastructure. And there are a lot more of us.

It is not surprising that human population "boomed" in the mild, stable climate of the last 10,000 years, and especially in the modern era, with its lower burden of infectious disease and new techniques of food production. But the long view tells us that the boom is part of a cycle. In the relatively brief agrarian era, each boom was typically followed by a plateau, in which population held steady at an upper limit constrained by resources, and that is what most people assume will happen eventually. However, our Paleolithic past warns that "bust" is another possibility, especially if our climate changes suddenly or returns to an unstable state. Our options for response to such a crisis, while expanded by technology, have become much more limited in other ways. For example, our most important prehistoric strategy, migration, became less frequent in the agrarian period, when land and other property were essential to survival and opportunities for migration were fewer.

Migration is potentially much easier in the globalized economy because goods, money, and information travel easily around the world. But the international migration of *people* is more restricted: today's states limit or exclude migrants under policies developed in the early twentieth century, when the beginnings of a global population boom prompted nationalist and racist responses with which not everyone agreed at the time. Today—despite our long and essential nomadic heritage—when people are displaced by some disaster, we call them "refugees"; nations place severe restrictions and quotas on people with this status and argue about what to do with them.

Climate change will require many people to move. Sea level will rise, either gradually or, and this would be much more disruptive, abruptly. An extreme but highly regarded study predicts that sea level rise will double in speed every 5 to 20 years to reach 5 meters (about 16 feet) within 40 to 145 years. This would displace hundreds of millions of people—and predicted final values for sea level rise, given the current and projected future carbon composition of the atmosphere, are much higher than five meters. While there is debate about how humans will respond to these changes, it is hard to imagine that large-scale migrations will not be part of the answer. It is an oversimplification, but still true, to say that this recourse will be easier if there are fewer people in the regions that we need to leave and also in the regions to which we need to go.[3] That is, reproductive strategy will be a vital part of whatever future modern society may have.

What we know about climate history and what we guess about our own prehistory cautions us that the brief 10,000-year agrarian era is not necessarily a good guide to the future. Though human population fluctuated around the limits of food supply throughout much of the agrarian period, we may be returning to an era of more brutal external checks. Preventing those disasters from claiming large numbers of lives in the first place, and from returning the world to a traditional demographic regime of high mortality (as might happen if modern political and economic systems collapse), should be very high priorities—allowing nature to destroy huge percentages of our population, as most likely happened in our prehistoric past, is an unacceptable alternative today. Predictions that climate change will cause a great wastage of human life such as we have not seen in recorded history are not implausible, and even a small chance of this outcome should deserve serious consideration. Hope of forestalling it lies partly in our talent for non-reproduction.

POPULATION THEORY: NON-REPRODUCTION AND THE FUTURE

Ever since Malthus, people have thought deeply about the idea that fertility control might be a painless and humane alternative to more violent checks on human population. Since then people have debated the point that, on the global level, it may be possible or necessary to ask the same kinds of questions that families faced throughout the agrarian period: How many people can we afford to maintain, and at what level of well-being? The breadth of this type of thought grew as the current population explosion became more obvious in the late nineteenth and early twentieth centuries. (Malthus himself was more interested in explaining why the English population had remained relatively stable than in raising the alarm about a rise he did not yet perceive as dangerous.)

Twentieth-century thought about population was wide-ranging and fascinating, and often politicized in many unfortunate ways that I have no wish to minimize. One of the best-known intellectual trends of this era was the eugenics movement, mostly developed in the United States in the early twentieth century, that advocated reproduction among those believed to carry desirable traits and non-reproduction for everyone else. As applied in practice, eugenicist policies victimized the poor, the nonwhite, social nonconformists, political dissidents, and members of minority groups. In the United States, the number of people who are known to have been subjected to coerced sterilization (that is, sterilization by main force or by strong threat or pressure)—many of them powerless inmates of prisons and psychiatric hospitals—as punishment, as therapy for perceived sexual deviance, or for other reasons, is a shocking 63,000, and this is probably an underestimate that does not account for many cases that left no record, or in which the procedure was falsely reported as voluntary. Victims of punitive castration or vasectomy were mostly African Americans and gay men. Coerced sterilization of the poor and of those with perceived mental defects targeted more women than men, and disproportionately affected African Americans, Native Americans, and Latin American immigrants. This practice peaked in the mid-twentieth century and was most frequent in California, where at least 20,000 forced sterilizations were performed; at different points over the course of the twentieth century, 32 states had compulsory sterilization laws in effect. Eugenicist politics and forced sterilization were not, of course, practiced only in the United States; they were pursued especially aggressively in Nazi Germany, where about 400,000 people were forcibly sterilized and 70,000 killed through the government's euthanasia program, quite apart from the atrocities of the death camps.[4]

When European and white North American populations were booming in the nineteenth century, the idea that high-reproducing peoples were entitled to extra land rationalized the conquest and colonization of regions that they perceived as underused or sparsely inhabited. By the mid-twentieth century, when fertility rates had already fallen below replacement level in many Western countries—nations that had also suffered huge losses in World War I and in the influenza pandemic that followed—Europeans and North Americans had become more concerned about the idea that other "races" were reproducing faster and might eventually overwhelm them, a notion that justified the Immigration Act of 1924 in the United States, and other restrictive immigration policies. Anxieties about "race suicide" led to the pronatalist ("pro-birth") propaganda of Hitler and Mussolini, to family support programs in France and Sweden, and to stricter laws against birth control and abortion in the United States and elsewhere, at the same time that eugenicists were targeting marginal populations for forced sterilization and (in Germany) "euthanasia" by carbon monoxide poisoning.

As mortality declined and population boomed in decolonized nations following World War II, Malthusian ideas took on a new dimension, and the term "overpopulation" came into common use; not surprisingly, updated versions of the racism of times past tainted the family-planning aid projects of the 1970s and 1980s that originated in the West and targeted the ex-colonial world. Governments sometimes tried to enact population control by coercion, with results that were disastrously incompatible with modern ideas of individual rights, as during the Emergency Period in India under Indira Gandhi (1975–1977) and the early years of China's one-child policy, when millions of people were sterilized by coercive methods, or even brute force.[5] This history of coerced sterilization is the reason that many colonized or oppressed people in North America and around the world are suspicious of feminist talk about contraception and abortion.

One of population theory's most important results was the "Green Revolution"—the development of the artificial fertilizers, pesticides, herbicides, and hybridized (or, more recently, genetically engineered) crops that have transformed the food supply in the modern era. The Green Revolution happened because an enormous amount of agricultural science and public policy were brought to bear on the Malthusian problem of hunger and food supply—and not only for humanitarian reasons, but in order to reap the profits generated by the new methods, and to support the national security policies and geopolitical strategies of the Cold War era and beyond. Hunger has indeed declined in recent decades, although about 795 million people—almost 11 percent of the world's population—still suffer from undernourishment today. As hunger came to seem a solvable problem, the emphasis in population theory shifted from concerns about food supply to the more complex questions of sustainability addressed by the science of ecology. Nevertheless, some of the methods of the Green Revolution probably cannot be used for long because of the damage they cause to the environment, and some have already been abandoned; while the race to match the world's food supply to its population continues at breakneck pace, a single stumble could return food to its old position at the center of population theory.[6]

While historians have rightly condemned coercive population policies, and governments and organizations were right to abandon the idea of population control, we have seen that long before Malthus and long before population theory, reproduction was coerced and limited in many unfair ways. Infants were killed, abandoned, or sold into adoption; extra sons, sisters, and poor parishioners were prevented from marrying; brothers, husbands, and mothers-in-law bullied women into bearing children or giving them up; single mothers were shamed and ostracized; and women who wanted to avoid giving birth had few means under their own control of doing so. In the modern world, reproductive coercion, while atrocious and sometimes widespread, has been the exception rather than the rule. Fertility has declined sharply in modern populations without coercion, sometimes to levels below replacement. Because very low fertility rates are, so far, a pattern across all modernized societies, most demographers agree that population will stabilize some time in the future. It is theoretically possible that it will stabilize at, or decline to, a level low enough to be sustainable at a high level of well-being for everyone, or to adapt to severe and abrupt climate changes without catastrophic loss of life.

But population stability has not occurred yet, and it is not on the horizon—United Nations projections show world population reaching 11 billion by 2100 and continuing to grow thereafter, though at a slower rate. Because of the low rate of mortality that is one of modernization's main benefits, population stability can occur only at extremely low rates of fertility. Worldwide, the crude death rate is about 8 per 1,000; that is, even if no children at all were born next year, population would decline by less than 1 percent.

Mortality is so low not only because of the advances that caused the Epidemiologic Transition, discussed later in this chapter, but also because of the rapid population growth of recent decades—worldwide, there are many more young people of reproductive age than old people, who were born when population was much smaller; since most people now die in old age, mortality is lower than it would be in a stationary population that was not growing. Eventually, if fertility declines below a theoretical replacement level, as it has done in some societies, population "momentum" will work in the opposite direction. In Japan, which has one of the lowest Total Fertility Rates in the world at 1.4, population is declining at a rate of about 0.2 percent per year, discounting the effects of immigration. If this rate were to persist over the long term, it would take 350 years for the population to decrease by half, but because changes in age structure should cause the rate of decline to accelerate, the United Nations predicts a decline of about 35 percent by 2100.[7] Thus population stability or decline can be achieved through behavior alone, without catastrophic suffering or a return to the high mortality of the foraging and agrarian periods.

The idea that humans might, as the natural result of billions of individual decisions, voluntarily limit their numbers to a level at which they can adapt to crises without collapsing may not have much support in our history, but it is not unimaginable. In one utopian alternative, humans might aim for an egalitarian system that guarantees everyone the main benefits of modern life, discourages too much individual consumption, and also reserves space and capacity so that large populations can move when they need to. They might design flexible infrastructures that can easily expand, contract, and move as external conditions change—we could become a people of hydroponic farms, yurts, fuel cells, cell phone towers, and vast tracts of wilderness. The idea that humans might return to some version of the nomadic, communal economy of our prehistoric past may seem unrealistic in today's intellectual climate, and economists, should I be unlucky enough to have any of these among my readers, will laugh at the scenario described here. But we should not forget that modern technology and modern communications have made many things possible that were unthinkable even a short time ago, and should make new—or, depending on how we see it, old—forms of social organization possible too.

Population decline is much feared by economists, because in theory it should lead to smaller markets, higher wages, less efficiency as economies of scale are lost, low investor confidence, lower aggregate surpluses to invest in research, and changes in the age structure that increase the ratio of retirees to workers, increasing pressure on the tax base. Historically, nations have been eager for population growth in order to maintain large armies and dominate other states; today, the size of the Gross Domestic Product (GDP) is another measure of international clout, and, to the extent that more people means more product, states with large populations are more powerful. More philosophically, a lack of "population pressure," per Boserupian theory, could result in less technological innovation. For many of these reasons, businesses and governments usually assume that population growth is good and decline is bad, despite the staggering implications of extrapolating annual growth rates over time.[8]

But some of these problems have proved less serious in reality than they are in theory—so far, for example, science and technology remain robust in the small, low-fertility countries of western Europe. National GDP may be important to politicians and military strategists, but it correlates poorly with well-being. The problem that gets the most publicity, and the one with which people seem to be most fascinated, is that of age structure, but this factor is also more complicated than usually described. In most modernized countries, governments offer support for people over a retirement age, and unless policies change, this support is expected to increase in the coming decades. But within families, net transfers of wealth are still strongly downward—that is, rather than being a burden on their families, old people in modernized societies are, on balance, helping to support their younger family members, just as they have done historically. (I am embarrassed to confirm this finding from personal experience, as, in my 50s and raising two teenagers, I still receive help from my 70-something parents.)[9] As age structures move toward stability at low, modern levels of fertility and mortality, with a higher proportion of old people, it is likely that public policies on health and pensions will be adjusted until society's overall flow of wealth, now slightly upward in most rich countries if public and private flows are combined, moves downward again through the generations as it has done for most of human history.

Retirement is largely an invention of the modern world in the lucky era that followed fertility decline; few new children needed public education, but the proportion of old people in the population was still low, because the population boom that followed mortality decline was still in the recent past. Most economists expect that the age of retirement will rise (and many governments, including the United States, are increasing the age at which retirees are eligible for public benefits), and if employers change their practices to make postponing retirement more appealing, all workers could benefit. Relaxing immigration policies could also temporarily increase the ratio of workers to retirees in low-fertility countries, although there is debate about how effective this measure would be. The age-structure problem has straightforward solutions, even if some are politically unpopular. Indeed, if people worked later in life and the healthcare costs of the last stages of old age could be controlled, the aging of the population would be a great benefit to society, because valuable skills and experience are conserved, and because older people in modern economies continue to support their children and grandchildren as they have done throughout our history.

Population decline has other advantages. Productivity is higher with an older workforce. Infrastructure, being costly to build, often lags behind demand by a substantial margin (where I live, the transportation system around Atlanta is the obvious example), and would better serve smaller populations, though costing more to maintain per capita in the long run. Historically, population decline has benefited the poor and working classes,[10] and that is likely to be the case in the future too: when labor is scarce, wages rise, unemployment falls, employers become more willing to invest in training employees, housing costs fall, and inequality decreases. All of these factors, of course, could eventually lead to rising fertility rates and another cycle of population growth. In the meantime, a smaller population would make all current and projected ecological problems, including carbon emissions, much easier to solve.[11]

In any case, my main point, which is perhaps obvious enough to go without saying, is this: low mortality is unsustainable without low fertility. While scholars rightly criticize the history of coercive population control, few would disagree that without the decline in fertility attributable to voluntary (even enthusiastic) behavior changes in modern societies, our future would be much more grim. The Malthusian mathematics of reproduction, simplistic as they may be, are unassailable on this point—no matter how much is done to control consumption in the richest countries, to increase food supplies through science, or to redistribute resources more fairly, and all of these are good and necessary policies, a population that is increasing at more than a small fraction of a percent per year will nevertheless, by doubling every few generations, eclipse all gains and reach some hard limit, with the impact of change accelerating rapidly as numbers get larger. Worldwide, the current rate of population increase is over 1 percent per year, even with widespread fertility control and a global TFR that, at 2.33, is shockingly low for our species or for any species. To update a famous calculation often performed by demographers, at that rate of growth, humans will cover the entire land surface area of the Earth at a density of one square foot per person in 1,100 years, and that population will double again in less than 70 years. Most likely this outcome is impossible—that is, in our near future, something will happen to increase mortality dramatically among humans, or behavior will change so that fertility declines and population stops growing. For as long as the Earth is a closed system—that is, unless an easy means of colonizing other planets is discovered—any significant rate of population growth is sustainable only for a very short time. This is true even if no catastrophic change in the environment reduces the number of people that can be supported; that we can foresee a good chance of this type of catastrophe—abrupt climate change—exacerbates the problem.

If modern society has a future, it is because of our capacity for non-reproduction. There is no merit to the silly idea, commonly expressed in popular media, that menopause is obsolete. As both a transition to the post-reproductive phase of life and a method of controlling the ratio of reproducers to helpers in a cooperatively breeding population, menopause is more relevant than ever—and we work hard to make the benefits of non-reproduction, bestowed naturally by menopause, available to a broader section of the population, to younger women and to men. Reduced rates of reproduction in industrialized societies are hard to explain except insofar as our reproductive strategies have always included high investment in children and a high adult-to-child ratio. Now as in the past, our capacity for non-reproduction is essential to our survival. The modern age is the age of menopause—if we did not already have it, we would have to go to a great deal of trouble to invent it.

THE DEMOGRAPHIC TRANSITION

The aspects of modernity most relevant to the subject of this book are two closely related phenomena sometimes called the Demographic Transition and the Epidemiologic Transition. While some scholars trace the origins of these two events as far back as the sixteenth century, for most of the world, they happened in the twentieth century, and among some populations they are still ongoing. These two shifts refer to profound changes in mortality, fertility, population, and the epidemiology of diseases.

The term "Demographic Transition" was coined by sociologist Kingsley Davis in 1945. In a famous article, he described how world population grew slowly and, to contemporaries, almost imperceptibly in the centuries before 1700, and then began to rise when (in his view) mortality declined during and after the Industrial Revolution. This happened first in Europe and then around the world as modernization spread. But as mortality declined, fertility also declined—only not as fast as mortality, so that the lag between the two changes caused what he called, in a famous metaphor, a population explosion with a long, European fuse. The end point of the Demographic Transition, he argued, was a more efficient, less wasteful equilibrium at much lower rates of fertility and mortality—because humans live longer and have fewer children, less energy is consumed by reproduction—and at a much higher and denser level of population than before. Davis's theory was an optimistic one: people, and women in particular, need no longer be burdened by bearing and caring for children who will not survive to adulthood.

At the time he was writing, Davis estimated world population at just over 2 billion and its rate of growth at about 0.75 percent per year. He expected it to rise to 4 billion by the year 2000, a guess that was too low, and considered an increase to the level of 10 billion wildly alarmist and unrealistic. Birth rates, he argued, were declining in Europe and were not rising worldwide.

Davis located the origins of the Demographic Transition in Europe, whose population, he estimated, had increased from 100 million in 1650 to 720 million in 1933. Europeans colonized the world, displacing, as he saw it, "sparse native populations" and filling whole continents with industrialized, modern progeny. But in some regions—especially Asia—Europeans dominated large, complex civilizations without displacing them. Because (as he argued) these subject populations had benefited from some Western techniques of "death control" without sharing in the other changes of industrialization, they had begun to grow without the corresponding fertility declines that would inevitably occur once they became Westernized. There was no need for European populations to fear being overwhelmed by Asian immigrants, because for Asian civilizations to sustain large populations they would have to modernize and inevitably become just like Europeans, with similar low fertility rates.

Davis's thoughts on colonization and on what he called the "Asiatic hordes" seem racist today (though at the time, he was arguing against even more noxious forms of racism in population research), and whether he was correct about the causes of population rise in Asia is open to doubt. In 1945 he could not have foreseen that the greatest gains in "death control" would follow the wave of decolonization that was one of the main results of World War II. Also, most modern historians argue that mortality did not begin to decline significantly, even in Europe, until the later nineteenth century—before that, population rose for other reasons. But the general outlines of his theory, and his term "Demographic Transition," have been accepted. Both mortality and fertility are dramatically lower in modern populations than they were in traditional ones. The disparity in these rates, as mortality fell earlier than fertility, caused the runaway population increase of recent decades.

Here it is important to emphasize again that what seems like a small growth rate of 1 percent per year causes a population to double within a single human lifetime, and that the effects of this kind of growth escalate as numbers become large. This is exactly what happened in the modern period, in which world population growth peaked at an annual rate of just over 2 percent in the 1960s. A process with roots a few centuries old looks like a mid-twentieth-century "explosion" of world population growth, and Davis was correct to point to its long fuse.

POPULATION GROWTH

The population explosion is a large effect, but it is caused by a relatively subtle disequilibrium of one to two percentage points between births and deaths, sustained over a few decades or centuries. For this reason it can be difficult to identify causes with precision. Some factors in the Demographic Transition were unique to each region, making it hard to generalize. In Europe, population began to rise before mortality fell, because fertility was higher. By the time Malthus wrote his *Essay*, rates of marriage in England were increasing, age at marriage was decreasing, fertility was rising, and population was rising, but mortality did not start to fall significantly until the late nineteenth century. At that point fertility started falling too, but it did not keep up with declines in mortality, and population grew even faster.

Why did England's population start to grow? Something must have changed that increased the amount of food and energy available. Historians have identified several possibilities. New plants from the Americas increased the food supply in parts of the Old World, including Europe (and especially Ireland, which was transformed by the introduction of the potato). Natural resources were imported from the overseas colonies that England began to acquire in the late sixteenth century. The beginnings of industrialization created more nonagricultural, urban jobs that allowed people to marry earlier. New agricultural methods increased yields and produced economies of scale even as some of them displaced small peasant farmers and shunted them into the urban labor force. Finally, temperatures were warmer after the end of the "Little Ice Age" around 1700.[12]

In China also, population rose very dramatically in the eighteenth century, from about 160 million in 1700 to about 350 million by 1800. As in the case of England, the extra food and energy came from various and interconnected sources. China's frontiers expanded under the Manchu Qing dynasty, opening new markets and new opportunities for migration. The Qing also lowered taxes and demands for requisitioned labor. More intensive agricultural practices spread as people adopted early-ripening varieties of rice and New World crops, including peanuts, sweet potatoes, and maize; they also used more manure and irrigation, and worked more days of the year. These economic changes were so effective that standards of living in China rose in the eighteenth century despite staggering population growth.[13] Finally, smallpox inoculation, the practice of deliberately infecting people by having them inhale the ground-up scabs of recent victims, became prevalent in this period. For people in the Qing lineage, inoculation was mandatory from the late eighteenth century, and their records show a dramatic decline in child mortality.[14]

In the Americas, European populations were booming in the wake of a catastrophic decline among indigenous peoples in the centuries after contact, as native peoples decimated by disease were further suppressed by exploitation, displacement, and persecution by Europeans.[15] In North America, it is notoriously difficult to estimate the size of the indigenous population before 1800, but this number declined from a fifteenth-century high, before contact, of perhaps 7 million to only 600,000 by 1800, and further to a nadir of about 228,000 in 1890, before rebounding in the twentieth century.[16]

For European colonists, on the other hand, the New World was a frontier region into which they expanded quickly. The population of Britain's North American colonies grew from about 240,000 in 1700 to over 3 million by 1800. Some of this rapid growth—about 14 percent in the 1700s—was due to immigration, but mostly it reflected a high rate of natural increase. At the time, Benjamin Franklin speculated that because land was abundant in the New World, colonists married earlier and had more children than their counterparts in England; modern demographers think he was correct on those points. Mortality was also lower in North America than in England, at least in the northern colonies, where average life expectancy at birth reached the 50s in some regions. The rate of natural increase among the British colonies of North America was about 2.5 percent per year in the eighteenth century—rates were highest in the northern and middle colonies and lowest in the South, where mortality was higher. Franklin's estimate that the population of the colonies was doubling every 25 years was not far wrong and influenced Malthus, as well as economic theorist Adam Smith and other writers.[17]

In colonial New England, the African American population was small and mostly free, but in the southern colonies, African Americans made up about one-third of the population, and almost all were enslaved. The forced migration of over 12 million Africans to the Americas over the course of the eighteenth and nineteenth centuries was one of the most important demographic events in recorded history and had a net negative effect on world population, as many died in the passage to the Americas and mortality was high among enslaved peoples. Most populations of enslaved Africans did not reproduce themselves and had to be replenished with new captives.[18] The situation in the British colonies, where about 270,000 Africans arrived over the course of the eighteenth century, was the main exception to this trend. Even with crude death rates higher than those of the white population (but much lower than those of enslaved populations in the Caribbean and Brazil, where most Africans were sent), over the course of this century most enslaved people were born in the colonies, rather than in Africa. A study of the enslaved populations of Virginia and Maryland in the eighteenth century shows very high rates of natural increase after 1730 or so—over 2.5 percent—and a Total Fertility Rate of about eight children. Despite the extreme adversity of their circumstances and the many ways in which slavery disrupted families, African Americans in the new United States numbered over 750,000 in 1790.[19]

This scattershot picture of population in the eighteenth century is intended to give a general idea of the kinds of factors that caused population to begin rising. These were different around the world, and not all populations grew in this period; some declined sharply. But overall, because of colonial conquests, new foods from the Americas, the early stages of industrialization, and perhaps the first effective measures to control infectious disease, world population expanded from an estimated 680 million in 1700 to 954 million in 1800, an annual growth rate of about 0.34 percent.[20] Then it began to rise faster. By 1900, world population had topped 1.6 billion, and by 2000 it exceeded 6 billion.

As industrialization progressed, age-old economic constraints on food and energy were relaxed, fossil fuel sources replaced wood and other organic materials, machines replaced human and animal labor, and advances in agricultural science increased the productivity of farms. The modern economy now supports a population much larger than what would be possible with peasant agriculture (to grasp the profundity of the change, we must also consider that a person living a modern lifestyle consumes about 20 times as much energy per day as a peasant farmer). Population would have boomed in the modern period even without sharp changes in mortality, because of the extra food and energy available, and this effect is exactly what we see in the eighteenth century for some populations. However, mortality also dropped precipitously in the modern period. This is the great achievement of the last century, and perhaps of human history more generally.[21]

"DEATH CONTROL"

Mortality decline happened earliest in England and western Europe and expanded rapidly in the mid-twentieth century, sweeping much of the world in the decades after World War II. In China, average life expectancy at birth rose by 20 years, from about 40 to about 60, in a single generation between 1950 and 1970 (and reached about 75 by the year 2000). At the same time, China's population boomed, rising from 600 million in 1954 to 900 million in 1974 as fertility remained high (with a steep drop during the devastating Great Leap Forward famine of 1959–1961, followed by a rebound).[22] Today, average life expectancy at birth worldwide is between 70 and 71, up from 47 in 1950. In the rich countries of Europe, North America, and Australia, as well as in New Zealand and Japan, the average is higher at about 78 (it is over 83 in Japan, which has the highest average life expectancy at birth of any nation); in poor countries it is about 69, up from 42 in 1950.[23]

Mortality decline came last to sub-Saharan Africa, where average life expectancy at birth was 36 in 1950 and is 57 today; all nations with an average life expectancy at birth below 60 are in this region, with the exception of Afghanistan. The reasons for this are complex, but one factor is that sub-Saharan Africa was decolonized last, mostly in the 1960s, which also delayed other parts of the transition to modernity. Another factor has been the persistence of malaria, and especially of the most deadly strain, *P. falciparum*, endemic to this region, where most people have genetic resistance to the more benign *P. vivax*. In the mid-twentieth century, malaria was eliminated from much of the modernized world, including the southeastern United States, by means of expensive public works and public health measures, and eventually, more cheaply, with the insecticide DDT. But DDT was very damaging to ecosystems, and its use was discontinued before it was widely deployed in Africa. No substitute as effective was ever found, and hopes of eradication have been abandoned.

A third factor affecting mortality decline in this region is the epidemic of HIV/AIDS, devastating to a few countries, including South Africa, where life expectancy had risen to about 63 before dropping sharply into the low 50s in the early twenty-first century (it is now starting to climb again). Unfortunately and paradoxically, we know now that HIV first became established in the Congo region by colonial efforts to prevent or treat sleeping sickness, malaria, syphilis, and other diseases, as medical teams spread the virus through unsterile needles—a reminder that public health measures, which have been so critical to achieving low mortality in the modern period, also have the potential to do a lot of harm if not executed carefully.[24]

The reasons that mortality has declined in the modern era are debated and not fully understood, and they have varied across time and place, but it is fair to say that changes that occurred piecemeal in Europe and resulted in a longer mortality transition combined to create a faster one in other parts of the world. One way to think about the mortality transition is as a change in the causes of death, and particularly in the diseases that kill people. Before the modern period, most people died of infectious diseases, many of which disproportionately affected infants and children; in modern societies, most people die of chronic, noncommunicable diseases, such as cancer and cardiovascular disease, that claim people later in life. In this way of thinking, the control of infectious disease is the main cause of the mortality transition.

But how did that control come about? This is actually not so easy to explain. Consensus agrees on a combination of several factors—public health measures, medical advances, better nutrition in modernizing economies, the spread of education and literacy (especially among women), higher GDPs, higher standards of living, and more stable modern states that can organize and implement public health programs and research.[25]

In the 1971 article in which he coined the term "Epidemiologic Transition," Abdel Omran described the first stage of this transition as one in which pandemic outbreaks of disease became less frequent and less lethal. In the late seventeenth century, bubonic plague, which had returned to Europe many times since its original catastrophic visitation in 1348, virtually disappeared from that region—we do not know exactly why. It is not clear that the disappearance of the plague affected mortality in Europe very much, but even a small change can be important for reasons emphasized in this chapter. Some scholars have credited quarantining and other practices learned as a result of the plague with reducing mortality in the early stages of this trend.

Medical science also played a role, though scholars dispute exactly how to describe it, especially in the early stages of mortality decline in Europe. In 1798, Edward Jenner publicized his discovery of vaccination; after that date, people could acquire immunity to smallpox by being deliberately infected with cowpox (*Vaccinia virus*, from *vacca*, the Latin word for "cow"), without having to get the disease themselves and without transmitting it to others. This was a big improvement over inoculation, a controversial practice for those reasons. While Malthus had speculated that controlling one disease would not make any difference to overall mortality—people would just die of something else—that is not what happened; instead, mortality declined as the practice of vaccination spread.

In the late nineteenth century, mortality in Europe began to fall more steeply. The discovery that cholera was transmitted by water contaminated with feces—after John Snow identified the famous Broad Street Pump as the source of the 1854 outbreak in London—led to improvements in sanitation and hygiene. In the late 1850s, Louis Pasteur and Robert Koch began a series of experiments that proved that infectious diseases were caused by microorganisms, and that these germs could be identified, distinguished from one another, cultured, modified, and killed. These findings led to an explosion of work in immunization and sterilization, and at the time, many believed that all infectious diseases would be wiped out by these methods. That hope proved too optimistic, but even diseases that could not be immunized against or cured, like tuberculosis, could be affected by public health measures designed to reduce their spread, such as isolating patients, disinfecting sick rooms, and inspecting cattle. By the time mortality decline came to the non-European world after World War II, other important breakthroughs had occurred—most importantly, the discovery of antibiotics.

Understanding how diseases worked was important, but it was not enough to cause mortality decline—ideas had to be circulated, research had to be funded, public health programs had to be implemented and paid for, and people had to be made aware of what to do and persuaded to do it, all tasks that were much easier in the presence of strong and stable governments. At the same time, it became clear that low mortality could be achieved even in poor countries that did not have fully modernized economies or high GDPs. The story of mortality "overachievers" such as Costa Rica, China, Cuba, and the state of Kerala in India, which attained average life expectancies just as long as—or even longer than—those in rich countries, has been some of the best news of the modern era. While it has not been so easy to explain or replicate these successes, certain lessons from them are now widely accepted: beyond a fairly low level, increasing GDP does not necessarily lead to decreases in mortality; it is more important to offer basic healthcare to everyone than to invest in expensive hospitals (China's "barefoot doctors" are a famous example of this approach); clean piped water is more important than expensive sewer systems, and low mortality can be achieved without sewers; and income inequality is an obstacle to low mortality. Education and literacy, especially for women, are related to mortality decline, and although the reasons for this correlation are disputed, women's dominant role in caring for children and making healthcare decisions for the family is likely part of the picture. If women can read and understand health literature, if education makes them more accepting of modern ideas, and if their decisions are more respected because they are educated, all of these factors may reduce infant and child mortality, and all have been proposed as explanations of the connection between women's education and low mortality.

Infants were the last age group to benefit from mortality decline. Before about 1900, most people had pessimistically assumed that human infants were doomed by nature to die in large numbers, and up to this point mortality had mostly declined among children and young adults. But around that time, infant mortality began to fall steeply in England and Europe. Pasteurized milk and sterile birthing conditions—two products of the new revolution in germ theory—were early causes; a vaccine against diphtheria, first introduced in 1890, took aim at one of the main killers of infants, but early versions were not especially effective. Also around this time, people and their governments changed their minds about the fragility of infants, partly from concern about falling birth rates. Governments and organizations tried to boost infant survival rates by educating mothers, placing a new, much expanded responsibility on mothers to follow the latest advice on hygiene and nutrition for infants. Combined, these factors proved surprisingly effective, and infant mortality plummeted in industrialized countries. In wealthy countries today, the rate of infant mortality is very low, under 1 percent; worldwide, it is about 4 or 5 percent. While adult mortality and old-age mortality have also improved in the modern era, this transformation of infant and child mortality has been more profound.

People resisted some public health measures, especially those that were most invasive, like smallpox vaccination, the chlorination of water, or requirements to notify public health departments about patients with tuberculosis, as well as, for other reasons, the education of women. This is not surprising, and indeed such resistance is only to be expected when states impose these types of interventions on their citizens *en masse*, no matter how good the result may ultimately be. In addition, government policies have not always been proven right. The same conflicts and arguments about individual liberty and the common good continue to arise, as for example in the AIDS epidemic of the late twentieth century, the SARS outbreak of 2006, and the Ebola outbreak of 2014. Modern societies could be doing a better job of establishing trust and developing the ethics of public health interventions, or at least of taking these issues seriously, though they are starting to get more attention now.[26]

FERTILITY DECLINE

The last and most mysterious stage of the Demographic Transition is the plunge in fertility that has reduced the world's Total Fertility Rate to 2.33 today and might, according to many demographers, cause it to fall below replacement level.[27] Most of the material discussed in this second part of my book suggests that as resources become more abundant, people should have more children—but this is not what happened. Fertility is falling in the modernized world, even as standards of living rise. The fertility transition is a recent phenomenon that we do not fully understand, and its future trajectory may be hard to predict—the transition in sub-Saharan Africa is proceeding more slowly than it has in other regions, for example, perhaps because modernization in that region has also been slower. But low fertility is the overwhelming trend in modernized nations. Several theories have been proposed to explain the fertility transition, and while none of them alone—and perhaps not all of them together—can account for it, two factors, education and mortality, have emerged as more important than others.

First, modern societies have high rates of education because children must learn skills like reading and writing that are important for nonagricultural jobs. Rates are higher for both girls and boys, although the education, political participation, and wages of girls and women have lagged compared to those of males. For these reasons, raising children in modernized societies is relatively more costly than in agricultural societies. Parents invest in their children's education, health, and development—creating "human capital"—because of the future payoff in income and well-being for the child, and they can invest more in each individual child if they have fewer children. The payoff is greater when low mortality makes the investment more secure, and because education tends to lower mortality, the two leading factors in fertility decline reinforce one another in a virtuous circle.

Some researchers have argued that education for women in particular has additional effects on fertility. Educated women have more opportunities for wage-earning jobs; this increases the cost of having children if women must stay home to care for them. Also, since women bear most of the burdens of reproduction, they typically want smaller families than men do (research shows that this is usually true, but not always), and educated women may have more access to modern contraceptives and more power to make their own decisions about reproduction. Educated women may be more economically independent, and therefore less motivated to please a husband by producing many children, or to bear the sons who, as adults, will protect and support them.

Besides education, mortality is the most important influence on fertility, as fertility decline usually follows a drop in mortality. In the second half of the twentieth century, the time lag between mortality decline and fertility decline was normally about 10 years. People respond not only to child mortality, but to mortality at all ages and to average life expectancy at birth. That is, once they see that their children have excellent chances of long life, parents usually prefer to raise only a few—in many societies, less than two on average.[28] The era of large families of six or eight or more surviving children that characterized my grandparents' generation here in the United States was short-lived. Even so, this lag between mortality decline and fertility decline was enough to create the modern population explosion.

Other factors that researchers have linked to fertility decline are urbanization (when more people live in cities, fertility is lower); political freedom (which has opposite effects in rich and poor populations); government programs, whether voluntary or heavy-handed; and a culture that sees small families as normal. Modern contraception and abortion have made it much easier to limit the size of families, and are easier for women to control than most traditional contraceptive practices. Most scholars agree that good access to modern contraception contributes to fertility decline but does not cause it—people decide to limit families for other reasons, and technology makes it easier to achieve this goal. Finally, as in the case of mortality decline, the most recent studies conclude that GDP per capita—a measure of a society's wealth—is not an independently important influence on fertility. This means that poorer countries with good education and low mortality have about the same fertility as richer countries in which those conditions are similar.

Again, the encouraging lesson is that we do not have to be rich to have nice things like low fertility and low mortality.

It is also important to note that the trajectory of fertility has not been straight down since the start of modernization—there has been one major reversal, in the Western world in the 1950s. In the aftermath of World War II, average age at marriage dropped (to about age 20 for women in the United States), the percentage of people marrying rose, and the crude birth rate rose steeply in the era now called the "baby boom" before fertility began falling again in the late 1960s. This event can be explained in hindsight but was not predicted; even now, we do not understand all the factors at work in fertility decline, and it is possible that more surprises await us in the future. Thus, although demographers commonly speak of both mortality decline and fertility decline as irreversible, and although both trends are resilient, it is not really true that they never reverse. Mortality soared in Russia, eastern Europe, and the former republics of the Soviet Union after its breakup, especially for men; it also rose in parts of late twentieth-century sub-Saharan Africa as a result of HIV, alone or in combination with political upheaval; and in many countries because of war, political instability, or genocide, such as Vietnam, Cambodia, Bangladesh, Bosnia, Uganda, Rwanda, Iran, Iraq, and Syria, all places where average life expectancy at birth dropped significantly in the later twentieth or early twenty-first centuries, after it had begun to climb with modernization.[29]

Low fertility is one of the defining features of the modern world, one of the foundations of hope for the future, and the most important factor in the changed (and still changing) status of women in modern society. For most of our history, our species, like all species, was tightly bound by the constraints of subsistence and reproduction—we spent almost all our energy trying to get enough food to survive and trying to bear enough children to ensure our families' future. The modern era has liberated large numbers of us from these constraints. Lower costs of reproduction have created an energy surplus for young women that can be used for other activities, a surplus once available mainly to men. Women past menopause, who historically used their energy surplus to help their families survive, can now use it in other ways too. This is one reason modern societies are so explosively productive, both economically and culturally—low fertility has liberated the energies of women, while lower adult mortality conserves the skills of educated and experienced people of all sexes. The modern life cycle is much more efficient than the wasteful conditions of high mortality and high fertility that preceded it. The runaway material *productivity* unleashed by our lower energy expenditure on *reproductivity* has transformed our world and threatens to overwhelm it.

None of the accepted explanations for fertility decline is consistent with evolutionary fitness—people who have two children will almost certainly have fewer descendants than those who have more. Although children are more costly to raise in modern societies than in traditional ones, modern economies are so much more productive that most people in today's wealthy nations could support many children at a level of health and comfort far above that of a peasant farmer. But although evolution can be an instrument of fine precision, it is never more precise than it has to be, and is often quite blunt. Evolution is probably the reason that sex is fun and babies are cute, motivations that for most of our history were enough to guarantee high fertility as long as additional children were not too much of a cost, but they are not enough to drive people to have large families in modern conditions.

However, it is likely that our evolved strategy of high investment in children—a strategy that includes menopause—is part of the explanation for fertility decline. Humans are willing to pour almost unlimited resources into raising one or two children who are very likely to repay the investment by surviving to old age. We also have a long ancestral history of cooperative breeding, so that investing in children who are not our own, those of our daughters and other relatives, or even children unrelated to us, can feel very rewarding. While having few children is not the best strategy in terms of evolutionary fitness, humans take great satisfaction in maximizing the well-being of offspring. Most explanations of fertility decline depend on the background assumption that humans are altruistic toward their children and will sacrifice to achieve the best outcomes for their descendants.

It is not an exaggeration to say that without our flexible reproductive strategy, which includes high investment in children, a high ratio of adults to children, and a large number of non-reproductive people, modern life would not be possible. Beyond this, it is likely that our future depends on harnessing our ancient and unusual ability to cooperate and share with people we are unrelated to. We should not assume that the motives and behavior that drove people in the agrarian period and its aftermath, to the extent that these are understood at all, necessarily apply in the modern world. Many of the assumptions about human psychology and behavior that underlie modern economic and political thought are plain wrong, and should not be allowed to dictate a short future of greed, exploitation, and spiraling consumption leading to catastrophe.[30] A species with grandmothers can do better.

PART III

•

Culture

●

CHAPTER 9

Women's Hell

MENOPAUSE AND MODERN MEDICINE

When I floated the idea of katakori as a distinctly local disease, citing my own experience in America and the evidence of words—the lack of equivalents in local vernaculars—some friends still resisted. People everywhere *must* get katakori, they protested. Its absence is surely only apparent.

As if to think otherwise were to think the impossible.

Perhaps people elsewhere *just don't realize* that they have katakori. . . . After all, a person pointed out, a person can have arteriosclerosis or even cancer, and for a long time be unaware, still feel "normal." When pressed, though, about whether one could still speak of katakori in the absence of any felt discomfort, my friend wavered. Perceived pain was definitely not all there was to katakori, but it seemed essential.

Then there was this possibility: perhaps other cultures knew katakori, but in a different guise. Perhaps they organized their sufferings differently. Perhaps, instead of identifying katakori as a discrete entity, they encompassed its symptoms within another, more comprehensive disease. Or, alternatively, they parcelled out katakori into a few disparate complaints. Perhaps.[1]

HISA KURIYAMA'S EXPERIENCE debating with his friends about the nature of Japan's most ubiquitous physical complaint—the deep ache between the shoulders called *katakori*—is like some conversations I have had about menopausal symptoms. Surely, his interlocutors argue, something that is so real and so fundamental to their lived experience must be universal; it must have some simple organic cause, like arteriosclerosis; it could not have part of its origin in the higher functions of the brain, in how we think, or in the influence of culture. When I mention that the idea of menopause, with symptoms and a name, is absent from most medical traditions outside of Western medicine,[2] many people respond with similar arguments—people *must have* experienced symptoms, but their complaints and their therapies are lost or hidden from us. Or they misinterpreted their symptoms. Or they did not notice them. Some of this may be true, as I will explain in this chapter. But it is exactly the experience and understanding of symptoms that make menopause what it is in modern European and North American cultures today—just as *katakori* is defined by the pain that is its essential element, and the concept does not make sense among people who do not feel it, menopause is defined by its symptoms. I've opened with this quotation from Kuriyama in the hope that my readers, if they find that this chapter challenges some ideas that seem profoundly true to them, will keep an open mind.

Before I begin the complicated discussion that forms the rest of this chapter, let me try to reduce confusion by summing up the message of this book. Menopause is a developmental transition that has been important to the success of our species in its long prehistory, and also in the shorter, more recent agrarian and modern periods. How women experience this transition depends largely, though not entirely, on what their beliefs and expectations are. Since about 1700, a strong tradition in western European medicine has perceived menopause as a dangerous, pathological condition with many symptoms, some of them dire. This fundamentally wrong view of menopause acquired an extra dimension when the new science of endocrinology—the study of hormones— came to understand it as a pathological deficiency of estrogen, and this Western idea of menopause has influenced its perception around the world. This chapter explores the influence of culture—the sea of values, beliefs, institutions, and practices in which we swim, unaware for the most part that it surrounds us—on the experience of menopause.

When people I know use the word "menopause," for example, when they say someone is "going through menopause," they don't only mean the end of menstruation and reproductive life. They mean something more—that is, they mean a collection of physical and psychological symptoms thought to be associated with this transition. Modern medicine uses the term "menopausal syndrome," where "syndrome" means a number of symptoms that seem to occur together but whose relationship is not really understood. A "syndrome" implies an underlying, unknown or poorly understood disease or disorder. In the case of menopause, most researchers believe in a vague sort of way that the hormonal fluctuations of menopause cause these symptoms, even if we don't know exactly how that works. But despite an overwhelming trend in recent decades to view and speak of menopause as a permanent and pathological deficiency condition, most physicians shy away from calling it a disease that needs to be cured. This means that the symptoms women experience at menopause—and not, for the most part, the underlying condition—are the problem, and this in turn is part of the reason for the "syndrome" label. The study of menopausal syndrome is the study of the deep and cryptic subject of symptoms.

When did menopause become a medical condition with symptoms and a name? One of the reasons I wrote this book was to answer that question. I knew from my research that medical sources in ancient Greek and Latin barely mention menopause and do not have a word for it. There is plenty of gynecology in the ancient medical tradition—no one could accuse the ancient Greeks of not being interested in women's medicine. Puberty, menstruation, and childbirth are all imagined as difficult and dangerous, or as requiring intensive management. But they have little to say about menopause.[3]

We know that menopause has been part of the human life course for a long time, either since the origin of our species, *Homo sapiens*, or even earlier. The natural human lifespan has always included a long postreproductive period for women. But it has not always been experienced the same way. The experience of menopause as a medical disorder—a deficiency of estrogen, a constellation of symptoms, a problem—is typical of modern societies but not traditional ones, and seems to be an idea that spread along with modern medicine. It is possible that low fertility and perhaps other effects of modernization, such as better nutrition, more sedentary lifestyles, and living indoors, have changed the physiology of menopause. Moreover, within modern populations, differences in diet, lifestyle, climate, or even genetics may affect the experience of menopause. That is, factors other than cognition and culture most likely play some role in menopausal syndrome and its variations, although efforts to prove these connections have been frustrating.

Culture, and cognitive functions like attention and expectation, also play a large role in menopausal syndrome. I will not try to argue that women in traditional societies never experience symptoms of menopause or even that they always lack a cultural concept of menopause. But I will argue that modern medicine includes a concept of menopause that is specific to it and that deeply influences experiences of menopause in modern societies, though with variation across cultures. I will also suggest that menopausal syndrome is similar to other "cultural syndromes" that challenge the distinctions modern science tends to draw between mind and body. That some syndromes have part of their origin in culture and cognition is widely accepted, and there is a good case to be made that menopausal syndrome has this quality.

While I see today's concept of menopause as a product of modern medicine, I don't think that modern medicine invented it. Some time in the late Renaissance in Europe, menopause entered popular consciousness as a critical time of life during which women might experience upsetting symptoms; the signal of this idea becomes strong after 1700 but is traceable earlier. Modern medicine, arising in Europe over the following century, picked up the signal and amplified it, so to speak.

Also, I do not think it is the case that modern medicine has made no valuable contribution to women's experience of menopause—in particular, it offers an explanation for sensations that may have troubled women in traditional societies, but that they did not know how to interpret. Among the Maya studied by modern researchers and discussed in this chapter, for example, women were not told in advance to expect either the onset of menstruation at puberty or, as one study found, its end at menopause; these events therefore caused a lot of anxiety when they happened.[4] Younger people who learned about the reproductive cycle in school were grateful for the information, even if elders thought the taboos should be maintained. So there is value to naming the concept of menopause and ascribing symptoms to it, within reason and in perspective. Traditional societies are usually more patriarchal than modern ones, and there is a sense in which attention to women's life stages and experiences grants women more equality.

On the other hand, the changing status of women in the modern world has naturally generated resistance, and dominant groups can be very creative in inventing new ways of oppressing people who might otherwise gain more status than they are willing to grant. In the "medicalization" of adolescence in ancient and Renaissance sources, it is easy to see anxiety about controlling girls at a stage of life that was potentially dangerous to their families and to society—a stage when they were fertile but as yet unmarried, and not sexually monopolized by a husband—in a time and place when that stage was long, because women were marrying late, in their mid-20s. Modern menopause encourages a view of middle-aged women as weak, vulnerable, and different from men, in a world in which women have more potential to challenge men for money and power, and this connection between social change and ideas of menopause is sometimes very obvious. Behavior perceived as aberrant or inappropriate—not doing housework, fighting with husbands, engaging in enthusiastic or promiscuous sex—has been blamed on menopause, often by women themselves. In the early twentieth century, when some women were disrupting patriarchy by demanding the vote, delaying marriage, attending college, and having fewer children, physicians and other experts warned that single women, lesbians, and masculine-seeming women would suffer most at menopause.[5]

It is reasonable to identify two broad types of medicine for our purposes. One is the type I am calling "traditional" medicine—those medical systems attested in every horticultural and agricultural society that I am aware of (though not as much in foraging populations), the roots of which mostly predate a society's exposure to modern medicine. "Traditional" does not mean "unchanging"—traditional medical systems evolve on their own, and almost all have evolved faster in recent decades, under the influence of one another and of modern medicine.

Most traditional medical systems, and especially those most highly standardized and widely practiced today, have changed fundamentally in the modern period, as the process of standardization rationalizes and streamlines enormously complex and variable traditions. For similar reasons, modern medicine, while recognizable everywhere, is not everywhere the same—it has adapted to different social and cultural environments, and incorporates local traditions and ideas.

Some agricultural societies developed written intellectual traditions with theories of physiology and disease; these survive in ancient Greek, Chinese, Sanskrit, Arabic, and other "classical" (that is, scholarly) languages and some vernacular languages. Virtually all agricultural societies have relied on traditions about which plants, substances, rituals, objects, or incantations can treat the conditions and diseases they recognize and describe. Some traditional systems, such as Traditional Chinese Medicine (often abbreviated as TCM), Ayurveda, or Unani (Greco-Arab) medicine, are regulated, taught in professional schools, and widely practiced around the world, but countless other systems not as globally influential are practiced by traditional healers or by grandmothers who serve as the expert healers for their own families.

In many rural parts of the world, people have limited access to modern medicine and rely mostly on traditional practitioners; also, it is common for people in highly modernized societies like the United States to use practices like acupuncture for certain problems. While practitioners of modern medicine are often hostile to traditional practitioners and vice versa, the trend today is to recognize the value of traditional medicine. Because of their deep roots and local origins, traditional systems often do a good job of addressing social and psychological factors in illness. Imposing modern medical technologies can cause many unexpected problems if it is done without sensitivity to the cultural factors that traditional medical systems reflect so exquisitely. And although efficacy is not the main strength of traditional medicine, some traditional remedies do work; in recent years modern medicine has devoted a great deal of research to identifying traditional materials and practices that might have broader (that is, cross-cultural) use. An important example of such a crossover remedy is artemisinin, a first-line malaria drug derived from the plant *Artemisia annua* and used in classical Chinese medicine to treat intermittent fevers. Traditional practices can also improve the efficacy of modern medicine by reaching more people, by making patients less resistant to treatment, or by reducing social and psychological stress. For chronic or painful conditions that modern medicine is not very effective at treating, traditional medicine can provide some relief.

A 2012 study of the use of traditional medicine by doctors in Japan illustrates some of these points.[6] Although Japan is one of the most modernized nations in the world, researchers found that 84 percent of the doctors they surveyed used *kampo*—Japanese traditional medicine—in their practice, and 80 percent of patients used both *kampo* and modern medicine. National health insurance covered 148 *kampo* prescriptions. Traditional medicine was especially important in obstetrics and gynecology; almost 45 percent of these patients used only *kampo* medicine. In private practice, most of the conditions treated with *kampo* were common, non-life-threatening, and not especially responsive to modern medicine; the common cold was the top-ranked condition in this category, and "menopausal syndrome" was also among these conditions. (A great deal of high-quality research on menopause in Japanese culture has been conducted and is discussed further in chapter 11.)

Traditional medicine has strong support from a number of governments and organizations worldwide. The government of the People's Republic of China, continuing Maoist policies, promotes both traditional and modern medicine, partly because bringing modern medicine to its large population has been a Herculean task, and partly because of its goal of encouraging a strong sense of national identity. For other reasons, the official policy of the World Health Organization is "to support Member States in harnessing the potential contribution of T&CM [Traditional and Complementary Medicine] to health, wellness and people-centred health care and to promote the safe and effective use of T&CM through the regulation of products, practices and practitioners."[7]

Although traditional systems differ widely from one another, the system I am calling modern medicine, which developed in western Europe in the late nineteenth century, diverges in important ways from all of them, and today it is at least equal to all the others combined in influence and in the resources dedicated to it. While it has clearly played a part in the transformation that created the modern era, it is actually not that easy to define what modern medicine is.[8] Its roots in the natural sciences are one of its defining features; practitioners of modern medicine learn systems of biology, biochemistry, physiology, and pathology that are quite new, having been invented or totally transformed in the last 100 years or so by drastic "paradigm shifts" (to use the famous phrase coined by Thomas Kuhn in *The Structure of Scientific Revolutions*, first published in 1962). Modern medicine includes a tendency to seek cause and explanation inside the body (and not, for the most part, in society, the environment, or the supernatural world) and at the microscopic level—in the world of microbes and cells, of molecules like DNA and hormones, and of other entities that we can see only with advanced technologies. It tends to assume that these microbiological forces affect all bodies in standardized ways according to a set of natural laws, and to rely on statistics and, increasingly, on what is now called "big data."

Modern medicine is mostly new, but it inherited certain very old features from the Mediterranean background in which it arose. Ancient Greek medicine included a unique tradition of anatomy and dissection that was transmitted and developed over centuries by mainly Islamic scholars in the medieval period, and then revived in western Europe in the Renaissance. That tradition has influenced modern medicine's tendency to look inside the body using technology and to view the body as a collection of assembled parts working together.

With regard to disease, modern medicine's overwhelming tendency is to identify and name a huge number of discrete conditions and pathologies, differentiating them, when it can, by their microscopic causes—a system that works well for conditions caused by germs and parasites, some cancers, and some diseases with genetic causes, and less well for chronic conditions like heart disease, most mental illnesses, and a huge variety of problems whose micro-level causes are complicated or poorly understood (osteopenia, which simply means "too little bone," and hypertension, which means "high [blood] pressure," are examples of unhelpful modern names for conditions). Modern medicine's much greater efficacy against infectious disease compared to traditional medicine is responsible for some of the transformations that created the modern era.

My point in trying to distinguish traditional medicine from modern medicine is that modern medicine is implicated in the answer to the question I posed at the beginning of this chapter: When did menopause become a medical condition with symptoms and a name? Researchers sometimes assume that traditional medical systems must have treated menopause and that these traditions have been lost or suppressed along with the voices of the women who developed them. I disagree with this view. Many traditional medical systems did not treat menopause, and many languages did not have a word for it until exposed to modern culture and modern medicine. Some traditional cultures have attributed physical changes or problems to menopause, or at least have developed theories of why menopause happens, but these explanations are not necessarily very similar to modern menopausal syndrome. Most traditional cultures acknowledge role changes in midlife, when a son marries or a grandchild is born, but this is different from modern culture's emphasis on hormonal changes that occur at the end of menstruation.

THE MODERN HISTORY OF MENOPAUSE

Before the eighteenth century, it is hard to find references to an idea of menopausal syndrome in European sources. Sources from Greek and Roman antiquity assume that women's ability to have children ends around the age of 40 or 50; their references are impressionistic and not based on rigorous observation, but this range is consistent with the average age at last birth and the upper limit of fertility in documented societies.[9] In the Hippocratic Corpus, the classic collection of treatises that was the foundation of ancient Greek medicine, fertility lasts until age 42; this may reflect a popular view of the importance of seven-year periods in the stages of human life, also attested in the treatise *On*

Sevens. Aristotle wrote that "menstruation ends for most women around age 40, and if it persists longer, it lasts until age 50, and some still give birth [at that age], but no one beyond it." Centuries later, Pliny the Elder, a Roman aristocrat writing in Latin, agreed that "a woman does not have children after age 50, and for most menstruation ends at age 40." Soranus, a physician of the early second century CE whose treatise on gynecology survives, thought that women were likely to be fertile between the ages of 15 and 40. He believed that most women stop menstruating between the ages of 40 and 50, although some might continue as late as age 60. Regulations for the Roman prefecture of Egypt discouraged marriage between women over age 50 and men under age 60 and vice versa, probably because those were perceived as the upper age limits of procreation.[10]

The people of ancient Greece and Rome knew that women's fertility ends in midlife, but they had little more to say about it. Today, the standard reference work in English on women's medicine in the Roman era does not mention menopause.[11] The author of *Diseases of Women* in the Hippocratic Corpus thought that women past menopause might be susceptible to the peculiarly Greek problem of displacement of the uterus—whose movements around the body were thought to cause many afflictions, some of them serious, a theme that remained strong in Greek medicine and popular culture through the Roman period and far beyond. In this view, because an older woman's uterus was dry and light, it was more likely to wander around the body.[12] But most ancient authorities did not associate "hysterical suffocation" (*hystera* meant "uterus" in Greek) specifically with menopause. They disagreed about who was most susceptible to it, but most often named young widows (that is, women who were once married and then forced to give up sexual intercourse), and women who were sexually mature but not married, who were married but childless, or who had suffered miscarriage.[13]

Soranus thought that it was dangerous for women to cease menstruating suddenly, and that women who did should be treated with emmenagogues (medicine to promote menstruation), but I believe this is the only passage in antiquity that suggests that menopause is a condition that might require medical intervention. Soranus's emphasis is on the suddenness of the change, not on the transition itself. He thought menstruation was generally unhealthy for women, and later in the same treatise, he writes that the end of menstruation does not affect women's health except maybe to make them stronger: "That they no longer menstruate does not at all harm the health of those women who are past their prime, but on the contrary, the drawing out of blood renders many women more delicate."[14]

On the other hand, a large number of problems plague reproductive-age women in much of the Greek medical tradition, beginning at puberty. In the humoral theory that dominated most medical thought from the Hippocratics onward, women were believed to be more cold and moist than men by nature, and were also assumed to live more sedentary lives. This meant that they had to be purged of the excess blood that in men was used for exercise, muscle, and hair. A Hippocratic treatise entitled *On the Problems of Virgins* describes the dire consequences of delayed or suppressed menstruation at the onset of puberty— lethargy, psychotic hallucinations, and suicide.[15] In this theory, better flow through the female body was achieved by sexual intercourse (that is, marriage) and childbirth, which were assumed to be healthful in the Hippocratic tradition. Pregnancy was also thought to make use of the blood otherwise purged through menstruation.[16]

Given the critical importance of menstruation in ancient Greek gynecology, it is at first surprising that the tradition has so little to say about menopause. But the same humoral theory that contributed to the preoccupation with menstruation also suggested a reason for menopause—bodily constitution changed with age, and Greek physicians may have assumed that women became dryer and more like men as they got older, so that they did not need to purge blood anymore. In this view much of the female life cycle was spent in a state of humoral imbalance that had to be managed, but childhood and post-menopausal life presented no special challenges and thus are largely absent from medical treatises.

Most of what survives in Greek and Roman medical texts represents an elite classical tradition. But this tradition was not isolated from folk practices, and many people see the influence of popular culture especially in collections of "herbal" remedies that survive in the Hippocratic Corpus, in Dioscorides's famous collection of recipes, in Galen's vast pharmacological works, and in the encyclopedic *Natural History* of Pliny the Elder. These collections do not contain any recipes meant to treat physical symptoms of menopause, or at least they do not say so.

Ancient drug recipes address problems of conception, pregnancy, childbirth, lactation, and breast health; a large number of them are intended to induce or regulate menstruation. Regular menstruation was thought important to the health and fertility of reproductive-age women, and perhaps, as many scholars have speculated, emmenagogues to induce menstruation were meant to cause abortions (though ancient sources are usually specific about which emmenagogues also cause abortions). A typical entry in Dioscorides—this one discussing a substance he calls *kankamon*—reads like this:

> Kankamon is the sap of an Arabian tree, similar in quantity to myrrh, offensive to the taste, which they use as incense. They burn it to fumigate their clothes, together with myrrh and storax. An amount of three obols [about one-thirteenth of an ounce], drunk with water or oxymel over many days, is said to have a reducing power on those who are fat. It is given to those who are splenetic, epileptic, and asthmatic, and it brings on menstruation when taken with honey and milk. It quickly clears sand [? *oulai*, barley-groats] from the eyes and heals dim-sightedness when put in wine, and for diseased gums and toothaches it helps as nothing else can do.[17]

It is possible that some remedies, like this one, that treated irregular menstruation, or other remedies treating fever or headache, for example, were used by women for symptoms they associated with menopause in a tradition of which no trace survives. While that kind of guess is not totally implausible, it would be pure speculation.

Through the medieval and early modern period, Western medicine retained its focus on fluids, on the idea of disease as "plethora" (a buildup of noxious fluids, especially blood), on bloodletting, and on menstruation. Medieval European sources that mention the age at which menstruation stops usually name the age of 50. Some medieval authors comment that the age at menopause is variable, from 35 or 40 to 60 or even later.[18] St. Hildegard of Bingen, the twelfth-century ab-bess, visionary, philosopher, musical composer, and polymath whose works include the exhaustive *Causes and Cures*, thought that menstrua-tion ceased for most women around age 50 but could continue as late as 70 in some healthy women, and that mothers had occasionally given birth as late as 80. (She also believed that men menstruated through age 80—early European writers often equated the purgation of blood through hemorrhoids and nosebleeds with menstruation.) Hildegard thought that menstruation was accompanied by eye problems, head-aches, and weakness, and that retained menstruation in young women could cause dire problems, but her writings do not mention symptoms associated with the end of menstruation.[19] The *Trotula*, a handbook on women's medicine composed in the twelfth century that circulated widely through the 1400s and was translated into several vernacular lan-guages, advises on menstrual problems, pregnancy, childbirth, sterility, vaginal itching, contraception, infant care, wet nursing, and displace-ments of the womb (including a chapter on "suffocation of the womb," the famous hysterical suffocation of antiquity), but although it informs us that menstruation usually ends around age 50, it does not mention symptoms or offer treatments for the transition.[20]

The word "climacteric"—an old-fashioned synonym for menopause—traces back to medieval theories that divided life into pe-riods of seven or nine years, punctuated by dangerous *climacteres* (the word, borrowed from the ancient Greek term meaning "rung on a lad-der," also had many other uses). These were also called "crises" or "criti-cal" times (from the Greek *krisis*, which means "turning point" or "de-cisive moment"). By far the most dangerous of the climacterics in Western belief was the "grand climacteric" at age 63, much feared by both men and women, though stage-of-life schemes mostly addressed men.[21] The words "critical" and, eventually, "climacteric" began to be applied to menopause in the eighteenth century.

For the Renaissance and early modern period, the most thorough study of menopause in European sources was published in 1999 by Mi-chael Stolberg, whose work I summarize here.[22] References to menopause before the eighteenth century are rare but not entirely absent. Giovanni Marinello, writing in Italian in 1563, associates a large number of symptoms with irregular menstruation in women of reproductive age, from weak vision to fever, vomiting, raging libido, difficulty breathing, and death.[23] Although the idea that inadequate menstruation is dangerous for young women drew on a long tradition, Marinello also mentions that women who stop menstruating due to their age have the same problems. Jean Liébault, writing his own version of Marinello's work in French in 1598, the *Three Books on the Health, Fertility and Diseases of Women* (which was close enough to Marinello's work that subsequent scholars identified it as a translation of the Italian, though Liébault did not credit Marinello), also associates problems with menopause, though not such dramatic ones:

> This flow ceases when women can no longer conceive, which is around the end of the seventh seven-year period of their age [age 49], when their nature begins to weaken; thus the parts of the body retain for their use and support all the blood that is brought to them; for some women, it ceases earlier, around thirty-five, forty, or forty-five years; for others later at fifty-five, and for the latest at sixty years. . . . [T]he signs when it is about to cease . . . are pains and heaviness of the loins, legs, and thighs, many little flushings which appear in the face mostly after eating, and which end profusely in sweats, less appetite than usual, migraines, dizziness, hardness of hearing, ringing in the ears: this flux diminishes little by little each month.[24]

I note here for future reference this early description of what a modern physician would call hot flashes ("many little flushings . . . which end profusely in sweats").

Aside from these references, the idea of a syndrome relating to menopause is missing from European works on medicine and gynecology before 1700, whether herbal recipe books, formal theoretical treatises, or works of any other genre, including those written by women; the slang expressions for menopause that are well attested for the eighteenth century and later do not appear before that, nor was there a medical term for it, nor sections in medical treatises or handbooks devoted to it. For example, the 1653 *Choice Manual, or Rare and Select Secrets in Physick*, a collection of herbal remedies by Elizabeth Grey, Countess of Kent, published shortly after her death, lists no disorders of menopause among the conditions for which she offers advice. These conditions range from warts to bubonic plague, and include gynecological problems like miscarriage, heavy menstruation, difficult labor, and "the Mother" (short for the condition called "suffocation of the mother [uterus]" in that era). Sarah Jinner's almanacs of 1653 and 1659 offer remedies for menstrual problems, uterine suffocation, problems of pregnancy, childbirth, and lactation, hernias, uterine prolapse, miscarriage, watery or phlegmy fluxes, and white flux. Three of her remedies promise to expel a dead fetus and may have been more commonly used by women wanting an abortion. But she mentions no problems relating to menopause.[25] Likewise, the two books on women's diseases in Jane Sharpe's famous *Midwives Book* of 1671 mention no disorders of menopause. A survey of 66 printed advertisements by female medical practitioners in London in the late 1600s finds that they promised to treat miscarriage, green sickness, menstrual problems, sexually transmitted infections, and infertility; to abort unwanted pregnancies; to cure many common diseases, such as rickets and backache; to foretell whether a client was pregnant; and to dye grey hair and disguise smallpox scars—but they mention no problems of menopause.[26] The physician Johannes Storch, who practiced in the town of Eisenach in the early eighteenth century and left a collection of nearly 2,000 case histories of his female patients, the *Weiberkrankheiten* (*Illnesses of Women*), was so little concerned with menopause that in her fascinating study of these eight volumes Barbara Duden could still write that "menopause had not yet been invented."[27]

But this was not quite true. The first dissertation "On the end of menstruation as the time for the beginning of various diseases" was defended in 1710 by Simon David Titius in Halle, Prussia; many others followed.[28] After this, it was common for works on women's medicine to include chapters devoted to menopausal problems, in striking contrast to the previous century. The second volume of Jean Astruc's exhaustive six-volume *Treatise on the Diseases of Women* of 1761, written in French, included a section (chapter 11) on "The cessation of menstruation, and the problems that it may cause"; the most important of these were hemorrhage, white flux, hysterical seizures (that is, fits like those that occur in cases of uterine suffocation), and emaciation. In 1777 John Leake referred to menopause as "a critical change" in his *Medical Instructions Towards the Prevention and Cure of Chronic or Slow Diseases Peculiar to Women*, of which "Cessation of the Periodical Discharge, in the decline of life, and the Disorders arising from that critical change of Constitution" formed one section; Leake believed that mortality was highest for women at this stage of life. Almost all European works on gynecology from the mid-eighteenth century onward likewise contained sections on menopause. Handbooks on "Advice to Women of Forty Years" and "New Advice to Women on the Supposed Critical Age" were published and devoured. That is, today's popular literature on menopause has a long history. So does that dreaded chore, the annual gynecological exam: physicians were recommending them for all women as early as 1785, to check for the uterine cancers they believed were associated with menopause.[29]

Thus menopausal syndrome came into being suddenly and with some drama—it hardly appeared at all in European literature before 1700, and became a popular concept after that. Most eighteenth-century writing on menopause was in French, and it was a French author who finally gave it its modern name: in 1816 Charles de Gardanne invented the word "ménespausie" to replace the many colloquial expressions that were then in vogue, including "women's hell," "green old age," and "death of sex" (he switched to the simpler "ménopause" in the second edition of the book, published in 1820). His treatise, *Advice to Women Entering the Critical Time of Life*, ran to over 400 pages and listed more than 50 related conditions, from erysipelas and scurvy to epilepsy, nymphomania, gout, hysterical fits, and cancer.

Although there is little direct evidence for women's thoughts about menopause in this period, it is likely that physicians were at least partly reflecting popular anxieties rather than only manufacturing them. Even as they described the dozens of dire symptoms they attributed to menopause, it was also a common theme to reassure readers that it wasn't, after all, as bad as they might believe. Eighteenth-century physicians thought that their patients nurtured an old-fashioned idea that menstrual blood was toxic and that its retention at menopause caused serious problems, and they denounced the tonics, pills, purgatives, and tinctures that their patients took for disorders relating to the end of menstruation; advertisements for such medicines survive from this era. A few passages from eighteenth-century women's correspondence confirm that they expected dire consequences when their periods stopped (and were surprised when nothing happened). For example, one Madame Viard d'Arnoy wrote to the Swiss physician Tissot complaining of irregular periods and frequent episodes of what would today be called hot flashes, occurring 10 to 15 times each day; she was sure she was plethoric—that is, suffering from a buildup of too much blood—and feared a slow and painful death. (This patient does not mention her age, however, which complicates the picture.)[30]

In the first half of the nineteenth century, some authors began to doubt whether all the symptoms attributed to menopause were really caused by it, and whether women really had higher mortality in this stage. Theories of health and disease centering on the nerves became popular starting around 1700, and especially after the late eighteenth-century discovery that nerves conducted electricity. Physicians began to attribute menstruation and menopausal symptoms not to plethora, but to the nervous irritation of the uterus and ovaries, thought to be highly sensitive organs. Some argued that women were healthier after menopause, once relieved of what they portrayed as the damaging effects of menstruation—provided their patients safely traversed the critical stage at which the irritation of the uterus was at its height, causing years of intense psychic and physiological disruption.[31]

In a common trope, physicians writing treatises on menopause blamed its symptoms on the urban lifestyles of their upper-class, literate audiences—women in the countryside who worked hard, spent most of their time outdoors, did not indulge in romance, and had lots of children, they argued, did not suffer at menopause. They prescribed strict regimens for their patients, insisting that only if they ignored the advice of old women (that is, lay female healers) and their drastic cures, and submitted entirely to the doctors' own prescriptions for food, drink, exercise, reading material, and everything else, could they hope to achieve a peaceful transition to old age. They advised against love and sex, and prescribed drugs and bloodletting, often with leeches applied to the vulva. In retrospect, it is easy to see in these exhortations a desire of male doctors to impose their authority on female patients, to establish themselves as the only legitimate practitioners of women's medicine (or of any medicine), and to enforce the patriarchal values in which they believed. On the other hand, in this model, once past menopause women might enjoy many years of renewed health and productivity.

Stolberg argues that physicians and laypeople of the early modern period believed so strongly in the damaging effects of menopause because of the perception that menstrual blood was poisonous—a perception old-fashioned by the eighteenth century, and one that physicians tried to overcome, but not without substituting their own theories about why menopause caused problems. The theory of plethora—the buildup of too much blood (rather than toxic blood)—that became dominant among professional physicians in the late seventeenth century also influenced the idea of menopause in different ways.[32] Episodes such as Liébault's "little flushes"—which foreshadowed the "hot flash" of later decades and centuries—were consistent with the theory of plethora, and were interpreted by patients and doctors as signs of it. The hemorrhaging of blood or white fluid from the uterus, and from other orifices and organs, was another symptom commonly mentioned by early writers on menopause.[33] In their view, normally harmless blood could become spoiled or corrupted when retained, and this could cause a huge variety of problems, from "hysterical" attacks or seizures to arthritic swellings, fever, heat, rashes, itching, abscesses, cysts, scurvy, headaches, anxiety leading to apoplectic attacks, and death.[34]

While the link between ideas of plethora or of the toxicity of menstrual blood and the idea that menopause causes health problems may seem a natural one, and may have been important in early modern Europe, not all cultures that perceive menstrual blood as poisonous or dangerous, or the purgation of blood as a beneficial release of excess fluid, have a concept of menopausal syndrome. These beliefs about menstruation are common in traditional cultures, but the idea of a menopausal syndrome of symptoms or problems is not. Plethora was a concept foundational to ancient Greek medicine, in which the excess of one humor or another was thought to cause diseases of all kinds, but menopause as a medical condition was absent from ancient Greek medicine. Similarly, menopause has only a small presence in classical Chinese medicine, even though blood and menstrual regularity are central to women's health in that system. The list could go on—I'll cite several examples in chapter 10 of cultures that believe menstrual blood is toxic and polluting without attributing significance or symptoms to menopause. To offer another example here, among the Lusi, a horticultural society of Papua New Guinea studied by Dorothy Ayers Counts in the 1970s and 1980s, menstrual blood was thought to be a powerful poison, but the Lusi did not think that menopause had symptoms or that it was important except as the end of fertility, which they generally welcomed.[35] Nevertheless, Stolberg is correct to emphasize that European ideas about menstrual blood are part of the background for the rise of menopausal syndrome in the West, even if they are not sufficient to explain it, and he is certainly correct that the idea of menopausal syndrome is older than modern medicine in the West.

Eighteenth-century theories of plethora and nervous tension were not only an intellectual phenomenon; ordinary people thought of health and disease in these terms, in some places until quite recently. When anthropologist Dona Lee Davis went to Newfoundland in 1978 hoping to study ideas of menopause in a traditional society, she found no communities that met that definition. She decided instead to focus on a small fishing village, with a population of under 800 people, that had only recently become accessible by road (and had previously been reached only by boat). Although the people of the village used modern medicine and Davis's study does not mention traditional practitioners other than two retired midwives, they explained health and disease, including menopause, which they usually called "the change," in traditional ways. The quality and consistency of blood was a central concept: blood could be too thick or thin, too high or low, or poisonous. Ideas of "high" and "low" blood, "the blood," or "too much blood" mixed older concepts of plethora with modern medical discourse about blood pressure—problems doctors attributed to high blood pressure, such as a husband's stroke, were interpreted as resulting from too much blood. Menstruation was thought to purge women of excess blood, but could also leave women with too little blood. At menopause, women were believed to purge all the extra blood that might have otherwise accumulated for the rest of their lives, and heavy bleeding at menopause was considered healthy for that reason; hot flashes were also thought to purge extra blood.

Women understood nerves to be strings holding the body together; they could be thick or thin, strong or weak. Thin or weak nerves could cause psychological and behavioral problems. Anxiety, irritability, depression, forgetfulness, social avoidance, and hot flashes were all problems they associated with a weakening of the nerves, which could happen in midlife. The women in Davis's study also complained of developing a lower tolerance for small children. Hormone replacement therapy and other medicines for menopause were popularly understood to be pills for the nerves. Davis emphasizes, however, that menopause played a relatively minor role in her subjects' thoughts about health and illness; problems with blood and nerves were considered typical of every life stage.[36]

Thus a concept with an earlier history—represented especially in the thread of influence that followed from Marinello's treatise published in 1563—became much more popular around 1700, and after that can be found in most European writing on women's medicine. It seems a fair conclusion that at some point in the early modern period—between the sixteenth and eighteenth centuries, but mostly in the eighteenth—menopause became a condition with symptoms and, eventually, a name.

These roots of menopause in western Europe antedate modern medicine by the definition I am using in this book. But as modern medicine developed in western Europe, the idea of menopause was not abandoned; it became a part of the new system and was exported around the world.

The social reasons behind the surge of interest in menopause among physicians and their patients in the eighteenth century are hard to pinpoint; none of the guesses offered by scholars has been very convincing. Was menopause more of a problem because women were living longer, or perhaps because fertility was lower? Did male professional physicians suddenly take an interest in women's problems they had long ignored, as part of a program of wresting control of medicine from illiterate female healers whose practices have left no record? Was menopausal syndrome the product of a general atmosphere of anxiety in the time of the French Revolution? Or did it arise because the doctors who produced the new type of professional treatise in which we hear so much about menopause lived in cities, treating upper-class urban patients—that is, were they partly right that affluent urban living caused menopausal symptoms?[37]

Some of these guesses are more plausible than others, and none is easy to prove. The hostile takeover of medicine by male professionals in early modern Europe was a real and striking phenomenon, described by Barbara Ehrenreich and Deirdre English in 1973 in their classic book *Witches, Midwives, and Nurses.* But arguments that menopause is absent from published literature because women's voices were suppressed, or because medicine of the time was slavishly following ancient sources, or because a separate tradition of informal folk medicine left no traces, are unconvincing for the seventeenth century in light of the quantity and variety of the material that survives from that period. I think it is possible that the subject of menopause became popular among academically trained male physicians precisely *because* it was a new condition not treated by traditional female healers, but that speculation would be hard to prove.

The explanation that menopause became a popular subject as women began living longer has the weakness that in 1700, the beginnings of the mortality shift were more than a century in the future, and the further weakness that even before that shift, post-reproductive life was not an anomaly among humans but part of our normal life history. Similarly, menopause was a popular and well-established concept long before the French Revolution of 1789.

We may be overthinking the problem. Syndromes are historical phenomena in their own right, and in Europe, the period from around 1600 to 1900 was the great era of syndromes—disorders characterized by mostly nonspecific symptoms, with disputed and shifting explanations and a prominent role for emotional and behavioral problems like psychosis, seizures, or sexual promiscuity. Some of the most popular syndromes of the time were melancholia, or ancient "black bile disease"; hysterical suffocation, called "suffocation of the mother," "fits of the mother," "mother fits," and several other names in this period; uterine fury, sometimes called nymphomania, a disease causing behavior so salacious that Jean Astruc, for example, wrote this section of his *Treatise on the Diseases of Women* in Latin instead of French; and green sickness, a disease of adolescent girls for which the closest analog today is *anorexia nervosa*. Melancholia and uterine suffocation were both conditions with old roots, linked by an uninterrupted chain of references over centuries of medieval literature back to sources of the Greco-Roman period. Both of these conditions gained hugely in popularity in the seventeenth and eighteenth centuries, and both were reinterpreted over the course of the eighteenth and nineteenth centuries as nervous disorders, and eventually as psychiatric disorders.

Thus around 1800 physicians began to speak of "hysteria" rather than of a condition—hysterical suffocation, suffocation of the mother— defined by its index symptom of choking. Some writers viewed hysterical suffocation, with its traditional origins in the uterus, as a female form of melancholia, distinct from the male hypochondriac form, which was thought to originate in the upper abdomen; some argued that hysterical and hypochondriacal disorders were the same thing and that both originated in the brain. In the nineteenth century, because women were perceived as more delicate and weak-nerved than men, they were thought to be more susceptible to these nervous syndromes, including a popular new syndrome, neurasthenia ("weak nerves")—although one study shows that in a hospital treating working-class patients in London, this condition was diagnosed with about the same frequency in both women and men, and especially in soldiers returning from World War I.[38]

Green sickness was more of a new invention of the late Renaissance—a condition of unmarried adolescent girls characterized by suppressed menstruation, lack of appetite, pale or greenish skin color, and a number of other symptoms that varied over time. The best cure was marriage. Green sickness is first attested in the mid-1500s and was believed to be a new disease at the time, although in the earliest reference to it, a letter dated to 1554, the Silesian physician Johannes Lange argued that it was the same condition described in the Hippocratic treatise *On the Problems of Virgins* dating to the fifth century BCE (this work had not been preserved in the West and became available in Latin translation only in 1525). Lange referred to the condition he describes in his letter, a response to an inquiry from the father of a long-distance patient named Anna, as "the disease of virgins," but it was more commonly called chlorosis or green sickness. Green sickness was not the continuation of a classical idea in the same way that hysterical suffocation was—there is no chain of tradition linking the Hippocratic treatise that Lange cites to the sixteenth century. But it became popular very quickly, and is well attested in the European medical literature of the later sixteenth through nineteenth centuries; over time it was reinterpreted as iron deficiency anemia and disappeared from medical literature around 1930.[39]

It is not surprising that a new syndrome, menopause, arose in Europe some time in the centuries around 1700, an era that was such a creative source of other syndromes, including some that were specific to women. Early writing on menopause often linked it to other women's syndromes. For example, Astruc considered "hysterical vapors" one of the most prominent symptoms of menopause. These sound like what are called hot flashes today—"blushings, heat sensations that frequently rise suddenly to the face, and that end in sudden sweats"[40]—but he also mentions other symptoms then associated with hysterical attacks, such as choking and strangling sensations, "swarming of entrails," and involuntary laughing and crying. In the 1816 treatise in which he gave menopause a name, de Gardanne listed "hysteria, or nervous affection of the uterus" and "nymphomania, or uterine furor" among its symptoms. In the mid-nineteenth century, Edward Tilt associated menopause with chlorosis—for he saw menopause as the mirror of puberty and closely connected to it—and with nymphomania and hysterical symptoms. He considered melancholia, mania, alcoholism, uncontrollable murderous instincts (in his view menopause could be cited as an insanity defense for murder), and other types of mental derangement typical of meno-pause, although few of his own patients had these conditions.

Today, the diagnoses of hysteria, melancholia, chlorosis, hypochon-driasis, and other early modern syndromes do not survive, except as ghosts haunting some modern diagnoses among psychiatry's mood disorders, dissociative disorders, and eating disorders. It is not accurate to say that these syndromes were imaginary or only literary traditions— real patients suffered real problems, as attested in diaries, letters, and doc-tors' case records. But culture was likely the most important factor shap-ing how they were experienced over the centuries. Today, neurasthenia remains a diagnosis in the European-based *International Classification of Diseases* and in the current, third edition of the *Chinese Classification of Mental Disorders*, and has a particularly fascinating history. But for the most part, of the syndromes so popular in European medicine between about 1600 and 1900, only menopausal syndrome remains a modern di-agnosis, and I wonder whether we have paid too little attention to its similarities, in its origins and nature, with other syndromes of that era.

To return to the story of its history, in the late nineteenth century, menopausal syndrome was quantified—that is, reduced to numbers— as part of a general trend in the transition to modern medicine. Edward Tilt, who wrote the first full-length book on menopause in English, *The Change of Life in Health and Disease*, was an early pioneer in the statis-tics of menopause. (This book is best known in its second edition of 1857 and eventually ran to four editions, the last published in 1882.) His database was 500 menopausal or post-menopausal patients from his clinical practice at the Farringdon General Dispensary and Lying-In Charity in London, where he treated women from all social classes. Tables in every chapter tally the number of women who experienced each of the many dozens of symptoms and conditions that Tilt associ-ated with menopause: for example, 244 experienced "an increased pro-duction of heat . . . with that irregular distribution of it which is called 'flushes'" (elsewhere he writes that 287 experienced flushes); 146 had regular leucorrhea (white flux); 208 experienced hemorrhaging, most frequently in the form of a heavy period, but some bled from hemor-rhoids or coughed, vomited, or urinated blood, and 4 bled from the ears. "Nervous irritability," headache, abdominal pain, back pain, "epigastric faintness," "hysterical state," and "pseudo-narcotism" (a sort of combination of dizziness, lethargy, disorientation, and memory problems—some patients forgot their own names) all affected more than 100 patients. The fingernails of 5 patients peeled off; 10 temporarily lost their hearing and 3 suffered "aphonia," inability to talk, both classical hysterical symptoms; 16 had heart palpitations; 16 suffered from insanity; 4 had uterine cancer; and 1 had breast cancer. Tilt also cites similar data compiled by his contemporaries; in the second half of the nineteenth century, this type of clinical sampling was a standard feature of the study of menopause in Europe.[41]

Tilt also includes a large number of case histories—stories about his patients—in his book. An example illustrates his general idea of menopause as a disorder that affects the nerves and brain, caused by a buildup of fluid that may be purged by hemorrhoids, abscesses, or other means.

Mary S., of average size, but thin, with a sallow complexion, dark hair, and hazel eyes, was 47 when, in October, 1852, she came to the Farringdon Dispensary. The m[enstrual] flow appeared at 13, and she had always been particularly free from morbid symptoms. Though twice married, she never conceived. About 15 months ago the m. flow became irregular, and she was much troubled with headache and abdominal pains. One night she went to bed as usual, and awoke delirious. She ran down the street in her night-gown, required 3 men to hold her, was taken to the Bristol Infirmary, and in 3 days, the m. flow having come on, her senses returned. Several abscesses appeared in both arm-pits, some broke, others were lanced, and, when better, she came to London, and applied at the Dispensary for relief from headache, nervousness, and lightness of head. There had been no m. flow for the last 10 months. I gave her a scruple of Dover's powder every night. . . . The singularity of this case is, that no nervous symptoms presented themselves until the attack of delirium. The patient was in tolerable circumstances, had a kind husband, and nothing to trouble her, so I cannot attribute the delirium to anything but the c[hange] of life.[42]

In Tilt's view, menopause was a critical point in the female life cycle when cancers of the breast or uterus might easily develop and when women were vulnerable to over 300 pages' worth of symptoms, including—and this was his main focus and interest—a long list of psychiatric symptoms he associated with the brain and nervous system. He considered menopause a nervous condition above all else, even though he also retained the idea of plethora, as in the case history just cited. Tilt believed that women past menopause were practically immune to disease, and his view of post-menopausal women was quite positive, but menopause itself was, in his work, a crisis of the most profound order.

By the late nineteenth century, statistics and tables had become a standard way to present research on menopause and on all topics in medicine. In the first American book on menopause, Andrew Currier's 1897 *The Menopause: A Consideration of the Phenomena Which Occur to Women at the Close of the Child-Bearing Period* cites data compiled by a score of physicians in France, Germany, Russia, England, Denmark, Norway, and other European countries, including Edward Tilt; Currier also cites data from his own clinical practice in New York City, and a study of Native American populations that he had conducted himself. He stressed the problem, also acknowledged by Tilt, that the data came from clinical populations—women who came to the doctor because they were sick—and did not represent the normal range of experience of women at menopause. He thought that Tilt's idea that menopause was a dangerous period followed by a golden era of "joy and felicity" was silly, and argued that "the typical menopause [was] a colorless, uneventful experience." Around this time, it was a trend in medical literature to argue that previous generations of physicians had overdramatized the transition. Nevertheless, Currier thought cancer of the uterus and breast occurred most frequently—though rarely overall—in the decade between 40 and 50, and that women predisposed to cancer might well contract it at menopause.[43]

Currier adopted the theory, then dominant, that menopausal disorders were caused by irritation of the nervous system, and thought that menopause could cause "profound vascular and nervous derangements." The slightest stimulus might cause "an explosion of nerve force, a flash of heat about the head and neck, and a minute or two later a profuse perspiration," episodes he compared to epilepsy, hysterical attacks, and the paroxysms of malaria. Currier was especially interested in these "heat flashes," which, he argued, might take several forms and involve other symptoms as well, including melancholia, headache, cold extremities, diarrhea, "abundant discharge of urine," mania, and "intense sexual ardor."[44]

It is hard to generalize about symptoms in early writing on menopause—or, indeed, in later writing—because they are so many and so varied. Physicians disagreed about whether menopause was dangerous: some thought it was a critical time of life with high mortality; others emphasized that it was normally well-tolerated; still others, such as Tilt, believed that menopause was followed by a period of rejuvenation and good health as women were freed from the burdens of reproductive life. Still, most attributed to menopause dire symptoms such as psychosis, convulsions, and cancer. Episodes that sound like some modern descriptions of hot flashes, with ascending sensations of heat and flushing followed by sweating, sometimes occurring several times per day, are described in early sources from the eighteenth and nineteenth centuries (and I have quoted from some of these descriptions). These "flushes" were the second most common symptom, after "nervous irritability," in Tilt's "Table XIX: Showing the Frequency of Morbid Liabilities at the c[hange] of Life in 500 Women."

Sexuality is a theme in much early (eighteenth- and nineteenth-century) writing on menopause, as it is now. Afflictions of the genitals—inflammations, polyps, ulcerations, itching, and cancers of the womb, vagina, cervix, and labia—are frequently mentioned, but the vaginal dryness and pain during intercourse often associated with menopause today do not appear in these early discussions. Many writers thought that nymphomania, or "uterine furor," with its symptoms of unhinged sexual promiscuity, were dangers of the menopausal transition. Some argued, and others may have assumed, that sexual feeling disappeared after menopause; they described atrophy of the uterus, ovaries, vagina, breasts, and vulva beginning with menopause. Among other austerity measures they recommended for midlife women, they often advised abstinence from sex and love. French physicians were more interested in sexuality at menopause than those writing in English or German, and it is mainly among the French that we find expressions such as the "death of sex" and florid, empathetic passages on the loss of physical attractiveness at menopause.[45]

Finally, it is interesting to notice the origin of the condition of "irritability," so often linked with menopause in the modern imagination. Early writers on menopause saw it as a disorder of the nerves, and linked irritation of the nerves of the uterus or other parts of the nervous system at menopause to physical and behavioral symptoms—women became oversensitive, overreactive, even psychotic. Tilt's patients "could not bear the slightest noise"; they found other people's conversations intolerable; one patient had to stop going to church because of the crowd. These symptoms Tilt labeled "nervous irritability," and he was careful to distinguish them from the emotional lability of "hysteria," a label his patients resisted. Today, despite changes in how we perceive the causes of menopausal syndrome, irritability remains at its center. In a review of eight studies by one researcher on the psychological symptoms of menopause in 2011, irritability and sadness or depressed mood were the only symptoms investigated in all of them.[46]

Menopause took the form familiar to modern readers in the early twentieth century, with the discovery of hormones—chemicals produced in the ovaries, thyroid, and other glands, which act on remote parts of the body by circulating through the blood.[47] Researchers had long speculated that the ovaries, as well as the testicles, secreted substances related to sex and reproduction (that these glands were the source of many secondary distinctions between the sexes had been known since antiquity). In 1929, Edward Doisy and Adolph Butenaldt, working separately in different countries, isolated the first estrogen, then called estrone, an accomplishment for which they shared a Nobel Prize. After that, menopause was understood as a condition of the endocrine system—hormones—and not one of nerves or blood.

With the new science of endocrinology, a very different view of menopause might have developed. Nineteenth-century physicians had little choice but to explain menopause in terms of aging and deterioration—the withering or stiffening of the uterus or other genital organs, the narrowing of blood vessels. There was now the possibility of seeing menopause as something different, as the cessation of a system programmed to shut down long before the rest of the body for some sort of adaptive reason. But this is not what happened. Instead, menopause came to be understood as a deficiency of the chemical estrogen, which in turn came to represent the essence of femininity—a formulation that restored to ideas of menopause the drama, tension, and flights of imagination that had begun to flag in the late nineteenth century and that would reach new heights in the later twentieth century. The new understanding of menopause also placed a great deal of money at stake, once estrogen replacement therapy became common.

In 1938, biochemists at the University of Oxford created the first synthetic estrogen, DES (diethylstilbestrol), which later became notorious for causing cancer and other problems in the children of women who took it to prevent miscarriage. A few years later, the Canadian pharmaceutical company Ayerst, McKenna, and Harrison developed a conjugated estrogen derived from horse urine, sold under the name Premarin.

At first, doctors publishing in medical journals were conservative about the use of estrogen replacement therapy (often abbreviated ERT today). Many suspected that ERT might cause cancer; others argued that symptoms of menopause arose from a period of adjustment, which might be prolonged by the use of hormones. Through the 1960s, the dominant discourse in medical journals in the United States—where most research on hormonal therapies took place in the twentieth century—saw menopause as a natural transition that needed no drug therapy in most women, although a few patients might experience severe suffering requiring treatment. Doctors advised listening to patients, educating them, and reassuring them, and argued for a strong psychosocial component to menopause, blaming what they perceived to be the psychological symptoms of menopause on the social stresses of midlife or on fears and misinformation picked up from gossipy friends.

This mostly sensible, if sometimes dismissive and patronizing, discourse in medical journals masked other trends. Pharmaceutical companies marketed their drugs directly to physicians (following regulations imposed by the Food, Drug, and Cosmetic Act of 1938), promising miraculous benefits for the huge variety of symptoms that had, by now, been associated with menopause for more than two centuries—headaches, hot flashes, nervous irritability, and so forth. Patients also heard about the drugs from books and magazine articles that popularized the new therapies, and demanded them; doctors, under pressure from both sides, largely complied. Husbands were the target of publicity that urged them to send their "irritable" wives to the doctor for estrogen. A large percentage of middle-class women going through menopause, perhaps about a third, were taking hormones or sedatives in the 1950s.[48] "Untreated" menopause came to seem old-fashioned and barbaric in some circles around this time.

In another trend, the hormonal model of menopause brought with it a new emphasis on the long-term health consequences of the reduction of estrogen, and not just on the symptoms of transition—an approach that became more prominent in the 1950s and eventually came to dominate menopause research. Doctors hoped that with estrogen replacement, women might be freed not only from transitory symptoms, but from osteoporosis, heart disease, and other chronic conditions they attributed to a lack of estrogen. Conversely, they shuddered at the cost to society of maintaining a growing population of old, unhealthy, estrogen-deprived women, a theme still common in scientific writing about menopause today. A corollary of this line of reasoning was that all women could benefit from hormone therapy and should be treated by default, unless they had some condition that contraindicated it. Thus some physicians began to argue that ERT was not just for a few women with severe symptoms, but a basic maintenance drug for everyone.

Pharmaceutical companies and physicians began to make claims that estrogen could preserve youth, attractiveness, and sexuality in older women who would otherwise become dour, shriveled shells of their former selves, destined to lose their husbands to younger women. (My language here is not nearly as florid as that in some mid-twentieth-century medical journals.) This focus on sexuality reflected changing ideas of marriage, as old patriarchal traditions became obsolete and un-appealing to women living urban, modern lives. The concept of "com-panionate marriage" arose in the 1930s, and by the 1950s white, middle-class marriage in the United States was a highly eroticized institution based on sex.[49]

At the same time, the dominant ideology in North America discour-aged women from living independently and reinforced a strict division of roles in both household and society. This was partly a reaction to the surge of women who had entered the workforce during World War II when many men were overseas, and who had to be put back in their place when the soldiers returned, a move many resisted. In this context, the theme of sexuality in menopause changed and became more impor-tant. Vaginal dryness and low libido began to be discussed as problems, even as much of the advice literature assured women that they could still have a good sex life after menopause. Sexual promiscuity, a theme so prominent in the discourse on menopause in previous decades and centuries, got less attention in this era, although it was still an issue (and twentieth-century writers added homosexuality to the list of problem-atic behaviors to which menopausal women were prone). Much ink was spilled warning women about the many ways in which they might drive their husbands away at menopause, and practically none, as other schol-ars have noted, on the ways in which an aging husband might arouse less sexual enthusiasm in his wife. Menopausal "irritability" was over-whelmingly construed as a marital problem—women being unpleasant to their husbands and sometimes to their children and in-laws as well.

Meanwhile, sex research with real women—including Kinsey's famous reports—became more common in this period. These data are limited by small sample sizes and narrow demographic representation, but they show a picture different from the one painted in popular litera-ture and medical journals of the 1950s. Most women reported little change in their sex lives at menopause and did not feel they had lost attractiveness; some reported that they felt liberated from worry about pregnancy; others used menopause as an excuse to end a sex life that they had not enjoyed in the first place. Some reported unsatisfactory sex lives but blamed their husbands' reduced sexual function. These surveys did not support the idea of a uniform effect of menopause on sexuality.

Race and class were always part of the idea of menopause—nineteenth-century physicians argued that their European patients of the leisured classes suffered more at menopause than peasants, the poor, or people of other races they deemed primitive in a sort of "noble savage" theme. Twentieth-century menopause was construed as white, middle- or upper-middle-class, and suburban. The typical menopausal woman of mid-twentieth-century medical and popular literature had the time and money for doctors and maybe drugs. She spread gossip at bridge parties, kept house for a husband, did not have a paying job, and was in danger of becoming bored and self-absorbed unless she dedicated her free time to worthy volunteer projects. As the twentieth century progressed, estrogen therapy was touted to professional women who wanted to remain competent at their jobs rather than deteriorating into the emotionally volatile or semi-vegetative mental states thought to be associated with menopause. Robert Wilson in particular, whose work is discussed further later in this chapter, and who had a keen instinct for branding, construed the audience for his book *Feminine Forever* as a class of "prosperous, upper-middle-class women in their middle years" who took up high-paying jobs after their children left home. The idea of menopause as a condition to be managed, about which complex decisions had to be made by a highly educated consumer, in consultation with a physician—an idea that was mainstream by the late twentieth century—also assumed a middle- or upper-middle-class type of patient.[50] That the twentieth-century idea of menopause was quite consciously developed for an affluent population in a highly modernized society is one reason that, as we will see, it might not serve the majority of the world's women in all the ways that practitioners of modern medicine tend to assume that it does.

The moralizing trend of the early modern period also continued in twentieth-century advice literature on menopause—although the example of the hard-living, hardworking peasant woman was now rarely invoked, experts urged women who wished to avoid problems at menopause to busy themselves with charity work and service to others, and to avoid complaining about their symptoms or focusing on themselves. Some construed menopause as a reward for decades of selfless devotion to housework and childrearing, reinforcing patriarchal values even as they partially exempted middle-aged women from them. Some writers emphasized the achievements of women past menopause and their important functions in society; others vilified older women (sometimes described as "fishwives") who spread what they deemed misleading and terrifying information about their experiences. Everyone, it seemed, had a strong opinion about what post-reproductive women ought to be doing with their lives.

The popularity of estrogen replacement therapy exploded after 1963 with the publication of "The Fate of the Nontreated Postmenopausal Woman: A Plea for the Maintenance of Adequate Estrogen from Puberty to the Grave" in the *Journal of the American Geriatrics Society*. This influential article was written by the physician Robert Wilson, mentioned earlier, and his wife Thelma Wilson, a nurse. Three years later, in 1966, Robert Wilson published *Feminine Forever*, bringing his ideas to a mass audience. Wilson's research and the foundation he established were funded almost entirely by Ayerst, McKenna, and Harrison, the makers of Premarin, and by other manufacturers of hormonal drugs.

Wilson cast himself in *Feminine Forever* as a champion of women, courageously battling "a serious, painful, and often crippling disease"— that is, menopause—in a world where mostly male physicians did not pay attention to women's complaints. He promised that women who followed his recommended therapy—or rather, demanded it from their physicians—would avoid menopause entirely and preserve their youth, their sex appeal, and their marriages. In hyperbolic language that put nineteenth-century French writers on menopause to shame, he painted a ghastly portrait of the withered, unnatural existence of women who had outlived their reproductive function and their supply of ovarian estrogen. "No woman," he insisted in the book's best-known passage, "can be sure of escaping the horror of this living decay. Every woman faces the threat of extreme suffering and incapacity."[51]

Wilson aggressively promoted the estrogen deficiency model of menopause and the image of estrogen as a miraculous cure, comparing it to insulin for diabetes. Because evidence connecting estrogen therapy to endometrial cancer (that is, cancer of the lining of the uterus) was growing, however, he advised a combined regimen of estrogen and progestin (usually abbreviated HRT, for hormone replacement therapy, today). Adding progestin plus a five-day hiatus in the drug cycle caused a woman's periods, and some of the related problems like bloating and cramps, to come back. Wilson tried to spin this outcome as positive, but women much preferred estrogen-only therapy, and it remained the dominant regimen until 1975, when two major studies independently found a strong connection between ERT and endometrial cancer.

Wilson thought that menopause, like diabetes, was a diseased state—a deficiency condition—that should be eliminated, not just managed or ignored. He argued that women should take hormones through their reproductive years and beyond—"from puberty to the grave"—by which method, he believed, they could avoid menopause altogether. Among the huge number of problems that he claimed estrogen might alleviate, sexual dysfunction and the appearance of aging were especially prominent. In lurid language he described the shriveling, drying, and shrinking of breasts and genitals and warned that women past menopause would find sexual intercourse painful or impossible. By now, sex was not only considered central to a good marriage; with the sexual revolution, it was thought indispensable to a healthy, fulfilled life. Wilson, and the drug manufacturers who backed him, capitalized on older women's desire to participate in this revolution, and also on their fears of losing their husbands to what seemed like an ever-expanding supply of young, sexually liberated women.

Wilson's views were extreme, and not all physicians agreed with him—the medical community became divided among those who thought of estrogen as a cure that should be made available to all women, and those with concerns about the risks of stroke, some cancers, and possibly cardiovascular problems among women on hormonal therapy. Hopes that estrogen would prevent heart disease in older women were proving elusive (and estrogen therapy for men at risk for this condition was abandoned soon after the first trials began in the late 1960s).[52]

But *Feminine Forever* was a bestseller, and Wilson, who was a tireless promoter of his views in popular media, transformed public discourse about menopause. Menopause was increasingly construed not just as a transitional period of crazy symptoms, but as a deficient and even diseased state in which women could not be healthy, attractive, competent, or reach their full potential without drugs. The sexual revolution, the sexualization of marriage, the increasing participation of women in the paid labor force, the various stages of the movement for women's rights, and the increasing profitability of drugs were all part of the background for this trend—in these ways and others, modern menopause is a product of the modern world. At the same time, certain older themes in Western thinking about menopause were remarkably resilient and remain so—these include the enduring position of the hot flash, first described in the literature of the sixteenth century, at the center of modern menopause's litany of symptoms, and the association of menopause with psychological problems, from "irritability" to more profound disorders.

Although hormonal chaos and estrogen deficiency became the new way to explain menopause, replacing the language of nervous irritation, medical and popular literature continued to link menopause to psychiatric symptoms. Psychoanalysis, the new type of psychology pioneered by Sigmund Freud, added a new dimension to this view by separating mental conditions from the physiology of the rest of the body—now the psychological problems thought to be associated with menopause could be blamed not only on fluctuating hormones, but on unconscious griefs and desires combined with a lengthy list of unpleasant-sounding personality traits. Freud himself had little to say about menopause; the most influential Freudian discussion of menopause was that of his American colleague, Helene Deutsch, who was born in Poland and fled Germany for the United States in 1935. Her *Psychology of Women*, published in two volumes in 1944–1945, mostly focuses on "Girlhood" and "Motherhood" (its two main sections) but contains an epilogue on "The Climacterium," which starts unpropitiously by declaring that "with the cessation of this function [menstruation], she [woman] ends her service to the species." "The mastering of the psychologic reactions to the organic decline," Deutsch goes on to state, "is one of the most difficult tasks of a woman's life." Menopause is, in her view, an unrelieved disaster of deterioration and mortification—caused partly because "the endocrine system is deranged in its functioning" at menopause, but also by women's reaction to the organic changes they are experiencing.

> The changes that take place in the body of a menopausal woman have the character not only of the cessation of physiologic production but of general dissolution. Woman's biologic fate manifests itself in the disappearance of her individual feminine qualities at the same time that her service to the species ceases. . . . [W]ith the lapse of her reproductive service, her beauty vanishes, and usually the warm, vital flow of feminine emotional life as well.[53]

I can offer no speculation on the source of Deutsch's idea that women become emotionally cold at menopause, but with her contention—and this is the main focus of her discussion—that women going through menopause are sexually promiscuous and make themselves ridiculous by dressing and acting like young girls, she continues a long tradition. Deutsch theorizes that women at menopause may seek creative outlets ("with little success," however, for despite their ambition and manic energy menopausal women are "emotionally impoverished" and "intellectually unproductive"); indulge in erotic fantasies of rape, prostitution, and old lovers; grow bored with their husbands and divorce them; have repressed sexual feelings for their female friends, sons, and young men; become anxious, overexcited, or depressed ("almost every woman in the climacterium goes through a shorter or longer phase of depression"); become emotionally unstable; and suffer psychotic delusions and morbid fears of cancer. Feminine women with satisfying erotic marriages, in her view, suffer less at menopause than masculine women or those who are "frigid" or "spinsterish." Deutsch's advice to women going through menopause is to resign themselves to the tragic reality, to focus on "what they can still enjoy" despite the catastrophe happening to them, and to try to be a good grandmother.[54]

Deutsch's central idea that women going through menopause are engaged in a desperate pitched battle against "partial death" continued a tradition that viewed menopause as a time of crisis, even if critical-age theory was long buried in the past. As the first comprehensive work on the psychoanalysis of women, *The Psychology of Women* was very influential, but Deutsch's ideas also reached a wide audience through their influence on Simone de Beauvoir's *The Second Sex*, first published in French in 1949 and in English translation in 1953. De Beauvoir's fatalistic thoughts on aging and menopause in one of the most widely read feminist works of its time helped to set the lugubrious tone that characterizes so much popular discourse on menopause from the late 1950s onward (despite the fact that de Beauvoir herself was only 41 when *The Second Sex* was published). The idea that women feel grief and desolation for the loss of their femininity at menopause became a common theme of this era—a theme fuel-injected and turbo-charged in the mid-1960s by Robert Wilson, who one-upped Deutsch by beginning his and Thelma Wilson's essay of 1963 with the declaration that "the unpalatable truth must be faced that all postmenopausal women are castrates."

In the psychoanalytic model, grief and despair were the main contributors to mental problems at menopause, and so depression was the kind of mental illness most associated with menopause from the mid-twentieth century onward. "Involutional melancholia" or "involutional psychosis" became the main psychiatric syndrome of menopause beginning in the 1930s, replacing the nervous conditions of the 1900s—hysteria, neurasthenia, and so forth. The term was first used by the great German psychiatrist Emil Kraepelin to signify a kind of mental disorder thought to be specific to middle-aged women and men; "involution" referred to physical decline.[55] Kraepelin himself lost faith in this diagnosis and abandoned it in the later editions of his textbook, the *Lehrbuch der Psychiatrie*. But it remained popular among U.S. psychiatrists, and through the 1970s it was applied largely to women going through menopause (although psychiatrists also diagnosed it in some men, in whom it was thought to occur later). Symptoms included profound depression, anxiety, nervousness, confusion, memory problems, erratic behavior, and paranoid delusions, as well as several other kinds of psychotic delusions. Involutional psychosis was treated with hormone therapy, despite limited success, and in the mid-twentieth century, like other psychiatric disorders, with electroshock therapy and the other shock therapies in use at the time, insulin coma and drug-induced convulsions.

Psychiatry changed profoundly in the later twentieth century with the increasing use of drug therapies—more and more, the poetic diagnoses of decades and centuries past yielded to lumped-together diagnoses for people with similar symptoms who responded to the same drugs. Depression came to the forefront as the premier psychiatric diagnosis, treated first with tricyclic drugs and monoamine oxidase inhibitors and then, after 1987, with fluoxetine, the blockbuster drug Prozac. Also, while research through the early 1960s was conducted in psychiatric hospitals on patients who had been committed involuntarily and were quite ill, with the rise of drug therapy and the deinstitutionalization movement that closed many psychiatric hospitals, more and more research was carried out on general patient populations with less serious problems. Psychiatry's focus changed from florid, mainly psychotic disorders to common conditions like anxiety and depression. Major Depressive Disorder and other mood disorders became the main preoccupation of scientists researching the mental aberrations of menopause.

By 1980, most psychiatrists had abandoned the diagnosis of involutional psychosis, and it did not appear in the revolutionary third edition of the *Diagnostic and Statistical Manual of Mental Disorders*, published that year. The *DSM-III* reflected new trends in psychiatry by redefining diagnoses based on symptoms and criteria, rather than on psychodynamic theories (like Freudian psychoanalysis), traditional classifications, and patient characteristics like age, sex, or menopausal status.

However, gender remained an important part of the understanding of depression, which was, and still is, believed to be more common in women and possibly linked to their reproductive physiology (that is, hormones).[56] While the *DSM-III* defined disorders by symptoms and criteria, these criteria were themselves, by necessity, based on a long history of previous research in which depression was assumed to be mainly a women's disorder. Laura Hirshbein, investigating this story in her highly recommended book *American Melancholy*, found that, in fact, about the same number of men and women were hospitalized in the era when hospital populations were the basis for psychiatric studies. But men had more diagnoses of syphilis, alcoholism, and other conditions that psychiatrists thought complicating factors, and were excluded from studies and trials in greater numbers. For this reason, studies of depressive disorders overwhelmingly contained many more women than men in the sample population; researchers accepted the imbalance because they thought these disorders were more common in women, and then they generalized the results of their studies to all patients even if men were poorly represented in the sample. (This is exactly the reverse of the historical problem with the study of heart disease.) The association of depression with women persisted through the twentieth century because of the boundaries psychiatrists drew around their diagnoses, despite the protests of a few researchers who thought that alcoholism might often be comorbid with depression or a common symptom or expression of depression in men.

After the publication of the *DSM-III*, it remained the case that more women were diagnosed with depression than men, partly for the circular reason that they better met criteria developed through work with mostly female populations. In particular, *DSM-III* criteria emphasized moods and feelings, which women were more likely than men to describe or acknowledge. But many physicians continued to speculate, as they had for centuries, that something about women's bodies caused this mental illness. Despite the fact that researchers had tried using hormonal therapy on involutional melancholia and had found it ineffective, psychiatrists of the late twentieth century continued to investigate the relationship between mental illness—especially Major Depressive Disorder and other depressive disorders—and menopause.

On the other hand, in the 1970s, some researchers—including a few who were skeptical of pathologizing views of menopause—took up the more basic question of whether there really was a connection between menopause and depressive disorders or other mental disorders. The potential for self-fulfilling prophecy in this question was very great; menopause was widely assumed to cause mental illness, particularly involutional melancholia or depression, as well as a host of other problems, from the minor and cosmetic to the extremely serious. In popular literature, menopause turned women into shriveled, irrational harridans who drove their long-suffering husbands into the arms of younger women, or sexless castrates who had outlived their usefulness to the species. This context could affect how women answered questions about their mental state, how they expected to feel, and even how they actually felt—that is, it would not have been surprising if negative portrayals of menopause and of middle-aged women caused a certain amount of clinical depression, a problem some researchers acknowledged. It is all the more remarkable that most of the research conducted from the 1970s through the 1990s found no statistical relationship between menopause and depression (or any other mental disorder).[57]

More recent research has shifted from a focus on clinical depression—Major Depressive Disorder—toward the investigation of depressive symptoms, measured by standardized psychiatric questionnaires. A series of recent studies with large sample sizes has supported a connection between menopause and depressive symptoms.[58] The best publicized of these results have come from the Study of Women's Health Across the Nation (SWAN), which observed an ethnically diverse sample of over 3,000 middle-aged women in the United States once a year for seven years, from 1995 to 2002. A team led by Joyce Bromberger analyzed the data on depressive symptoms from that study and concluded that women who were transitioning through menopause, or who had completed the transition, were more likely to have a high score on the scale of depressive symptoms they used, by about 30 to 70 percent. After controlling for other factors such as income, they found that ethnicity significantly affected scores; Chinese American women had the lowest scores, while African American, Japanese American, and Hispanic women all had higher scores than white women. However, several other factors—poverty, stressful life events, and negative attitudes toward menopause—correlated more strongly with high depressive symptoms than menopausal status or ethnicity. Women with hot flashes and night sweats were also about 70 percent more likely to have high depression scores than women who did not have those symptoms, and women who used hormone therapy were more (not less) likely than non-users to have high scores. The SWAN researchers also concluded that women going through menopause were more likely than pre-menopausal women to meet criteria for Major Depressive Disorder, and that women past menopause were more than twice as likely to meet these criteria than pre-menopausal women.[59]

SWAN researchers were cautious about concluding that hormonal fluctuations, and not other factors, were responsible for the pattern they observed, and efforts to prove a link between hormones and depression at menopause have had mixed and unconvincing results.[60] Also, while studies from the last decade or so have tended, on average, to confirm a relationship between menopause and depressive symptoms or depression, their results conflict in ways that are hard to explain. For example, like the SWAN study, a study published in 2004 found higher rates of depressive symptoms in women going through menopause, but unlike that study, it found much lower rates of depressive symptoms in post-menopausal women compared to pre-menopausal women or to women in transition. In the 2004 study, women going through menopause met criteria for Major Depressive Disorder at rates that were slightly lower than those of pre-menopausal women, even though they reported more depressive symptoms, and African American women were twice as likely as white women to have a high number of depressive symptoms and to meet criteria for Major Depressive Disorder.[61]

It is also hard to reconcile the SWAN findings with the results of the Women's International Study of Health and Sexuality, nicknamed WISHeS, which is the only major study of menopause symptoms to survey women of a broad age range, from 20 to 70 years. This study, published by Lorraine Dennerstein and colleagues in 2007, surveyed 4,517 women in five Western countries (the United States, United Kingdom, Germany, France, and Italy) to look for symptoms that peaked in frequency around age 50, reasoning that such a pattern suggests a correlation with menopause rather than, for example, with age. In this study, mood symptoms, including depressed mood, anxiety, mood swings, and impatience, peaked at around ages 35 to 40 and declined steeply thereafter, a result very different from the findings of the SWAN study.[62]

I am not convinced that there is a meaningful connection between menopause and depressive symptoms or depression. But in the twenty-first century, scientists are more likely to perceive a connection than they were a few decades ago; the idea that menopause causes mental problems is now making a comeback in medical literature after a period of decline. On the other hand, today's researchers are more reserved than those of past eras on the question of whether psychological problems at menopause are caused by women's physiology (that is, hormones), although some have made the argument that hormonal changes are the cause of these problems.

In 1975, two separate large studies showed a link between ERT and endometrial cancer, and although the results were debated, prescriptions for estrogen and HRT declined for several years after that. But this was only a temporary setback: HRT, which did not appear to cause endometrial cancer, soon became the standard treatment for women who had not had a hysterectomy, and the number of prescriptions rebounded, reaching new heights in the 1990s. In 1998 one survey found that more than a third of women over 50 were on HRT, frequently for the long-term prevention of osteoporosis and heart disease.[63] In 2002, 90 million prescriptions for hormone therapy were written in the United States. But in that year a new setback occurred, when the results of the Women's Health Initiative became public. This was a 15-year clinical trial of hormone therapies with a huge sample size of over 68,000 women, aged 50 to 79, who were past menopause at the time of enrollment in 1993–1998 (an additional 93,000 women were enrolled in the control group). Twenty-seven thousand women received either ERT or HRT (depending on whether they still had a uterus) or a placebo. The results showed that women on hormonal therapy had a lower risk of hip fractures, fractures in general, and diabetes, but more frequently suffered stroke and blood clots. Those taking HRT also had a lower risk of colorectal cancer, a slightly higher risk of breast cancer, and a not quite statistically significant elevation in their risk of heart disease. The latter was a shocking result—hormone therapy had been expected to prevent heart disease, but neither HRT nor ERT appeared to do so. (Another surprise was a not quite statistically significant reduction in rates of breast cancer for women on ERT, which had been expected to increase the risk of breast cancer.) Women on hormone therapy who were over

65 also had more risk of incontinence, gall bladder disease, and dementia, although overall mortality was about the same as for those not taking hormones. The HRT trial was halted in 2002, and the ERT trial in 2004, because of the elevated risks to the women in the study; the latter decision was contentious, as the risks to the ERT group were fewer than those to the HRT group. No significant quality-of-life differences were found between the groups taking hormones and those not taking them. Overall, the Women's Health Initiative study found both the risks and the benefits of hormone therapy to be small, with the balance more favorable for ERT than for HRT (but ERT was prescribed only to women without a uterus, eliminating the risk of endometrial cancer). Estrogen just didn't seem to make much difference to women's health.

These findings undermined estrogen's reputation as a wonder drug, and threw cold water on the hope that hormone replacement could prevent chronic health problems. In 2005, the National Institutes of Health, in a published statement of its State-of-the-Science Conference on the Management of Menopausal Symptoms, concluded that hormone therapy was an effective remedy for hot flashes but carried elevated risk of stroke, blood clots, pulmonary embolism, breast cancer, and heart disease, and recommended rather unhelpfully that women with severe symptoms balance the risks and benefits.[64] Today, most physicians are likely to agree that hormones are best prescribed over a short term and for severe symptoms. Others, however, still advocate hormone therapy for life, based on the "timing hypothesis"—re-analysis of the statistics from the Women's Health Initiative showed that most of the elevated risk for heart disease was concentrated in the group of women who started hormone therapy 10 or more years after menopause. Advocates of this hypothesis estimate that women who are placed on hormone therapy at menopause and stay on it for 5 to 30 years might enjoy, on average, an additional 1.5 "quality-adjusted life years" (but not actual life years, which would increase by only 0.12 according to their model).[65]

This emphasis on large trials, chronic conditions, long-term therapy, and preventive medicine potentially prescribed to all women reflects modern medicine's understanding of menopause as a permanent and continuous deficiency state. The language of estrogen deficiency still pervades gynecological textbooks and published research[66]—so much that it's important to point out the assumptions it contains. To speak of estrogen deficiency is to say that women past menopause do not have as much estrogen as they ideally should; in this view women of reproductive age set the standard for the "right" amount of estrogen. If we assumed that women past menopause had the right amount of estrogen for post-reproductive life, we would not say that they had an estrogen deficiency—just as we do not say that an eight-year-old girl is deficient in estrogen. The language of estrogen deficiency also carries the corollary that women past menopause have outlived their main (reproductive) function and that their lives have been prolonged past what they are naturally adapted to—and many research papers published today adopt this assumption implicitly or explicitly, despite the enormous quantity of research on the evolution of menopause discussed in part I of this book.

Because of what psychologists and philosophers of science call "confirmation bias," the assumption that the physiology of women's reproductive years is the standard from which menopause deviates tends to create an accumulation of data in its support. Confirmation bias can take many forms but, to oversimplify here, it means that we usually find what we're looking for and don't find what we're not looking for—for example, when assumptions about menopause shape symptom checklists, or affect the perceptions and expectations of the individuals who fill out the self-reports on which most menopause research is based. These assumptions can also influence researchers' choice of topics, an effect that is exacerbated by publication bias, as papers that confirm hypotheses are more likely to be published than those that show a negative result. That hormonal therapies have delivered underwhelming results for women's health despite all these factors suggests that the estrogen deficiency model might not, after all, be a very good way to think about menopause.

It is also important to remember that even menopause's most energetic critics, researchers who believe that all menopausal women should be placed on HRT for life, are not arguing that we should try to extend fertility through old age (I think most would be horrified by the thought). The debate is a more narrow one, about whether mimicking reproductive levels of certain hormones, especially estrogen, is a good idea. Calculating the risks and benefits of HRT for individual women is very different from evaluating the costs and advantages of menopause more generally, as it does not consider menopause's most important effect, the end of fertility. The larger question is addressed not by doctors, most of whom do not seem to be very interested in it, but by anthropologists and evolutionary biologists.

In the story of the origins of menopausal syndrome, it can be difficult to disentangle ideas from realities, and different ideas from one another. Theories about the nature of the menopausal transition and the causes of its symptoms have sometimes conflicted; medical discourse has sometimes diverged from popular literature; and both have often differed significantly from what women actually thought and experienced, which, until we reach very recent times, can be hard to recover. Nevertheless, a few important themes emerge. There is no evidence that menopausal syndrome was a significant concept in European medicine or culture before about 1700, after which it became very prominent. After that, physicians explained it first by the retention of blood, later by the irritation of the nerves of the female reproductive system, and finally, in the twentieth century, by fluctuating hormones and estrogen deficiency. While nineteenth-century physicians were likely to describe menopause as a period of crisis from which women might emerge strong and healthy, in the late twentieth century, they came to see menopause not only as an acute syndrome but also as a chronic state of deterioration, and even as an epidemiological disaster, as the rising number of post-reproductive women weakened by estrogen deficiency threatened to become a global health problem. As with other European syndromes of the early modern period, psychological and behavioral problems, sometimes dramatic ones, were thought central to menopausal syndrome, and the idea that menopause causes mental disorders is a "sticky" one that persists today despite decades of negative or conflicting results. Finally, sexual symptoms of menopause took center stage in the mid-twentieth century, after marriage was redefined as an institution based on sex.

•

CHAPTER 10

What Are You Talking About?

MENOPAUSE IN TRADITIONAL SOCIETIES

THE STORY OF modern menopausal syndrome begins in Europe and is longest and most complex there. Its history in Chinese medicine is very different, in ways that serve to illustrate how the idea of menopause spread around the world. Classical Chinese medicine is an especially massive and varied tradition, tracing back uninterrupted to at least the third century BCE. The most frequently cited passage relating to menopause comes from the *Inner Canon of the Yellow Emperor*, Chinese medicine's oldest foundational text, originally dating to some time between 200 BCE and 200 CE. The first section of that work describes the stages of life for women and men, divided into seven-year periods (for women) and eight-year periods (for men). For women, the aging process is complete at 49, whereas men age more slowly and do not reach total depletion until age 64: "At forty-nine, the conception vessel is depleted, the great penetrating vessel wanes, fertility is exhausted, menstruation ceases, the body has become old and she can no longer have children."[1] Here menopause—the end of menstruation—is a part of aging, a sign rather than a problem in itself. This is the passage that modernizing Chinese physicians would later cite in support of adding menopausal syndrome to the conditions treated by Traditional Chinese Medicine.

Because of the limits of my language skills, I must rely on the work of other scholars about the idea of menopause in imperial China, but my impression is that references are not very common. An interesting case from the memoir of the female doctor Tan Yunxian (1461–1554) is both an exception to this rule and, in another way, an example of it. The memoir contains 31 case histories of Tan's patients, all female.[2] Tan's own life is worthy of a short digression—she was raised by her grandmother, a physician's daughter; she published her memoir at age 50, and lived to be 96. Her grandmother had married uxorilocally, and Tan's grandfather Tan Fu had been trained in his in-laws' family profession of medicine and was a famous physician. But although it was unusual for women to learn or practice classical medicine, Tan's grandmother was also a doctor, and Tan began to study with her at an early age. While she practiced medicine all her life, she achieved professional status only after she had finished raising her children and her son was married.

Tan treated only women in her practice, and the 31 patients in the case histories her memoir describes are all female. They suffer from rashes and sores, coughing, vomiting, nerve damage, abdominal masses, vertigo, diarrhea, fever, parasites, and many other problems. Several suffer from gynecological conditions—uterine flux, difficulties related to pregnancy or birth, and infertility. One case concerns a 53-year-old woman with menstrual irregularity:

> Because the menstrual period of a 53 year old woman was irregular and her original qi was very weak, she suffered an ailment with deficiency of both qi and blood. I rechecked her pulse; her heart channel pulse was extremely floating and surging, and it would stop for six beats [irregular pulse]. The reason was that the woman was often taxed from toil which damaged her heart. The heart is the master of the whole body; when her heart fire stirred, menstruation did not come when expected, increasing her deficiency and weakness.... My opinion was that this woman [suffered from] unregulated blood and qi. Then I used *Guī Pò Wán*.... She took these formulas and then recovered; the woman became mentally sharp and strong as before.[3]

Despite the woman's age, Tan does not attribute either her patient's irregular periods or the other, unnamed symptoms that she was experiencing (perhaps psychological ones, in light of the case history's last sentence) to natural menopause. In her view a heart condition caused by overwork has resulted in the depletion of blood and *qi*, which has caused both her menstrual problems and her other problems. This conclusion is typical of Tan's memoir: she usually attributes her patients' illnesses to overwork, grief, or anger, sometimes identifying precipitating circumstances—the death of a husband, for example, or his decision to take a concubine.

The idea that menopause could cause health problems directly is attested at least once in the Chinese classical tradition. In 1793, the *New Book of Childbearing* described the stages of female life this way:

> At the time of her menses a young married woman of 18 to 21 years may ache all over, feel numbness in her hands and feet; she may be now hot, now cold, dizzy and faint. Women in their midtwenties are liable to irregular periods and intermittent fevers and are vulnerable to the serious wasting disease "bone-steaming." By 24 or 25, a woman's "sea of blood" is already in danger of depletion and attacks from cold. In addition to the usual symptoms of headache, dizziness, hot and cold sensations, and cramping, these women will often suffer from vaginal discharges, menstrual flooding or spotting, or protracted periods. By 28 or 29, childbearing has taken its toll. Her "*qi* has dispersed and her blood is depleted," and she suffers from "blood depletion with stomach heat, a condition of fatigue." At 35 internal weakness has made her so vulnerable to noxious invasions from without that her best defense when menstruating is to take to her bed. By her early forties her menstrual cycle has ceased, leaving her with a chronic depletion of *qi* and blood. She is now subject to disorders caused when old, bad blood is not dispersed, but congests and stagnates within.[4]

This physician imagines that women at menopause are depleted of blood as well as of vital *qi*, and also that they might suffer from the accumulation of stagnant blood because of its failure to disperse. Here, menopause is the last of many stages of life in which women experience diseases and syndromes arising from their reproductive biology.

This last reference shows that it is not quite correct to say that classical Chinese medicine did not consider menopause a medical condition; nevertheless, there seems to be little discussion of menopause in that enormous corpus, which is much more preoccupied with the problems of reproductive-aged women. Today, however, Traditional Chinese Medicine treats menopausal syndrome, and is an especially well-documented example of modern medicine's influence on the idea of menopause.[5] Traditional Chinese Medicine signifies a standardized form of traditional medicine originating in China but now taught in professional schools all over the world.[6] While it draws on the vast history of medical literature in Chinese that I am calling "classical" Chinese medicine, Traditional Chinese Medicine is not the same as classical Chinese medicine or even "traditional Chinese medicine" without capitals. It is a specific and evolving interpretation of a much more complicated system. The ancient, vast, and highly complex and variable literature that underlies it still survives, and is one source of variation and change in TCM—practitioners sometimes return to the classical tradition rather than relying on the modern textbook version of TCM. This is especially true in China, where more practitioners can read the classical texts.

Traditional Chinese Medicine emerged in 1959, when the first modern-style textbooks were published as part of the Maoist government's effort to modernize China. TCM has some of the features of modern medicine—for example, it has adopted ideals of standardization, regulation, and clinical research and testing. But TCM is usually understood as an alternative to modern medicine, a form of "complementary and alternative medicine" (often abbreviated as CAM).

Traditional Chinese Medicine identifies and treats an equivalent of modern medicine's "menopausal syndrome," called "symptoms and signs associated with the cessation of menstruation" in one formulation. There is a word for "menopause" in Chinese; practitioners of modern medicine call it *gengnianqi*, a loanword borrowed from the Japanese *konenki* (which was itself a new word introduced in the early twentieth century, to signify the European concept of menopause). Traditional Chinese Medicine avoids using the loanword *gengnianqi*, which of course does not appear in the classical literature on which it is supposed to be based. Classical Chinese medicine had no word for menopause, nor is the idea of "symptoms and signs associated with the cessation of menstruation"—menopausal syndrome—attested there, either as a topic or as a phrase. It appears for the first time in 1964, in the second edition of a textbook published as part of the Chinese Ministry of

Health's official series, called (in English translation) *Lecture Notes for Chinese Medical Gynecology,* which includes a section on "manifestation patterns associated with the cessation of menstruation." So it is possible to date the introduction of menopausal syndrome to Chinese medicine quite precisely (this section is not present in the first edition of the same textbook, published in 1960). It was inserted into TCM as part of a broader project of making Chinese medicine more modern. As one scholar of Chinese medicine puts it, "Menopause, which did not constitute a medical problem in the classical literature suddenly became one in 1964."[7] Pressed to find references in the classical literature to support the new concept of menopausal syndrome, the inventors of TCM cited the passage on stages in the female life cycle from the 2,000-year-old *Inner Canon of the Yellow Emperor* quoted at the beginning of this chapter, in which the end of menstruation is one result of the depletion that also causes other signs of old age.

This was an era of optimism that Chinese and modern medicine could be harmonized. Chinese physicians connected the Western idea of hormones with the kidneys, which in the Chinese system produce both yin and yang forces that regulate growth and development. In TCM textbooks today, menopausal symptoms are attributed to a decline of kidney essence that causes a deficiency of either yin or yang, resulting in hot flashes if yang predominates, and depression if yin predominates, along with other problems.

Menopausal syndrome, then, was written into Traditional Chinese Medicine in 1964 as part of a larger effort to modernize Chinese medicine and harmonize it with Western medicine. This is not to say that no traditional cultures outside of eighteenth-century Europe have associated menopause with symptoms and problems, although many or most have not. Evidence is difficult to collect, since most early traditions are not as well documented as those of ancient Greece and China, and few cultures today have not been profoundly influenced by modern medicine. But traditional equivalents of modern "menopausal syndrome" are, at least, not easy to find.

A great deal has been written about the evolution of menopause and the role of post-reproductive women in foraging populations, but despite their interest in the subject, researchers have published little on the *experience* of menopause among foragers. My impression is that this is because there is not much to say. Frank Marlowe, in his book-length study of the Hadza, writes that the women he interviewed reported no symptoms of menopause, and that most did not know when they entered menopause because they continued to give birth and nurse babies to the end of their reproductive lives: "When asked if they have reached menopause, they may simply shrug and say, 'We have to wait and see,' meaning they will know only if they do or do not resume regular menstruation or get pregnant again after weaning the current child."[8]

Marlowe cites a 1991 study that concluded from blood tests that the average age at menopause among the Hadza was about 43, but the study is obscure and appears to be unpublished, and age at menopause is notoriously difficult to determine by blood tests, or indeed by most other measures used by anthropologists.[9] Age at last birth is easier to measure, but has not been calculated among the Hadza. In their study of the Ache, Hill and Hurtado report that Ache women have a word for menopause and seem reluctant to acknowledge being menopausal (that is, post-reproductive) until they are old, but that is all the authors have to say about the subject.[10] Median age at last birth among the Ache was 43, among the highest for any documented population.

While evidence from the Hadza suggests that women in our foraging past experienced menopause several years earlier, on average, than they do in modern cultures for which age at menopause is well attested, the late age at last birth among the Ache casts doubt on this idea, and we do not really know whether it is true. It is also possible that Hadza women experienced symptoms of menopause but did not report them or consider them significant; Hill and Hurtado are silent about symptoms among the Ache, suggesting that the symptoms of menopause were also not very significant among them.

A few researchers have published studies on ideas and experiences of menopause in traditional, agrarian societies without much exposure to modern medicine or to modernity in general. As with foragers, these populations are dwindling, and research about them is challenging in many ways, so the number of these studies is not large. The Maya of

Yucatán (a state of Mexico) and Guatemala, a large group of indigenous Americans numbering over 8 million, are the traditional population whose experiences of menopause have been observed most closely. Researchers have studied a few separate groups of Maya, with conflicting results that have never been explained.

The first of these studies, and also the most thorough and sensitive, was conducted in 1981 by anthropologist Yewoubdar Beyene as her dissertation research. She was drawn to the subject through her work in a mental health clinic, where she was surprised by the reports of some older women that menopause was causing depression. "Coming from a non-western cultural background," she writes (Beyene is from Ethiopia originally), "I was not aware that menopause causes depression or any other emotional or physical illness. I only knew that menopause was a time when women in my culture felt free from the menstrual taboos; otherwise, nobody pays any attention to this life event."[11] Beyene's study is noteworthy because she does not confine it to the narrow focus on symptoms and attitudes that is typical of most transcultural research on menopause—partly because her subjects had no symptoms to report.

For a year Beyene lived in the village of Chichimila in Yucatán, where she convinced the women of the village to take her on as a student, teaching her the skills of daily life, and was able to gain trust and form relationships with them in that way. She interviewed 107 women from ages 33 to 57; she had originally planned to interview women over age 40, but it became obvious in the course of the study that menopause was occurring earlier than expected in this population. All families in Chichimila were farmers, practicing swidden agriculture on a relatively short (five- to six-year) rotation. Houses were single-room structures of mud and thatch or, for people with extra income, masonry. Married sons often lived in separate huts in the same courtyard as their parents, at least until they could afford to move out; widowed grandparents used the same arrangement. The population's diet consisted of corn (maize), beans, a few garden vegetables, and an occasional egg; the villagers raised pigs and chickens for cash but rarely ate them, and milk was very difficult to obtain. However, village women had a relatively high Body Mass Index and did not seem malnourished. On average, the women

Beyene interviewed had begun menstruating at age 13, had married between 18 and 19, and had borne 7 children, of which 4.7 survived the first year of life. The medical professionals who lived in the village were a male Maya traditional healer; two midwives, both in their 60s, one of whom also treated menstrual disorders with herbs; and a healthcare worker literate in Spanish, with a formal education through sixth grade and a few months' training in modern medicine, who was able to dispense some drugs.

Chichimila was not isolated from the modern world. A bus stopped there four times daily on its route between the larger towns of Valladolid and Chetumal. The village store sold a few commercial foods (like sardines and sugar), sewing supplies, and basic medical supplies (aspirin, tetracycline, and herbs). The younger generation were mostly educated through elementary school, could read and write, and were restless and resistant to a life of farming; some young people traveled during the week to work at resorts in Cancún. Chichimila had several bars where its men spent much of the cash they earned as migrant workers or from selling the corn crop, a habit that was a major source of stress for the women Beyene interviewed. Women scrambled to get by on the money they could earn from selling the products traditionally under their own control—tomatoes, squash, herbs, tamales, and handmade clothes and hammocks.

Modern medicine was represented in the healthcare worker's clinic, which promoted vaccination and family planning, and encouraged people to boil drinking water and dig latrines, though without a lot of success. People with serious illnesses sometimes sought treatment in Valladolid. One 70-year-old physician in Valladolid saw most of the Maya women who came there for treatment, and he was one of Beyene's interviewees; people also made the trip to town to see the itinerant *curandero* (Latin American traditional healer) who visited there once a month. But because the people of Chichimila used few consumer goods, practiced near-subsistence farming, lived on a typical peasant diet, and had fertility patterns typical of those before the Demographic Transition, Beyene's study is our best insight into the experience of menopause in a traditional population.

Mothers-in-law were powerful in Chichimila in some of the same ways discussed in part II of this book. They had authority over decisions about childrearing and especially the diagnosis and care of sick children, even when their sons did not live with them. (I think about this pattern a lot whenever I have to offer a snap diagnosis for someone in my family—even in a modernized, literate culture teeming with doctors and online advice, mothering experience plays a big role in medical care.) Because mothers-in-law had authority over their daughters-in-law, women's status increased when their sons married. They no longer had as much housework to do, and focused instead on grandchildren, church, visiting friends, weaving hammocks and embroidering dresses to sell, and on other occupations—both of the midwives in Chichimila, for example, were old enough to have daughters-in-law at home. As in Chinese culture, sons were expected to be loyal and attached to their mothers, even more than to their wives, and tensions were common.

The women of Chichimila worked hard and had many children without respite from their workload. Pregnancy and childbearing were sources of anxiety and stress, and Beyene believed that at least some of the village women were open to the idea of family planning, although their husbands, who feared losing control of their wives, resisted. But overall the people of the village valued large families and felt that the burdens of raising children were concentrated in the years when the children were very young—older children could help with the farm and take care of themselves and their siblings. The rate of infant and child mortality in Chichimila was higher than in modernized societies, and most women lost two or three children in the first year of life. Women continued to bear children to the end of their reproductive lives, and it was not thought unseemly (as in some traditional cultures) for mothers and daughters to have children at the same time.

Menstruation carried a lot of cultural baggage in Chichimila, as it does in many traditional and modern societies.[12] Women in Chichimila stayed home during their periods, and it was thought dangerous for menstruating women to visit newborn babies. Many health problems were believed to be caused by spoiled, dirty, or stagnant menstrual blood. This last idea is very common across cultures, attested in ancient

Greek medicine, in the rural Greek village that Beyene also visited as part of her study, in villages in The Gambia studied by anthropologist Caroline Bledsoe, in classical Chinese medicine, in Japanese *kampo*, among Hmong refugees in Australia, among women living in rural Thailand, and likely in many other traditional societies.[13] The women of Chichimila used a number of remedies for cramps, irregular periods, and other menstrual problems. Women found menstruation a hassle and welcomed the amenorrhea of pregnancy and of menopause.

The physician at Valladolid who treated the villagers' more urgent health problems told Beyene that his Maya patients rarely complained of hot flashes around menopause; sometimes they complained of irregular or heavy periods, but once they learned the reason for their experience they did not return. He said that he treated some older women for headaches and other symptoms, but neither he nor his patients connected the problems with menopause. The most frequent problem requiring modern medical treatment among Maya women, according to the Valladolid doctors, was complications from pregnancy, and malnutrition, which caused anemia and other deficiency disorders, was the most common health problem overall. All of the physicians at Valladolid agreed that menopause caused no physiological or psychological problems for their Maya patients.

This consensus of physicians in itself would not be so compelling, but the middle-aged Maya women of Chichimila agreed with them. The village's midwives knew of no problems associated with menopause other than, sometimes, heavy and frequent periods, and they had trouble understanding why Beyene was asking about symptoms. The other women and men of Chichimila all agreed that menstrual irregularities were the only symptom of menopause and that there were no special remedies or therapies related to the problems of menopause. When Beyene interviewed menopausal and post-menopausal women in depth, they affirmed that they associated no symptoms with menopause. They did not have a word for "hot flash," and when the concept was explained to them, they said that they did not experience them (in contrast, urban, educated women living in the city of Mérida knew what a hot flash was and had a word for it, *bochorno*). The village women also did not have a word for menopause, though they knew it was normal for women to stop menstruating in midlife. They were eloquent on the symptoms and problems of pregnancy and menstruation but puzzled by Beyene's questions about menopause. Some even wondered if she was asking about symptoms because menopause was an abnormal condition that happened only to Maya women.

The women of Chichimila believed that menopause occurred when a woman had used up all her blood giving birth to children. Women with many children were expected to enter menopause soonest. Village women welcomed menopause as a relief from the stresses of pregnancy and the hassles of menstruation. By the time they reached menopause, they had all borne many children and did not think they needed any more; most had grandchildren to whom they devoted their attention. When asked what their husbands thought, they responded that their husbands were indifferent and that their sex lives were better without the worry of pregnancy. Many felt "young and free" at menopause, as though returning to a stage of life before the heavy burdens and restrictions of adulthood.

When asked about the age at which they had experienced menopause, the average recalled age was 42, with ages clustered between 35 and 45, much younger than in most populations that have been studied. Although based on recollection, a method that is not very reliable, it is hard to dismiss this finding. Beyene had to adjust the age categories of her subjects downward when it became clear that women under 40 could not reliably be assigned to the "pre-menopausal" group. Women in Chichimila did not have many periods in their adult lives because they were usually pregnant or nursing babies, and they remembered the number of periods between pregnancies and whether their period had returned after their last pregnancy (even their husbands remembered these things). Beyene does not say how her subjects knew their ages, information that would be helpful in interpreting her findings, as it is common in traditional societies with low levels of literacy for people not to know or care about their exact age.

Beyene and her collaborator Mary Martin also investigated whether the biochemistry of menopause was different for Maya women than for those of North America.[14] For this study, they interviewed and tested 232 women from Chichimila and Valladolid. They tested for seven different hormones and bone density, asked detailed questions about the women's medical history, and looked at medical records. They found that the average age at menopause was around 44; blood levels of Follicle Stimulating Hormone, which are higher in women after menopause, seemed to confirm this finding. None of the women reported hot flashes or any other symptoms of menopause. Bone density was lower for Maya women than for North American women, and it began to decline at age 35, as it does in North American women, but the researchers found no history of fractures in any of the women they interviewed or in the medical records they reviewed. They speculated that because Maya women were short and stocky, their body structure protected them from the trauma of falling, or that because they were so active, their strong and flexible muscles protected them (the latter explanation is still widely accepted). However, because of the difficulty of getting equipment to Chichimila, the researchers could not perform CT scans or other imaging studies, which might have revealed hidden fractures. They also could not perform skin conductance tests, sleep studies, or other studies that might have revealed whether the women "really" experienced hot flashes or not. Their finding, however, that Maya women did not think they experienced hot flashes and did not appear to suffer negative effects of bone loss is quite important. Beyene's studies provide support for the idea that many factors, including culture, might influence how women experience menopause—even if menopause is biochemically the same across cultures, the experience of menopause can be different for complex reasons in which physiological and cultural factors are difficult to untangle.

About 20 years after Beyene first visited the village of Chichimila, Donna E. Stewart published a paper on another group of Maya living in the highlands of southern Guatemala. This was a "qualitative" study based on in-depth interviews with 27 women, 9 from each of three ethnolinguistic groups of Maya in that region. Three of the women were what Stewart calls "key informants," well-connected women who helped her find other subjects, and who translated her Spanish into their languages. The women she chose to study were undergoing menopause, were experiencing irregular periods, or had ceased menstruating in the last three years. Like the Maya of Chichimila, her subjects were subsistence farmers with a meager diet, a heavy physical workload, and high fertility (by modern standards). Stewart asked each woman, "What happens around the time when women finish having their menstrual periods at midlife?" She also asked "probe questions" regarding "any special feelings or experiences, special foods, herbs or practices that the women engaged in during the time when their periods were becoming irregular, different or after stopping." She was careful not to ask questions more leading than that and did not use the "checklist method" of collecting information about symptoms (described later). Each interview took one to two hours, and none of the interviewees allowed her to use a tape recorder or to take notes; she recorded what she could remember immediately after the conversations. The women who knew their ages were between 38 and 55. On average they had borne seven children, the first at age 18. The average recalled age at menopause among those who had not menstruated for a year was 48.[15]

Like the Maya of Chichimila, the women Stewart interviewed were relieved to have entered menopause, and for the same reasons—they no longer had to worry about becoming pregnant, they felt free to move around and visit, and they did not have to worry about staining their clothes. While 5 reported improved sex lives, 11 of the women said they had lost interest in sex or found it uncomfortable.

Unlike the women of Chichimila, most of the women of highland Guatemala described feeling hot, and many described sweating at night. It was common in one of the groups for women to switch to a lighter style of blouse in midlife. Feelings of heat could be episodic, but they could also be continual, and it was thought that women were generally hotter in midlife than previously. About half of the women described "increased irritability, moodiness, and anger," and some attributed these feelings to the arousal of the animal spirit thought to reside in everyone. Several reported vivid dreams. About half reported heavy or painful periods for which they used herbs, massage, and steam baths; some consulted traditional healers or midwives. Midwives and *curanderas* agreed that women in midlife commonly complained of the symptoms Stewart noted.

In 2004, Joanna Michel, an expert in traditional pharmacology interested in finding new herbal remedies for menopausal symptoms, visited a separate group of Maya in four villages near the town of Livingston, in the eastern lowlands of Guatemala, to study their "symptoms, attitudes, and treatment choices" regarding menopause.[16] She lived among the Q'eqchi Maya for eight months and, like Beyene, observed and participated in village life as much as she could. She interviewed 40 Maya women and men, including five midwives, five male traditional healers (*curanderos*), and eight women past menopause. The older women had not heard of menopause before they experienced it, because it was considered wrong to discuss it, but the younger generation knew about menstruation and menopause from school.

As among the Maya of Chichimila, menstruation was thought a dangerous condition, and menstruating women especially took care to avoid newborn babies and did not cook or serve food. On average, the older women Michel interviewed had begun menstruating at 13, had experienced menopause at age 46, and had given birth to almost nine children, the first at age 17. They had an average of less than two years of formal education. Their diet was similar to that of the Maya of Chichimila and of western highland Guatemala—mostly corn tortillas and black beans—and anemia and other deficiency problems were common. Most of these women's day was spent in hard physical farm work, such as grinding corn and hauling water. (Beyene estimated that her subjects hauled over 250 pounds of water up a typical 60-foot well every day, and more on laundry days![17])

Michel's findings are especially interesting because the different groups she interviewed gave very different answers to her questions about menopause. Like Stewart, she tried to avoid using specialized terms relating to menopause and asked whether her interlocutors "were familiar with any health problems related specifically to older women and if there are any symptoms associated with the cessation of menstruation." When she talked to lay community members, they seemed not to understand what she was asking and not to associate any symptoms or remedies with menopause, though they mentioned remedies for menstrual problems. But when she interviewed the midwives, their answers were different. Two of the midwives lived in Livingston and knew the term "menopause" from radio and TV. The other three midwives did not have a term for menopause or for a hot flash, but they did describe something else, using the Spanish term *baja presion*, "low [blood?] pressure." Michel reports that this term signified "the symptoms of profuse sweating followed by chills, heart palpitation and emotional instability associated with the end of menstruation." It would be good to know more here about how the midwives thought about *baja presion* and its connection with menopause—Michel, like other Maya researchers, was mostly speaking through an interpreter—and why the term did not come up in interviews with lay members of the community or the male *curanderos*, especially since *baja presion* is not one of the conventional Spanish equivalents of "hot flash." The midwives confirmed that because most women did not know about menopause before they experienced it, they came to the midwives about symptoms, but it is not clear what symptoms they had in mind besides menstrual irregularity, which must indeed have been alarming for women who did not expect their periods to stop.

The five male *curanderos* Michel interviewed had much more to say about symptoms and remedies. They all said they knew that menstruation stopped in midlife, and in answering Michel's questions, they mentioned "headaches, body aches, irregular menses, excessive worry and insomnia," but not hot flashes or *baja presion*. They thought the problems they described might be caused by age, witchcraft, or divine retribution, and they treated their patients with prayers, incantations, and several different herbs. Again, it would be helpful to know more about their perceptions of how these symptoms related to menopause, but certainly the male healers had some colorful ideas about the types of symptoms older women might experience.

Finally, Michel singled out eight women who were past menopause for in-depth interviews. In response to her questions, they mentioned experiencing a number of symptoms, of which headache, muscle aches, and anxiety were the most common; each of these were mentioned by seven of the eight women. One mentioned vaginal dryness, two mentioned sexual problems, three mentioned night sweats, four mentioned irritability, and five mentioned depression. Three mentioned "hot flashes"—it would be interesting to know here whether they used the term *baja presion* or whether they described feeling hot, as Michel indicates that they did not have a word or term for hot flash. (Michel also reports on several herbs the Q'eqchi used to treat hot flashes without explaining this point.)

Much has been made of the conflicting findings of the three qualitative studies, plus Beyene and Martin's more quantitative study of 2001, on menopause in Maya women. It is important to note, however, the limits of all of the studies. All were conducted by researchers living in highly modernized societies who did not speak Maya languages, and almost all interviews were done through interpreters (local "informants"). While all of the researchers took steps to gain acceptance and trust from the women they worked with, it is hard to know how open the Maya women really were with them, how well they understood what the researchers were asking, and how well the researchers understood their answers. All three qualitative studies involved small numbers of women. All were focused on issues important in Western culture and medicine but not necessarily in the culture of their subjects—menopause, attitudes to menopause, and symptoms of menopause; of the researchers, Beyene was the one most sensitive to this point. The studies of Stewart and Michel are published in shorter form, and it is harder to get a sense of what their subjects were thinking.

It is clear that all of the Maya groups studied lacked a word for menopause. They also lacked a word for hot flashes, though the women of highland Guatemala described feeling hot when queried about their experiences over the last few years, and the midwives of the Livingston area, but apparently not the lay population, used the Spanish term *baja presion* to describe an episodic experience that included sweating among other symptoms. Finally, none of the researchers interviewed men or younger women about their recent experiences or symptoms. This means that it is very difficult to tell whether the women interviewed by Stewart and Michel thought their symptoms were associated with menopause *or whether the symptoms were, in fact, associated with menopause.* We do not know how many younger women or men, if asked the same questions, would have complained of extremely common problems like headaches, irritability, or lack of interest in sex. The failure to investigate this question vitiates a huge amount of cross-cultural research on menopause.

Beyene's subjects did report irregular menstruation and sometimes heavy menstruation, which they assumed was normal as their periods came to an end. They also reported other symptoms: "Some women said that they had headaches and felt dizzy at times. However, they did not associate these with menopause because headaches and dizziness are common symptoms associated with other ailments."[18] That is, the women of Chichimila did not share the assumption, built into a great deal of cross-cultural research, that all complaints experienced by middle-aged women are automatically symptoms of menopause.

The most puzzling finding of the Maya studies is the young age at menopause of the women in Beyene's report. Although her findings were not fully borne out in studies of other groups of Maya, it is possible that environmental or genetic factors affect the age at which menopause occurs at the population level. Some studies suggest that women in less modernized societies experience menopause a few years earlier than women in industrialized societies. For example, based on their own recollections, women living in Bangladesh reached menopause at an average age of 46; among Bangladeshi immigrants to London, this age was 47.5, and among white women native to the United Kingdom, it was later still, at 49.[19] Research has linked an earlier age at menopause, if only tentatively, to lower levels of education, episodes of extreme nutritional stress in childhood (many of the Bangladeshi women had endured food shortages in the wake of catastrophic cyclones), and a higher prevalence of infectious diseases in childhood—all features of more traditional societies. Consumption of meat and alcohol, which are more likely in modern societies, may also raise age at menopause. On the other hand, women with children reach menopause later than women with no children (and in some studies, age at menopause increases with the number of children borne, though that result is less consistent), which would favor a later age at menopause in traditional societies compared to modern ones. The finding that smoking lowers age at menopause is the most consistent result in studies that look for these effects. Because age at menopause is partly inherited, genetic differences among populations may also account for some of this variation. Average ages at menopause as low as 42 (in Beyene's study) and as high as 52 have been documented, with highly modernized societies like the United States dominating the upper end of the scale.[20]

A complication is that average age at menopause is usually calculated by asking women to remember when their last period occurred, which is not a very reliable method. Some studies do not exclude women who reached menopause because they had hysterectomies, resulting in a lower average age at menopause. Some studies exclude cases in which women report a very early or very late age at menopause, which can have an important effect on averages; others do not. Some studies, instead of asking women to remember when they stopped menstruating, ask every woman in a sample whether or not she is still menstruating and then calculate a median using a method called probit analysis. That method usually arrives at a later age at menopause, with less variation across populations. However, it is usually only possible in highly modernized societies.

In Puebla, Mexico, researchers found that the average age at natural menopause, when measured by asking women to recall their last menstrual period, was 46.7; excluding outliers, that figure rose to 47.6, and median age calculated by probit analysis was 49.6 (with outliers included) or 50.1 (with outliers excluded). So it is actually not easy to know what the average age at menopause for a population is, or to compare that age across populations. It is not clear to what extent results showing a later average age in more modernized societies reflect different sources of information and methods of calculation, and how much of that difference is real.[21]

At a loss to explain why her finding that Maya women experienced symptoms of menopause differed from Beyene's results, Stewart suggested that Beyene's interviewees, whose Body Mass Index was "relatively high,"[22] were better nourished than the thin women she observed. I think it is possible, however, that the most important variable in the

Maya studies was the researchers themselves: the assumptions they brought to their studies and what they expected to find. Stewart and Michel were able to map their subjects' experiences onto a modern idea of menopause by asking questions about the things that interested them most—symptoms—whereas Beyene focused more on her subjects' perceptions of menopause and midlife, and was less inclined to interpret the symptoms and problems they mentioned as related to menopause. In Chichimila as Beyene describes it, the transition to becoming a mother-in-law and a grandmother was very significant, whereas the fact that women's periods stopped was not. We saw in part I that it is probably not a coincidence that these events happen at the same time, but cross-cultural research on menopause has tended more toward the approaches of Michel and Stewart than to that of Beyene. It also may be important that Beyene's study took place more than 20 years before those of Michel and Stewart, decades in which modernization vastly expanded its reach everywhere, including Central America. Without more detailed information, it is hard to tell how much this might have affected the experience of menopause in the rural regions they studied.

While many traditional and folk systems of medicine appear to lack an equivalent of menopausal syndrome, there are exceptions. In the early 1980s, Nancy Rosenberger visited an isolated village of subsistence farmers and fishermen in Japan, which had only recently become accessible by road, to interview women about menopause. These village women understood menopause as a phase of *chi no michi*, or "path of blood," an old-fashioned term for problems related to menstruation and childbirth.[23] They believed that women were especially vulnerable to these problems just after childbirth, when blood might be retained or spoiled. If women did not get enough rest—which was often the case, because of their lifestyle of hard work—they might suffer symptoms later in life, when menstruation ended. One could become quite sick—one woman developed a fever shortly after her periods ended, at 42, following the birth of her last child. The fever lasted months, during which she was confined to bed.

The term *chi no michi* is also used in urbanized areas of Japan, although people are more likely to speak of menopause using the modern terms *konenki* and *konenki shogai* ("problems of menopause"). Women from the city of Matsuyama in Ehime Prefecture, on the island of Shikoku in southern Japan, used all of these expressions when interviewed by anthropologist Jan Zeserson in the early 1990s.[24] Here too the idea of *chi no michi* connected menopausal problems to childbirth and the reproductive cycle more generally; the first 30 days after childbirth were critical, and failing to rest and care for oneself during that period might result in symptoms later, when menstruation ended. It was particularly important to avoid reading or any sort of eye strain, which might cause failing eyesight, or even blindness, later in life. (Rural women of Haryana, India, also associated menopause with weak vision and blind spots.[25])

Traditionally, Japanese brides left home at marriage to become daughters-in-law in their husbands' families, but it was and remains traditional for women to return to their parents' home to give birth, and to have the support of their mothers for the weeks after childbirth. As an alternative, the mother-in-law has a moral obligation to care for her daughter-in-law and to protect her health at this time. In this way, *chi no michi* links women to their natal families, children to their maternal grandmothers, and both to a value system that emphasizes the duty of potentially exploitative mothers-in-law to care for their daughters-in-law. One woman interviewed about her distressing symptoms at menopause discussed her difficult relationship with her mother-in-law, who made harsh demands on her for service and had not allowed her to return to her own family to give birth; conversely, a woman who had few menopausal symptoms attributed this to her good relationship with her mother-in-law, who had cared for her after childbirth.

Chi no michi is an old concept, attested in Japanese literature as early as the tenth century, though without research in a language beyond my competence I can't discuss here how it may have changed over time. It does seem likely that in Japan, modern ideas of menopause blended with, or displaced, a preexisting tradition that also associated symptoms and problems with the female reproductive life cycle, including menopause, even if these symptoms and problems were quite different from those emphasized in the European tradition.

Women living in rural northeastern Thailand in the late twentieth century also thought that menopause could cause health problems, even though they retained a traditional understanding of medicine and physiology. Anthropologists Siriporn Chirawatkul and Lenore Manderson studied villages in northeastern Thailand that had only recently begun to modernize, with the introduction of electricity and roads, mechanization, and a change to cash crops from subsistence farming. The village women's ideas of menstruation were similar to those of many traditional societies around the world—they believed menstrual blood was toxic and polluting, and that retaining it could cause problems like headaches and dizziness, which they often experienced as pre-menstrual symptoms. Women thought regular menstruation was important for health, but also found it disgusting, and restricted their movements and behavior during their periods. Their language included phrases equivalent to "menopause," and women were likely to explain the end of menstruation as the result of running out of menstrual blood. Women were generally happy to be free of menstruation and its burdens and looked forward to menopause. Some, however, experienced symptoms, especially dizziness, that they attributed to the retention of blood, assuming that the supply of menstrual blood in their bodies had not yet run out. Eventually, as they saw it, any retained blood would dissipate, and symptoms would go away. Village women mostly did not consider the symptoms of menopause to be serious, but some took medicines for headaches or insomnia, or went to herbalists for help inducing menstruation to purge bad blood.

In this culture, then, researchers found a belief that menopause could cause symptoms and a rationale for why that might happen. Interestingly, when they compiled a list of the symptoms mentioned most frequently in their work with the village women, they found that almost every symptom on their list—including dizziness, hot flashes, irritability, lack of libido, headache, tiredness, feeling depressed, poor concentration, and so on—was either more common in pre-menopausal women or about equally common in both groups. (Because of their small sample size, however, the researchers did not try to calculate statistical significance.) Only forgetfulness, blind spots before the eyes, and insomnia were more common in the post-menopausal women they interviewed, and village women interpreted these as normal experiences of aging. The two symptoms village women attributed specifically to menopause were dizziness and headache, although both of these were actually more common in the sample of pre-menopausal women. They attributed emotional symptoms to family or money problems. That is, like the women of Chichimila, Thai village women were not inclined to associate every problem they experienced in midlife with menopause. They had no term closely equivalent to "hot flash" or "hot flush" in English; the researchers explained that the phrase they were translating as "hot flush" signified a burning sensation like that of touching a hot chili, often affecting the ankle or waist, or radiating from the chest.

Senior status in this population was linked not to menstruation but to the marriage of a woman's children, or of her nieces and nephews if she had no children of her own, and to the birth of grandchildren. The role of grandmother in this society was changing as modern education and economic development eroded, as the Thai women saw it, the value of older people's experience and wisdom. As in many (but not all) traditional cultures, it was thought inappropriate for older women to have sex, and women past menopause said they did not enjoy sex very much, a development that many reported with relief, as they felt liberated to focus on religion, meditation, and peace of mind. They saw midlife and old age as a period of calmness, a quality they valued.

The researchers also interviewed psychiatrists and gynecologists trained in modern medicine in Thailand. These professionals shared the enthusiasm for HRT that was current in Western medicine at the time, as well as the view that menopause was a deficiency state that should be remedied with hormones. Unlike the village women, modern Thai medical professionals thought that menopause caused a wide variety of symptoms, especially emotional symptoms, and that it increased the risk of mental illness. Some were prescribing HRT to all their patients who could afford it, and one had opened a menopause clinic. Magazines and television were also promoting the idea of menopause as a time of crisis for which women should seek medical help, and HRT as a wonder drug that could preserve youth and health. Some village women had begun to absorb these messages about menopause—and it is hard to know how much their exposure to modern ideas might have affected the answers that they gave to the researchers' questions (that is, although I am including this study in my discussion of menopause in traditional societies, Thai village women did have exposure to modern culture and medicine). Modern medicine had a powerful reputation for efficacy among the villagers, and a few women had asked for medication for menopausal symptoms but could not afford the HRT that the doctors they visited would otherwise have prescribed for them. Chirawatkul and Manderson thought that their changing attitudes toward menopause were closely related to the changes that modernity was also bringing to the status of old age, as less value was placed on experience, and more on youth and productivity in the modern economy.

Among traditional explanations of menopause, the idea of depletion—the using up of a fixed quantity of something—stands out as a theme across cultures. The women of Chichimila believed that their periods stopped because they had used up all their blood bearing children. In a similar way, village women of The Gambia surveyed by Caroline Bledsoe in the early 1990s believed that God granted every woman a fixed endowment of potential children, and fertility ended once that endowment was used up. Because each potential child faced many risks and might be lost, and because each pregnancy also weakened and aged the mother, affecting her ability to bear and raise healthy children while using up some of her endowment, ideas of spacing, conservation, and depletion dominated thinking about reproduction and its end. Thus some women in the study used modern birth control methods not so much to limit fertility as to conserve it, by extending the time between pregnancies. Gambian women showed "almost no cultural interest in menopause" as it was understood in the West. Many retired from childbearing and sought co-wives for their husbands in their mid- or late 30s, before they reached menopause.[26]

When researchers interviewed Hmong refugees from southeast Asia living in Melbourne, Australia, in the early 1990s—this population was still unacculturated and non-English-speaking at the time, and most had lived in Australia for less than six years—they explained menopause in ways very similar to the women of Chichimila and The Gambia. The Hmong women had no word for menopause and perceived the end of menstruation as a part of growing old and achieving respected status. Menstruation stopped when they had had all the children they were destined to bear, whether 7 or 10 or some other number; one interviewee speculated that women run out of blood through childbearing. They believed that women who started menstruating early would stop early, perhaps because they used up their allotment faster. Although the Hmong valued large families, they were relieved when menopause lifted the burden of producing more children. By the time they reached menopause, a daughter or daughter-in-law was usually producing grandchildren that they helped raise. Menstruation was thought polluting and embarrassing, and most Hmong women were glad when it stopped and they became clean, like men. When asked about symptoms of menopause, whether physical or emotional, Hmong women could not think of anything besides light and irregular periods, and questions about whether their health had changed for the worse after menopause seemed silly to them. They were surprised to learn that many white Australian women experienced problems at menopause, and speculated that perhaps those women had not followed the right practices after childbirth, such as lying near a fire and eating special foods to restore their strength. Their thinking on this point recalls Japanese ideas of *chi no michi*; also like traditional Japanese women, the Hmong thought that lack of menstrual blood after menopause might weaken eyesight, making it hard to thread a needle, for example.[27]

If depletion is a common explanation for menopause—and this notion makes a lot of sense in societies with "natural fertility"—this may be one reason that menopause is perceived less often, and less pervasively, as a cause of medical problems in those societies. As the body's response to the exhaustion of its resources for reproduction, menopause is not a problem, but a solution. The women of Chichimila felt rejuvenated by menopause. Women of The Gambia enjoyed renewed strength in their post-reproductive years, when they were relieved from the burdens of childbearing and hard labor, and could finally achieve the status of respected elder after decades of sacrifice.[28]

While menopause can be hard to document among traditional populations and often does not seem to be too important among them, role change at midlife is typically very important in these cultures. In the usual pattern, women enjoy peak authority and well-being from the time they become grandmothers or mothers-in-law until they begin to become frail in old age. This pattern has been observed even in foraging societies, including the !Kung of the early 1960s, among whom women over 40 arranged marriages, adjudicated matters of kinship (determining who should use what kinship terms for whom), had the largest number of trading partners, and were uninhibited contributors to public discourse and sexual adventures. It was not unusual for women over 40 to have younger lovers or husbands (a finding that is interesting to me, as my own husband is much younger than I am, and this type of marriage is rare and widely considered unnatural in my own culture). Middle-aged !Kung women also had less work to do, because they did not have to carry or feed small children.[29]

In agrarian societies of the usual patriarchal pattern, middle-aged women typically have fewer constraints on their behavior—they are more free to leave the house, visit the market or religious shrines, go on pilgrimages, and generally appear in public. Where menstrual taboos are burdensome, these restrictions are lifted at menopause, and women feel clean. Middle-aged women typically have authority over daughters-in-law or younger co-wives whose labor they command and to whom they delegate the more exhausting and unpleasant household tasks. They control the distribution of resources, especially food, in the household, a power that can be used as a weapon. Through the loyalty of their adult sons, they can exercise considerable power even over male family members and the larger community. In cultures in which it is normal for husbands to be many years older, a middle-aged wife may take over most of the authority in the household while her husband, in old age, remains nominal head. Older women are sometimes eligible for special roles in the community as priestesses, matchmakers, or midwives. Many of these advantages come from having sons in a patriarchal social system, and so it is easy to see why women in many traditional societies prefer sons over daughters.[30]

It is hard to generalize about ideas of menopause in traditional societies, especially since we have very little of the right kind of information from cultures without exposure to modern medicine. Nevertheless, some themes emerge as typical. Many traditional cultures have perceived menstruation as polluting, dirty, and embarrassing, and retained menstrual blood as toxic and the cause of health problems. Freedom from menstruation and its social restrictions has been a welcome feature of menopause. Menopause has coincided roughly with a transition to becoming a mother-in-law or a grandmother, and thus with a higher and more respected status, and possibly less housework and hard labor; while this transition has been very significant, menopause itself has not been so significant. Women often have not associated physical or emotional symptoms with menopause or thought about it as a health issue, and many traditional societies do not have a word for menopause or for hot flashes, although in a few, there is evidence for the idea that residual menstrual blood may cause problems like headaches or weak vision at menopause. A typical explanation of menopause has been that it occurs when women run out of menstrual blood, or run through their predestined allotment of children. Women have usually borne many children before menopause, and although they have typically valued large families, they have welcomed the relief from further childbearing that it brings. No society that I have come across in researching this book has marked menopause with any type of ritual.

While this description does not fit every traditional society's ideas of menopause, I offer it as a common template. It's a view that makes a lot of sense in patriarchal, agrarian, high-fertility societies, and it is not surprising that it has altered under the influence of modernization.

In the agrarian period, menopause fit into a patriarchal social and economic system in which the household was the main economic unit, the labor of older children increased prosperity, control of reproduction was often advantageous because land and other resources were fixed, and mothers-in-law and grandmothers wielded authority over younger women. Work and responsibility were divided not only by sex but by age. In the modern period, all of that changed—wage labor mostly replaced peasant farms, the multigenerational family became obsolete, marriage was redefined as a companionate partnership, divisions of labor by sex came to seem arbitrary and unfair, experience became less important with universal education and literacy, women were included in new constructs of individual rights, and reproduction could be controlled much more effectively than before. It is no surprise that the modern world saw a reconfiguration of the role of older women, and this is likely one of the reasons for the persistence and spread of the new Western concept of menopause in medicine and popular culture; as old systems collapsed, the roles of post-reproductive women changed. Modern culture offered a new ideology of menopause to fill the void it had created with its erasure of age-specific roles, and it is not surprising that people bought into it, flawed as it may be.

The anxiety that seems to lie behind the modern concept of menopause—that is, anxiety about what older women should be doing with their lives once they are no longer bearing children and managing peasant farms—seems misplaced to me. Roles have changed for everyone, not just middle-aged women, and it is not clear why a group of people who seem, on balance, to have performed complex roles with a high level of competence throughout human history should now come under the kind of scrutiny that the construct of menopause encourages. That this construct is, in the big picture, part of an age-specific backlash against the challenge women pose to men as sex roles become less differentiated seems very likely. Menopause rhetoric took on new dimensions in the early twentieth century with women's suffrage and the emergence of the "New Woman," increased in volume in the mid-twentieth century as more women entered the paid workforce during and after World War II, and reached a fever pitch in the 1970s with the advent of the women's liberation movement. On the other hand, ideas of menopause and midlife problems may also reflect the vulnerability and confusion of women in modernizing societies who, without farms or children at home, find few roles prescribed for them.

CHAPTER 11

•

Symptoms

IT IS AN axiom of modern medicine that menopause has symptoms. The approach taken by much cross-cultural research on menopause focuses on symptoms and also on attitudes (and sometimes on the connection between these). One does not have to read far in this research to become frustrated with some of its methods. Even when studying populations with no word for menopause, researchers usually do not address the question of whether the phenomenon they are interested in—that is, menopause—is relevant to the women they are studying. Interviewees dutifully comment on a subject they are required to address ("What are you talking about?" is not an option on a typical checklist or questionnaire), and scientists then draw conclusions about their "attitudes to menopause." Researchers often ask themselves whether their subjects are "aware of" or "conversant with" a canonical list of symptoms, rather than whether their subjects believe that menopause has symptoms. Assumptions that menopause creates burdensome problems to society as the population ages, or that more attention to and treatment of menopause is urgently necessary in societies in which menopause is not medicalized, are frequent in this body of research.

In a common approach, researchers deploy standard checklists of menopausal symptoms, which they ask their subjects to fill out—I will call this the "checklist method." This method has flaws that are easy even for nonexperts to see. That recollection and self-report are not very accurate measures of experience, and that people trying to be helpful can overreport symptoms they had not previously noticed or thought important, are problems frequently acknowledged. Researchers using the checklist method usually do not study younger women or men for comparison, assuming that the many symptoms typically reported by women in midlife are related to menopause, when a search for correlations might show a weak connection or no connection between symptoms and menopausal status. Most do not ask their subjects whether they think their symptoms are related to menopause.

A doctor might say (and I am quoting a doctor here, from casual conversation), "A sign is something you see; a symptom is something the patient tells you." Symptoms are hard to measure by their very nature. They are subjective, individual experiences that can be detected only by asking someone about them.[1] The potential for circumstances, environment, beliefs, expectations, personal history, and even personality to influence both the experience and the reporting of symptoms is very great. But most researchers treat symptoms as objective, easily measurable facts that can be turned into statistics. They might argue that with a big enough sample, all the fuzzy factors will wash out and a valid result will emerge. A problem here is that for many menopause studies, samples are small, but even with large samples, we will see evidence that this argument is dubious—results that vary widely when studying the same populations, difficulty proving which if any symptoms correlate with menopause in most societies, repeated and frustrated calls for standardization in cross-cultural studies.[2]

The example of one well-known cross-cultural study may illustrate some of these problems. As part of their work with populations living in Israel in the 1970s, researchers read a list of symptoms to their interviewees (many of whom were illiterate), ranging from pounding heart, dizziness, and constipation to joint pains and worry about going crazy. They prefaced the list with the statement "Now I am going to read you a list of problems women sometimes have. Please tell me, for each one, if during the past year you were troubled by the problem often, sometimes, or not at all." Although they did not ask whether their interviewees thought the problems were related to menopause, or offer readers a justification for their choice of symptoms or evidence that they might correlate with menopause, or investigate whether women going through menopause reported the symptoms more frequently than other women (or men); although all the symptoms they asked about were common and nonspecific to menopause; and although recall over a period of a year is not a very accurate way to measure experiences, these researchers reported the results as rates of "menopausal symptomology." This is a fairly typical example of the way menopausal syndrome has been investigated in cross-cultural research. On the other hand, this is also one of a small number of studies that have asked women whether they thought menopause was important, even if only as an afterthought.[3]

There is thus a lot of cross-cultural research on menopause that is, bluntly speaking, not very good. In this part of the book I have not cited every study in the field, but have focused mostly on those studies that go beyond the checklist method, that are based on longer and more intensive field research, that include "qualitative" discussion to place the results in context (that is, they go beyond just numbers), that address issues of language and the translation of words and concepts (this is quite rare), and that make at least some effort to find out whether symptoms are related to menopause, or whether they are just common problems that a lot of people report when asked about them.

While some researchers invent their own checklists of menopausal symptoms, there are several different standardized lists in use: the Blatt Menopausal Index, the Menopause Symptom Checklist, the Midlife Symptom Index, the Greene Climacteric Scale, the Menopause Specific Quality of Life Questionnaire (and its variation, the MENQOL+), the Women's Health Questionnaire, and several others. Where do they come from?

As an example, let's consider the Midlife Symptom Index (MSI), first published in 2006, which was developed to better capture variation in menopausal symptoms across cultures.[4] The origins of this instrument trace back to the Cornell Medical Index, published in 1959, which screens for a large number of general symptoms not specific to midlife women. Researchers interested in menopause produced the Modified Cornell Medical Index (MCMI), an early version of the MSI, which added 14 "additional questions on menopause-specific symptoms reported in previous studies of menopause in Western and Asian populations." It was reviewed for validity by "two groups of experts in women's health."

Because the 164-item MCMI seemed too long, researchers produced a shorter version, the MSI. To do this, they used "recent literature" and "recently developed menopause-specific instruments." They subjected the resulting checklist to "reviews from a panel of eight experts in the area of women's health." The remaining list of symptoms contained 88 "physical, psychological, and psychosomatic" items. This list was further reviewed by "20 experts in women's health," who were asked, "How well do you think the MSI measures menopausal symptoms of diverse ethnic groups of middle-aged women?" It was shown to a convenience sample of 77 women between the ages of 40 and 60, who were asked to indicate whether they had or had not experienced the symptoms. These subjects were also given the 28-item Menopausal Symptom Checklist, first published in 1963 and commonly used in menopause research, and their scores on the two instruments were compared. Because the subjects' answers to similar questions on the two tests correlated well, the authors determined that their new Menopausal Symptom Index was valid. In its final form it lists 73 questions beginning with "Have you recently gained weight?" and ending with "Are you forgetful?"

While this process seems rigorous, it is important to note that this new checklist of menopausal symptoms mostly derives from, in the first place, previous checklists, and, in the second place, what women's health practitioners already believe about menopause and its symptoms. It would be hard to correct old traditions or assumptions about menopause using the MSI, even if its questions are embedded in an even longer checklist of more general symptoms (but most of the "symptoms of menopause" are very general), or if open-ended questions are added—two techniques researchers sometimes use to mitigate the methodological problems of checklists. My purpose is not to single out the MSI for criticism, but only to use it as an example; most checklists have been generated using similar methods. Checklists are self-reproducing and reflect the accumulated history of about 200 years of research on symptoms believed to be associated with menopause.

When validating the MSI, the researchers focused only on women of menopausal age and did not give the checklist to men or to women outside that age range. And while they asked their subjects to report on their menopausal status (whether they still had regular periods), they did not test whether the symptoms they reported correlated with menopausal status. Most of the symptoms included on menopause checklists are common complaints—joint pain, fatigue, "irritability," and so on. Some of these symptoms may also result from aging, and the effects of aging and menopause can be very difficult to disentangle. If a symptom is reported by a large number of middle-aged women, this does not necessarily make it a "symptom of menopause"—we need to know how many people not experiencing menopause have the same problem. For example, although the culture-specific shoulder pain known as *katakori* was the symptom most commonly reported, by a wide margin, in studies of middle-aged Japanese women, and although women *believed* that it was a symptom of menopause, when researchers tested for a statistical correlation between the two, they found no association. It was a symptom commonly reported by Japanese adults of both sexes and in all age groups, and was not more prevalent in women going through menopause.[5]

Although most studies of menopausal symptoms ignore these problems, more careful studies have tried to establish which symptoms really correlate with menopause, producing results that are often confusing and contradictory. I have discussed the example of depressive symptoms earlier, in chapter 9. Sometimes a symptom correlates with menopause in one study or one population, but not across cultures (sometimes, not even across multiple studies of the same society). This is true even though most menopause research focuses on highly modernized, Westernized societies with mostly white populations. The National Institutes of Health's "State of the Science Conference Statement on the Management of Menopause-Related Symptoms" of 2005 accepts only a small number of proven symptoms of menopause—"vasomotor" symptoms (those having to do with the control of our blood vessels and blood supply, that is, hot flashes and night sweats), vaginal dryness, and possibly sleep disturbance—concluding that there is not enough evidence to link menopause to other symptoms like mood disturbance, forgetfulness, stiff joints, or sexual dysfunction.[6]

A more recent review, published in 2011 by Melissa K. Melby and colleagues, includes studies from all over the world, with the intention of adding international and cross-cultural perspectives to conclusions based mostly on North American, European, and Australian populations.[7] The number of cross-cultural studies meeting the criteria for inclusion in this review was not large; only nine studies qualified. Four of the nine studied different ethnic groups within the United States, rather than comparing populations from different parts of the world, however, so the review's findings still disproportionately represent highly Westernized, industrialized societies, and the United States in particular.

Included in this review is the only major study of menopause-related symptoms to survey women of a broad age range, from 20 to 70 (most studies restrict participation to women between 40 and 60 or between 45 and 55). This is the study published by Dennerstein and colleagues in 2007 and nicknamed WISHeS.[8] Of the 36 symptoms these researchers asked about, 7—hot flashes, night sweats, sleeping difficulty, memory problems, vaginal dryness, problems with sexual arousal, and "aches in the head, neck, and shoulders"—showed a pattern that peaked around age 50. Of these, the first two—hot flashes and night sweats—showed a much stronger peak around age 50 than the other five. None of the remaining symptoms on Dennerstein's checklist showed this peak. A cluster of mood symptoms peaked around age 35 to 40 and declined steeply after that. Another group of symptoms, which included decreased strength and stamina, thinning hair, and skin changes, increased linearly with age but showed no peak at menopause. A third group that included weight gain, involuntary urination, increased facial hair, and fatigue showed a strong relationship to Body Mass Index but none to age. Among the symptoms associated with age 50 and, by inference, with menopause, vaginal dryness affected women in the United States and United Kingdom but did not show a relationship to menopause in Italy, France, or Germany.

Melby advocates including the seven symptoms that peaked around age 50 in the WISHeS study on all menopause checklists. She also advocates including other symptoms that do not have a proven correlation with menopause, either because they seem to be linked to symptoms that do, or because they have been commonly reported. I am not sure I follow all of the reasoning for the construction of Melby's recommended checklist, but it runs to 21 "core symptoms": the 7 already mentioned, plus depression, anxiety, stress, irritability, feeling nervous, difficulty concentrating, headaches, fatigue, palpitations, dizziness, breathing difficulties, numbness or tingling, gastrointestinal problems, and problems with sexual desire. I think it is important to emphasize that only a few of those 21 symptoms—to say nothing of the dozens of symptoms included on some menopause questionnaires—have a proven correlation with menopause. I note other points about the "core symptoms" here, for future reference—they are all general problems not specific to menopause, they are all common problems, and they are all symptoms that can be made worse by anxiety (and anxiety itself is one of Melby's core symptoms). Symptoms of anxiety and panic closely resemble lists of menopausal symptoms.

A limitation of the WISHeS study is that, like most other menopause research, it focused on modernized European and North American populations that are culturally very similar—its subjects lived in the United States, the United Kingdom, Germany, France, and Italy. Based on this study, it is not possible to say whether the correlations it found between symptoms and menopause are specific to modernized Western cultures or hold true across cultures—that is, whether they are really core symptoms in that sense. On this point, the most interesting of the nine studies reviewed by Melby is the Decisions at Menopause Study (nicknamed DAMES), which included two non-Western societies. This study surveyed 300 women between the ages of 45 and 55 in each of four countries—Morocco, Lebanon, Spain, and the United States (for a total of about 1,200 subjects)—with results published between 1998 and 2007. The synthetic report of 2007, by Obermeyer, Reher, and Saliba, is included in Melby's review. One of the populations studied, the women of Morocco, was less modernized than the others—although the sample came from the large city of Rabat, the women surveyed were less educated (only about half had any formal schooling), had married in their teens, and had more children on average than the women in the other groups; many did not know their exact age. Obermeyer and colleagues mention in passing that the women of Rabat whom they interviewed had no local word for menopause, though a few more educated women used the French term *ménopause*.[9]

The DAMES checklist contains 20 symptoms, and almost all women surveyed reported some symptoms. A few were frequently reported across three or all four countries: sleep problems, memory problems, impatience/nervousness, fatigue, and joint pain. When the researchers tested for a connection between symptoms and menopausal status—looking for symptoms that were statistically more frequent in women undergoing or past menopause than in women whose periods were still regular, independent of other factors—they found that only one, the "hot flash," was associated with menopause in all four countries. A few other symptoms were associated with menopause in some countries: vaginal dryness in Massachusetts, Madrid, and Rabat, but not Beirut; loss of sexual desire in Rabat, but not elsewhere; painful intercourse in Madrid, but not elsewhere; joint pain in Massachusetts, but not elsewhere; heart palpitations in Beirut, but not elsewhere. Overall, based on their pooled data, the authors of DAMES concluded that sleep disturbance and vasomotor symptoms (that is, hot flashes and night sweats), but not other symptoms, were significantly associated with menopause. In the case of women who had undergone hysterectomies—"surgical menopause"—two other symptoms, joint pain and depression, were also significantly associated.

Unlike most checklist studies of this kind, the authors of the DAMES study comment on some issues of language and translation. Regarding "emotional symptoms," it was difficult, they write, to "find . . . exact equivalents for emotional states across cultures and languages." The main item inspiring comment on language and understanding, however, was the hot flash.

Women in Massachusetts used only one or two terms for hot flashes—"sweats and flashes are the standard terms, and there is not much variability in women's descriptions"—but researchers found a much richer vocabulary in the other populations they studied. Spanish women used at least three terms—*sudores*, meaning "sweats"; *calores*, meaning "heat"; and *sofocos*, meaning "choking" or "suffocation." The women of Beirut also spoke of choking, using Arabic terms meaning "hot suffocations," "I am stifled," "I choke," and "I am suffocating," as well as "hot and cold flashes," "I am ablaze," and "My face is red." While

Obermeyer and her colleagues were confident that all their subjects were describing the same thing, there were regional variations in how women described the hot flash, and perhaps in how they experienced it, with sensations of choking and suffocation being especially prominent in Beirut.

Women in Rabat complained energetically and in colorful language about their symptoms, especially hot flashes and joint pains, although the researchers do not comment on the extent to which their subjects believed menopause caused these symptoms. In this population, 68 percent of perimenopausal women (women whose periods were irregular) and 70 percent of post-menopausal women (those who had not had a period in more than a year) reported hot flashes. Important context, however, is that nearly half (47 percent) of pre-menopausal women in the study also complained of hot flashes. The statistical association of hot flashes with menopause was significant in Rabat, but not as strong as one might think based on their high prevalence among menopausal women.

It's not unusual for studies of menopausal symptoms to get different results even when focusing on the same population. Using the example of the hot flash, five studies conducted in Thailand report very different rates of prevalence, ranging from 25 percent (the lowest result) to 80 percent (the highest). In South Asia, the range was 14 to 42 percent in four studies. In the United States, one recent large, well-regarded national study, the Four Major Ethnic Groups study, showed a prevalence rate of 84 percent in white non-Hispanic women, but another recent large, well-regarded study—the SWAN study—found a prevalence of only 24 percent for white non-Hispanic women. Some of this difference may result from different methods—the Four Major Ethnic Groups study asked women to recall symptoms over a period of six months (longer than what most researchers would recommend for reliable results), while SWAN's recall period was only two weeks. Still, the very high prevalence rates that researchers commonly cite or assume for U.S. women need qualification. (This is probably the best place to point out that sources reporting figures of 87 or 88 percent often cite Fredi Kronenberg's 1990 study, discussed further in the next section. The subjects of this study were not chosen at random; they were a group of women who wrote to Kronenberg because they were interested in her work on hot flashes, and Kronenberg emphasizes in the original publication that her prevalence rates are not meaningful.) It is harder to explain the range of results in studies from Thailand, where reviewers have found no major methodological differences that would account for it.[10]

Do women who have not entered menopause have hot flashes? Studies that ask this question have reported widely variable rates of prevalence, just like studies of hot flashes in women going through menopause or past menopause. Prevalence rates as high as 86 percent have been found (in Thailand), with a median prevalence of 21 percent. That is, many women whose periods are still regular report that they have hot flashes—when researchers ask. Men are hardly ever asked about hot flashes, but in a survey of 1,381 Swedish men aged 55, 65, and 75 published in 2003, about one-third reported that they experienced them.[11] When I tried this question on my then-32-year-old husband, he said he had hot flashes all the time. By contrast, I am one of the 16 to 76 percent of white non-Hispanic menopausal women in the United States who, depending on how you ask the question, does not have hot flashes.

HOT FLASH OR HOT MESS? RESEARCH ON VASOMOTOR SYMPTOMS

What is a hot flash, anyway? Typically, this item appears on menopause checklists without definition or explanation (and I did not explain it to my husband when I questioned him). I became quite interested in this question and kept track of the descriptions that I found in my research. Many of them have been quoted in this chapter and in preceding chapters.

A leading U.S. researcher of hot flashes, Robert R. Freedman, offers the following description: "Hot flashes (HFs) . . . are reported as feelings of intense warmth along with sweating, flushing, and chills. Sweating is generally reported in the face, neck and chest. HFs usually last for 1–5 min[utes]."[12] Freedman cites, in support of this definition, a classic study of Fredi Kronenberg, published in 1990, in which she surveyed a nonrandom sample of 501 women living in the United States about their experiences. The most commonly reported experiences in that study were feelings of heat, flushing, and sweating—all reported by large majorities of those surveyed—followed by anxiety, chills, embarrassment, change in heart rate, depression, change in breathing rate, and pressure in the head.[13] Western medicine's understanding of the hot flash, based on this influential publication, is quite specific—the quotation from Freedman is an example. That is, the hot flashes of Western medicine are episodic events—they come on suddenly and usually resolve quickly (though Freedman adds that they can last up to an hour). They involve not only feelings of heat, but sweating, flushing, chills, and sometimes psychological symptoms like anxiety and embarrassment (although these symptoms, commonly reported in Kronenberg's survey, tend not to be emphasized by later researchers). They are localized to the face, neck, and chest.

There is a long tradition in Western medical literature describing the hot flash in language similar to that used by Kronenberg and Freedman—as brief episodes of heat, flushing, and sweating. As early as 1598, Liébault, writing in French, described "many little flushings which appear in the face mostly after eating, and which end profusely in sweats." In 1761, Astruc wrote, also in French, of "blushings, heat sensations that frequently rise suddenly to the face, and that end in sudden sweats"—the "hysterical vapors" of menopause. For Tilt, "an increased production of heat . . . with that irregular distribution of it which is called 'flushes'" was one of the most common symptoms of menopause. Episodic fits of heat with flushing and sweating in the head and neck trace back to the earliest traditions of menopause in the West, though the addition of chills appears to be recent.

The authors of the DAMES study saw many similarities in the ways the women they interviewed described hot flashes, but the choking sensations reported by the women of Beirut—and implied also in one common Spanish translation of "hot flash," *sofoco*—are not part of the traditional definition in English literature. Choking was the classic symptom of the ancient syndrome of hysterical suffocation and its

Renaissance and early modern descendants, and I wonder whether choking symptoms of menopause reflect the continuing importance of that tradition in some parts of the Mediterranean world. The idea of hysterical suffocation was transferred to the Islamic world through an Arabic medical tradition deeply influenced by the ancient Greek one.[14]

In other studies, researchers have not been so confident about matching the heat sensations described by their subjects with the Western hot flash. Thai village women interviewed by Chirawatkul and Manderson described a burning sensation like touching a hot chili, often felt in the ankle or waist, though it could also radiate from the chest like a Western hot flash. The most illuminating research on this point is Melby's 2005 study of vasomotor symptoms (meaning hot flashes and night sweats) as reported by a sample of 140 women in Japan between the ages of 45 and 55, who were asked to recall their symptoms of the past two weeks.[15] Melby asked about four different words for heat sensations, including one term—*hotto furasshu*—borrowed from English that had recently been introduced in Japanese media. Interest in modern menopause and in HRT escalated in Japan in the late 1980s and 1990s, and *hotto furasshu* was coined to signify a kind of sensation specific to menopause that HRT might treat.[16] Hardly any women reported it on Melby's survey, however, and many did not seem familiar with the term. Twenty-four women, about 17 percent of the sample, said they experienced *hoteri*, which Melby translates as "feeling hot or flushed," while 15 women chose the term *kyu na nekkan* ("sudden feeling of heat") and 13 chose *nobose* ("rush of blood to the head"). Altogether, about 22 percent of the sample said they had experienced some type of heat sensation in the past two weeks. The percentage was nearly twice as high among those women Melby classified as "late perimenopausal" based on their menstrual patterns, but because of the small sample size, the difference was not statistically significant. About 16 percent of pre-menopausal women whose periods were still regular experienced heat sensations.

Asked to define each of the heat sensations they experienced, the women in Melby's study described *hoteri*, the term they chose most often, as a feeling of flushing or redness in the face, and sometimes in the hands and feet. This feeling was not necessarily of short duration;

hoteri could describe a state experienced over a long or medium period of time. *Nobose* was a feeling of heat with dizziness in the head and a sensation of "spacing out" that lasted for a long time. Many women were uncertain about the meaning of *kyu na nekkan*, although some selected it to describe their experience. None of the Japanese words for heat sensations, other than *hotto furasshu*, seemed to be specific to menopause in the same way that English speakers associate "hot flash" with menopause.

Melby's survey did not offer women the choice of phrases using the popular term *kaa to suru* (literally, "to be hot") or related expressions that signified, for the women in the study, a brief and sudden feeling of heat followed by chills and sweating, and accompanied by feelings of nervousness, anger, or irritability.[17] Although familiar to almost all of the women in Melby's sample, *kaa* expressions had not been surveyed by Margaret Lock in her pioneering 1993 study on menopause in Japan, and Melby wanted to be able to compare her results with the earlier data, so she left them out of her questionnaire but did ask her subjects what they thought these expressions meant. It is possible that if she had included *kaa* phrases among the heat sensations she asked about, more women would have reported them.

Although the percentage of women in Melby's study who said they had experienced heat sensations over the last two weeks was similar to the percentage of Caucasian women who reported hot flashes in the SWAN study over the same period of time, midlife Japanese women described different kinds of heat sensations that only partly overlap with descriptions of the Western hot flash. Some of their experiences did not have the sudden, episodic character so central to the Western concept. At the same time, the changing language of the hot flash in Japan illustrates how ideas of menopause are evolving in response to Western cultural influence.

It is not clear whether researchers studying the hot flash are measuring the same thing in Thailand or Japan as they are in the United States or Italy, but if we assume that all heat sensations reported by midlife women are related phenomena, they are the best candidate for a core symptom of menopause, as researchers have shown an association between them and menopause in many societies. On the other hand, many traditional societies do not recognize the concept of a hot flash or associate it with menopause, although this is not the same as saying that women in those societies do not experience what a Westerner would call hot flashes.

Some researchers have focused on measuring the hot flash as a sign, not a symptom—that is, on measuring hot flashes objectively. Tests of skin conductance, which is higher when we sweat, and which can be measured by electrodes attached to the chest, correspond fairly closely with women's own perceptions of hot flashes, when these tests are administered in laboratory conditions. Researchers have also invented portable monitors to measure skin conductance outside the laboratory, but in those conditions, there is much less agreement between the machine-measured hot flashes and the subjective experiences that women report. That is, women wearing hot flash monitors throughout their normal daily routines often fail to report hot flashes that the monitors detect, and they also report hot flashes that the monitors do not detect.[18]

Skin conductance tests were developed in the United States, but some researchers have taken monitors to other populations. In 2007, Lynette Leidy Sievert measured hot flashes using portable monitors in Hindu and Muslim women of Sylhet, Bangladesh.[19] Two-thirds of the 30 women in this study had been raised in rural villages, although all were living in modern apartments in the urban capital at the time of the study. They were mostly middle-class, they had received about 10 years of formal education on average, and their average number of children was 3.7. Nine of the 30 were Hindu, and the remaining 21 were Muslim. All were between the ages of 40 and 55. These 30 women were not a random sample; the research team interviewed many more, but the 30 who agreed to wear a hot flash monitor for the study were much more likely than others to have experienced hot flashes, or other problems like irritability or dizziness (the most common complaint), and to be bothered by them.

The term Sievert's team used to translate "hot flash" was the Bangla phrase *gorom vap laga*, "feeling steaming hot"; the women themselves used related terms in Sylhati dialect meaning "uncomfortably hot inside, and/or queasy or suffocating," "a sudden feeling of heat," or "smoke coming from the head."[20] When asked to indicate where they felt the hot flashes, the women indicated the top of the head and neck (especially Muslim women, who wore headscarves during the day), the chest, and in about half of cases, the hands and feet.

Although there were no important differences in wealth or status or, for the most part, lifestyle and activities between the Muslim and Hindu groups, there was a big difference between these groups in the test results. Monitors recorded hardly any "objective" hot flashes in the Hindu women, but they recorded an average of 1.5 hot flashes in the Muslim women over the eight-hour period for which they were worn (only eight women in total, however, experienced these "objective" hot flashes recorded by the monitor). Hindu women, however, reported hot flashes at a higher rate, an average of 3.5 compared to 2.5 for the Muslim women. Sievert's team guessed that different religious practices might have influenced the results, since Muslim prayers were more active, requiring repetitions of the cycle of standing, kneeling, and bowing (the *rakat*), which might elevate core temperatures.

Women in both groups reported many more hot flashes than were measured by the monitor, and more than half of the hot flashes measured by the monitor were not felt by the women. The rate of concordance—that is, the percentage of total hot flashes reported that were registered by both monitors and women—was unimpressive; machine and human agreed only about 19 percent of the time. Clearly, skin conductance monitors were not good measures of hot flashes in this population. In a later study, Sievert and her team examined the prevalence of subjective hot flashes in three groups—Bangladeshi immigrants to London, Bangladeshi women still living in Bangladesh, and white Londoners. In this study, concordance rates were also low, between 10 and 20 percent across all groups. The prevalence of subjective hot flashes correlated with menopause—only 29 percent of premenopausal women in the three groups said they had hot flashes, while 55 percent of perimenopausal women and 71 percent of postmenopausal women reported them—and with the number of children the women had borne, but not with other factors that the researchers studied, including ethnicity. Objective hot flashes, measured by monitors, did not correlate with anything.[21]

These complexities in measuring and talking about hot flashes are one reason that it has been so hard to generalize about them—why it is so hard to know, for example, what percentage of women going through menopause in a population experience hot flashes or to compare those rates of prevalence. For a long time, researchers have argued that hot flashes are less prevalent in East and Southeast Asian populations than they are in the West, and this idea is still widely accepted. One especially influential study by Boulet and colleagues, published in 1994, found prevalence rates over a four-week recall period of about 20 to 40 percent in Hong Kong, Indonesia, Korea, Singapore, and Taiwan, rates that seemed low to researchers studying Western populations (this same study found higher prevalences in Malaysia and the Philippines, however).[22] Researchers have speculated that East and Southeast Asian women might have fewer hot flashes because they consume more soy, which is high in plant estrogens, but that hypothesis has been hard to prove. On the other hand, some studies, including some of those cited earlier that surveyed Thai populations, have found a high prevalence of hot flashes in East and Southeast Asia, and recent research on hot flashes in the United States and other Western populations has shown prevalence rates similar to those in Boulet's study, using the same recall period.[23] So it is not very clear that patterns in rates of prevalence really exist, nor that there is an easy way to reduce those patterns to numbers and statistics.

Differences in the way the information about hot flashes is collected—"subjectively" or "objectively," by recall or by diary, by recall over a long period or over a short period, by ambulatory monitor or in the laboratory—can lead to big differences in reported prevalence rates. A well-known study of hot flashes in Japanese Americans and white Americans living in Hilo, Hawaii, published in 2009, illustrates this problem. Among women surveyed by mail, a greater percentage of white women than Japanese American women said they had experienced hot flashes over the last two weeks (47 percent and 27 percent, respectively, of the sample that was later monitored). During a subsequent 24-hour period in which they wore a monitor and were asked to record hot flashes by pressing a button, however, there was no statistically significant difference in the percentage of women who reported hot flashes (56 percent of white women and 51 percent of Japanese American women reported them). The monitor registered hot flashes in 78 percent of Japanese Americans and 72 percent of white women, again not a statistically significant difference. A likely explanation for the results is that Japanese American women did not normally notice hot flashes as much as white women did, but when asked to pay special attention to them, they reported about the same number.[24]

The question of whether hot flashes are more common in Western women than in Asian populations is also complicated by the progress of modernization and globalization, which might reduce any differences caused by diet, lifestyle, medicine, and beliefs, as all of these become more similar. One reason that Melby found a 22 percent prevalence of heat sensations in Japanese women in 2001–2003, while Lock found a much lower prevalence in the 1980s, is that between those dates HRT received a lot more publicity and acceptance in Japan. Melby's study can be interpreted in many ways; depending on how we read it, we could say that there was not much difference between her results in Japan and the SWAN study's results for white women in the United States; that Japanese women experience the heat sensations of menopause differently than Western women do; or that heat sensations are not correlated to a level of statistical significance with menopause in Japan, as they are in the United States. There is also the likelihood that the next study of hot flashes in Japan will show different results.

Despite all the problems with cross-cultural research on hot flashes, most researchers agree that for large numbers of women all around the world, hormonal changes at menopause interfere with the body's system of temperature regulation. In the theory that currently prevails, the body's thermoneutral zone—the range of temperatures in which we do not sweat to cool off or shiver to generate heat—contracts to nearly zero for some women, so that when core temperature rises (presumably because of some trigger like exercise), the body responds by flushing and sweating. Some researchers have argued that women in warm climates have fewer hot flashes, perhaps because their thermoneutral zone is set to a wider range.[25] Hot flashes respond well to treatment with hormones (although placebo response rates are also high when patients believe they are getting hormonal therapy), strengthening the case for an organic connection to menopause.[26] The heat sensation or hot flash is the best candidate for a "universal" symptom of menopause, meaning it is associated with menopause in many cultures (and not that all, or even most, women experience them).

But in an important way, hot flashes are indigenous to western Europe—they have been part of the Western *history* of menopause since its invention in 1700, when they were symptoms of the plethora, or buildup of excess blood, believed to affect menopausal women, and resembled the fits of uterine ("hysterical") suffocation whose career in Western culture was already long and influential by that time. Modern medical descriptions of the hot flash quite obviously descend from that tradition. In the rest of the world—including countries such as Japan and Bangladesh—the hot flash is a concept imported recently. And there are still many societies today that have no concept of the hot flash or do not associate heat sensations especially with menopause, though women may report heat sensations when questioned closely. Whether we should equate these sensations with the hot flash, a very specific phenomenon rooted in Western culture, seems questionable to me.

I think it is likely that a failure to appreciate the history and tradition of menopause explains some of the muddled results of many cross-cultural studies of the hot flash. Monitor studies show that "objective" experiences measured by skin conductance explain only a small part of what women report outside strictly controlled laboratory conditions, even when they are asked to pay attention to hot flashes and report them as they occur. When filling out checklists about experiences recalled over weeks or months, the influence of "objective" hot flashes must be even less, and that of other factors, including cultural traditions about menopause and hot flashes, much greater. Small changes in wording, slight shifts in the level of education or urbanization of the sample, or cues communicated by the researcher might have large effects in a culture without a deeply entrenched tradition of the hot flash. Some questions may not capture transmutations of the hot flash as it has been adapted by local cultures. It is, in fact, hard to say what we are measuring when we ask such questions.

These nuances are difficult to capture in an approach that favors checklists and statistics, but not all research has been so shallow. In 1993, Margaret Lock published a well-known book about menopause in Japan, one of the most modernized nations in the world then as now, that remains the most sensitive investigation of how and why the experience of menopause differs even among highly modernized populations.[27] She is a leading advocate of a complex "biocultural" understanding of menopause that attributes these differences to a large number of factors, including diet, environment, lifestyle, reproductive history, genetics, and culture itself.[28] Culture is the hardest influence to define or measure, and it resists the checklist approach. Fortunately, Lock's study goes beyond this method.

The modern Japanese equivalent of "menopause" is *konenki*, a term invented in the late twentieth century, after Japan ended its self-imposed isolation in 1868 and came into contact with Europe. Relations with Germany were especially close, and German medicine became influential in Japan. At that time in Europe, menopause was an important theme in both medical literature and popular culture, and *konenki* was coined to translate that concept. Japan already had the folk idea of *chi no michi*, as well as a concept of dangerous or critical years similar to medieval and Renaissance ideas of crisis or "climacteric" in Europe—in Japan, the critical years were 42 for men and 33 for women. But neither of these ideas was a close enough equivalent to European menopause, and so the term *konenki* was introduced.

In Europe in the late nineteenth and early twentieth centuries, the notion of the "autonomic" (that is, independent or self-regulating) nervous system—named in 1898—was developing rapidly as researchers explored the relationship between the central and ganglionic nervous systems, and discovered the function of the vagus nerve. This is the system that controls our involuntary functions and our state of physiological arousal (the "fight-or-flight" response, among other things).[29] At the same time, the new science of endocrinology was also developing, and researchers postulated deep connections between these two systems—the autonomic nervous system and the hormonal system—as they explored the effects of adrenaline, for example, on autonomic arousal. The ideological linkage between the autonomic nervous system and *konenki* became very strong and enduring in Japan; late twentieth-century Japanese doctors continued to describe *konenki* largely in terms of its effects on the autonomic nervous system.

Although *konenki* was introduced to Japan to assimilate a modern European medical concept—menopause—it is not quite the same thing. When Lock interviewed 105 Japanese women between the ages of 45 and 55 in the 1980s, they described *konenki* as a stage of life beginning in one's 40s and lasting 10 years or more. While most women thought the end of menstruation was part of *konenki*, it was not central to the idea, leading Lock to distinguish women's self-reported *konenki* status from their menopause status; the latter she defined, per Western practice, by whether a woman still had regular periods, irregular periods, or no periods. Because *konenki* was not closely related to menstruation, men could have it too, although it was commonly imagined as a women's condition. Also, although every woman expected to stop menstruating, it was thought possible to be years past menstruation without having yet entered *konenki*, or to avoid *konenki* altogether.

Medical ideas that Lock's Japanese interviewees associated with *konenki* were hormone imbalance and disruption of the autonomic nervous system. Some women thought *konenki* brought gray hair, fatigue, and weak vision, connecting it with aging and its signs. Others thought of it as a collection of symptoms, as in the West, but the symptoms were a little bit different. The problems they most closely associated with *konenki* were headaches, dizziness, *katakori*—the culture-specific shoulder pain discussed by Kuriyama in the passage that begins this part of my book—and feeling irritable. They did not talk about feeling hot, sweating, or irregular periods, the topics mentioned most often by Western women when asked about menopause, although a few described sensations very similar to the Western-style hot flash; none described sweating or night sweats. Women used the term *konenki shogai* ("problems of *konenki*") and *konenki shokogun* ("menopausal syndrome"), but popularly, *konenki* signified mainly the problems and symptoms associated with it, just as women in the West commonly speak of "menopause" rather than "menopausal syndrome."

When Lock surveyed over 1,300 Japanese women aged 45 to 55 with a checklist of symptoms, she found that *katakori* was the most frequently reported symptom by far—52 percent recalled experiencing it over the past two weeks, compared to 28 percent who recalled experiencing the next most common symptom, headache. While 16 percent reported chilliness, only 9.5 percent reported heat sensations. Sweating and sexual symptoms occurred at low rates (the only sexual symptom Lock asked about, lack of sexual desire, was reported by 4.4 percent of women, which seems low for any age group). Twelve percent reported irritability, and 8 percent reported depression (a further 4 percent reported *yuutsu*, "melancholy").

To Lock, neither *konenki* nor the end of menstruation seemed to be very important to Japanese women (and 63 percent of the women she surveyed rated it as not important, or only a little important). Women's ideas and experiences of midlife centered on social factors. It was the age at which they began to be perceived as mothers-in-law or grandmothers, a transition they were ambivalent and sometimes sad about. They worried about the burden of caring for aging in-laws in the near future, while relishing what many viewed as a period of liberation before that—free of the duties of motherhood, not yet encumbered with the most significant duties of a daughter-in-law. Most thought it unlikely that their own daughters-in-law would care for them in the same way, as society changed.

As in the West, Japanese women and their doctors moralized about *konenki* and its symptoms. Housewives with a lot of leisure time on their hands, they thought, were more likely to suffer; working women did not have time to worry about their problems. They were proud of what they perceived as a Japanese resilience to menopausal symptoms, and thought that women in the United States probably suffered more because they did not work as hard and were more self-indulgent. Lock's survey showed that in Japan, farming women and working women, many of whom worked in factories for low pay, reported more symptoms like shoulder stiffness, headaches, and lumbago than housewives, a finding that contradicted popular belief but was not, after all, surprising. Lock did not try to find out whether the most common symptoms that her sample of women reported and attributed to *konenki* had a statistical correlation with menopause, and suspected that they did not.

CHAPTER 12

•

A Cultural Syndrome?

IN MY DISCUSSION of menopausal symptoms, I worry that I may seem to be indulging in an old-fashioned, late twentieth-century Pyrrhonian skepticism—a demolition of obvious facts ("menopause has symptoms") for the sole purpose of showing how complicated everything necessarily is, a type of discourse I dislike. Putting this question to myself, I have two responses. The first is that some cultural constructs—some widely pervasive ideas—really do need this treatment. Not everything that is commonly or universally believed to be true is a fact, and it is the job of researchers in some disciplines, including my own profession of history, to point this out once in a while. Many things commonly believed about menopause are in this category. Second, my purpose is not to argue that menopausal syndrome does not exist. On the contrary, I think it is a very real and important phenomenon and product of the modern world. But I am not sure that most researchers, embedded as they are in the web of understanding about menopause that I have tried to describe, are approaching it in a useful way. I think menopause is best understood as a cultural syndrome, an idea I will explain next.

It is important to pay attention to the "syndrome" part of menopausal syndrome. A syndrome is a "running together" of symptoms believed to be related, but for which the ultimate cause or mechanism is unclear—when it is clear, doctors usually call the problem a disease rather than a syndrome. In the case of menopausal syndrome, researchers investigate symptoms and correlations but normally without trying to explain why menopause might cause them (and for none of the symptoms of menopausal syndrome is a cause well understood). Syndromes are deeply interesting and mysterious phenomena. Certain types of syndromes are closely linked to culture and cognition—that is, what we believe and how we think—and these have a number of names in scientific research. "Cultural syndrome" and "functional somatic syndrome" are most common; "functional" means that the disorder is debilitating (that is, it affects function) without having an obvious "organic," anatomical cause. Because the term "functional" now carries negative connotations—the idea that a problem is "all in the patient's head"—researchers have proposed substitutes for "functional somatic syndrome," such as "psychological and social factors affecting bodily condition."[1] Likewise, "cultural syndrome"—the *DSM-5*'s reformulation of the older term "culture-bound syndrome"—can convey an air of exoticism that has also prompted the use of substitutes, such as "distress syndrome."[2] A great deal of sophisticated cross-cultural research investigates the nature of syndromes that fall into this category, and once we know enough about them, it is hard to dismiss the possibility that menopause is one of them, or at least behaves a lot like them.

Menopausal syndrome is obviously a real phenomenon from which millions of women around the world suffer. This is true of cultural syndromes as well—many have high rates of prevalence. Unfortunately, the argument that some conditions have a cognitive element—that how we think plays a role in how we feel—is often accompanied by a moralizing stance that only self-centered, idle, weak, or irrational people will suffer from those conditions or that their suffering is trivial. It has been a theme in the modern idea of menopause since its invention in the early eighteenth century that hardworking, self-sacrificing, self-controlled women suffer less. That negative expectations of menopause can influence its course, and that attention to symptoms can make them worse, may have some basis in fact, as I shall argue. But I strongly disagree with the value judgments often embedded in this stance. Cultural syndromes exist and can be difficult to treat because they result from complex interactions among the web of cultural beliefs that everyone is exposed to, higher functions of the brain that psychologists struggle to understand, and the body's physiological systems. I empathize with the suffering of women for whom menopause is hard, and although I am lucky enough not to be in that category, I have suffered from other conditions that were greatly helped by my research into cultural syndromes and by understanding how they work. One reason I wrote this book was that I thought that menopausal symptoms might be easier for readers to bear if the nature and purpose of menopause—a developmental transition to an important stage of life—were better understood. It would also not surprise me if people with this understanding experienced fewer symptoms.

To say that something is a "cultural syndrome" is not to say that it has no physiological basis. On the contrary, it is the interaction between physiology, emotions such as anxiety, and cognition—ways of thinking—that creates a cultural syndrome. Part of the variation among syndromes may derive from environmental or genetic differences between groups of people—for example, some researchers speculate that a greater vulnerability to dizziness and nausea among East Asians accounts for the importance of those symptoms in many regional syndromes[3]; influences like diet and climate are other potential sources of difference. Even small differences attributable to these factors can be amplified, by a process described in this chapter, to create culture-specific syndromes.

This last argument of my book is complicated by the problem that menopausal syndrome is so hard to pin down. It seems that the closer we look for a cluster of symptoms related to the end of reproductive life, the more tenuous that phenomenon becomes. On the other hand, it is clear that in all modernized cultures—but not in some or most traditional cultures—menopausal syndrome is widely *believed* to exist; it is an important sociocultural phenomenon that supports a large industry of drugs, advice, and research. Individual women experience menopausal syndrome, even if those experiences tend to dissipate when aggregated into statistics, and even if the search for universal (that is, cross-cultural) experiences has not gone well. If we accept that women in the modern world experience menopause in ways that overlap to some extent but also vary across cultures, then Western menopausal syndrome is a good candidate for a cultural syndrome of a certain type. But first, let me explain in more depth what a cultural syndrome is.[4]

A symptom is a sensation or experience to which a certain kind of meaning has been assigned—it is more than just a transient feeling, it portends something. We are constantly feeling and processing what one researcher calls a "white noise" of bodily sensations—aches, pains, nausea, palpitations, and so on—which increases with age.[5] Trauma, illness, or stressful life events can also increase this background noise of sensations. If we think a sensation or experience is important, it becomes a symptom.

For many kinds of symptoms, the process of registering them, assigning significance to them, and paying attention to them causes us to feel them more strongly, which in turn prompts us to pay more attention to them. If they cause us worry, embarrassment, or other anxious emotions, the effects of these emotions on the body can also make the symptoms worse, and worsening symptoms then increase anxiety in a feedback loop.

Culture partly determines which sensations we interpret as symptoms. In American culture, chest pain is important; for Cambodians, neck pain; for Chinese, dizziness. Sometimes culture-specific disease labels—like *kyol goeu* among Cambodians, discussed later—arise that organize and assign meaning to our experience of symptoms. When a patient believes his or her symptoms signify one of these diseases, this belief may speed up the feedback loop by increasing anxiety, and it may also focus attention on the symptoms thought most characteristic of the disease, amplifying those particular sensations.

Cultural syndromes are therefore disorders that do not, on the surface, make a lot of sense outside the culture that has shaped them, but that nevertheless share broad patterns with other cultural syndromes. That is, cultural syndromes can be described as a class of phenomena characterized by the kind of feedback loop that amplifies symptoms, combined with culture-specific beliefs about how the body works and what diseases exist. These disorders can be quite common, affecting millions of people and large percentages of the subgroups to which they apply. It is a feature of cultural syndromes that they have few signs— they are not easy to diagnose or explain by laboratory tests, imaging, or any other objective method, and they do not, as a rule, kill the patient, although they can make him or her very miserable.

The role of higher brain functions—cognition, emotion—in generating the symptoms of cultural syndromes is not unique to them. In all kinds of diseases, including diabetes, heart disease, arthritis, and even respiratory infections like the common cold, symptoms vary widely in ways that are independent of how severe the problem is as measured by medical tests. Subjective symptoms and objective signs do not always match up very well.[6]

There are different types of cultural syndromes. In the United States, chronic pain syndromes dominate research and discourse on what are called, when they occur in Western cultures, not cultural syndromes, but "functional somatic syndromes" (reflecting the Eurocentrism of the discipline of psychiatry).[7] Another type of cultural syndrome, one that has been the subject of a great deal of research by Devon Hinton of Harvard University and others—and I think this type is more relevant to menopausal syndrome—is characterized by anxiety and panic.[8]

Psychologists of emotions identify anxiety as a sort of low-grade, chronic version of fear that concerns something in the more remote future rather than an immediate threat. It generates some of the same physiological responses as fear, involving our sympathetic nervous system—the part of our autonomic nervous system that keeps us keyed up and prepared to meet emergencies. Symptoms related to anxiety are very numerous and include different kinds of pain (headache, joint pain), muscle tension, fatigue, stomach problems, sweating, flushing, chills, shaking, dizziness, nausea, tinnitus, numbness or tingling of the extremities, shortness of breath, feelings of choking, sleep problems, blurry or spotty vision, loss of appetite, cardiac symptoms like palpitations or racing heart, and psychological symptoms like irritability and trouble concentrating.[9] These kinds of symptoms are especially likely to escalate in feedback loops. The term "anxiety sensitivity" denotes our tendency to worry about these symptoms, which then increase as our anxiety heightens. Almost all symptoms on menopause checklists— and on Melby's list of 21 core symptoms of menopause, described earlier—are of this type.

When sensations are thought to portend dire outcomes like death, "catastrophic cognitions" are common. People in the United States, for example, greatly fear chest pain; women of Greek antiquity afflicted with the condition they called hysterical suffocation were told that they could choke to death on their own uteruses at any moment. These thoughts can generate powerful responses that increase the force and speed of the feedback loop. Small, barely perceived triggers can sometimes escalate so fast—in a matter of minutes—that they cause attacks resembling seizures.[10] In Western populations and in modern psychiatry, panic attacks are characterized by racing or pounding heart, feelings of choking or suffocation, dizziness, sweating, shaking, feelings of heat or cold, chest pain, nausea, feelings of numbness or tingling, feelings of losing control, feelings of unreality or detachment from oneself, and an overpowering sense of impending doom (patients usually do not experience all of these, but a combination of four or more is required for a *DSM-5* diagnosis of Panic Attack). But similar attacks, spells, and fits— with some variations—are also typical of many non-Western cultural syndromes; Panic Disorder as diagnosed in the West is one form of a wider phenomenon.

Hinton has treated Southeast Asian refugees in his clinical practice over many years and is fluent in Khmer, the dominant language of Cambodia, and much of his work on cultural syndromes has focused on this population. In two lengthy, groundbreaking articles on the condition called, in Khmer, *kyol goeu,* he mapped out the most plausible theory of how cultural syndromes work—the theory described here—and set the standard for research into these conditions. Hinton has proposed a broad, 10-point approach to investigating cultural syndromes that not only describes their symptoms but also addresses the biological, social, and cultural factors that may influence their shape.[11]

Thirty-six percent of the Khmer patients Hinton saw in his psychiatry clinics said they had experienced a full-blown episode of the condition they called *kyol goeu,* and 60 percent said they had suffered an episode of near-*kyol goeu* in the past six months. *Kyol goeu* means "wind overload." In Khmer physiology, wind continually passes through the body's vessels, exiting through the hands and feet. If it becomes stuck in the joints or in the neck, so much pressure can build up that the wind explodes out of the vessels, killing the patient, or it can compress the heart so that circulation fails or even stops, causing even more wind blockage. Joint pain or neck pain can be a sign of stuck wind; dizziness, too, can indicate a buildup of wind in the head. Other signs of wind buildup are nausea, palpitations, sweating, and cold or numb hands and feet. Worry, poor appetite, and sleep problems can all cause sluggish or blocked wind circulation. Hinton's Khmer patients interpreted fainting and collapse during the Pol Pot period as *kyol goeu* events caused by the starvation and hard labor they endured. Sometimes people died during these episodes, which greatly increased the anxiety that Hinton's patients felt about *kyol goeu*.

Because of the horrific circumstances that brought them to the United States, almost all of Hinton's Cambodian patients met criteria for Post-Traumatic Stress Disorder, which increased the rate at which they experienced symptoms of autonomic nervous system arousal like pain, dizziness, and blurred vision; also, as a result of their experiences with starvation, malaria, and overwork, Hinton's patients were highly sensitive to cues like dizziness and abdominal pain. Dizziness upon standing up frequently triggered attacks of *kyol goeu*. In a typical attack, the patient feels a sense of wind rising from the abdomen, shakes, feels dizzy, feels unable to breathe, has blurred vision, feels wind shooting out of his or her ears, and may experience other sensations like palpitations or chilling of the hands and feet. Patients often feel faint and fall to the ground, unable to move or speak, and sometimes lose consciousness. Family members rush to stimulate wind through the blockage by pounding, massaging, biting ankles, and "coining," a practice of rolling coins down the arms and legs. If these measures are not taken right away, the belief is that the patient will die.

Kyol goeu is mainly a syndrome of sudden attacks, which is typical of many anxiety-type cultural syndromes. But other syndromes are perceived more as chronic states—for example, the syndrome *hwa-byung*, which is very common among middle-aged Korean women and typically lasts 10 years or longer. A recent survey found that 44 percent of a nonclinical sample of South Korean women over 40 met criteria for *hwa-byung*.[12] Originally a folk illness not recognized by modern medicine or classical Korean medicine—which is deeply influenced by classical Chinese medicine—it was first formally described by Korean psychiatrists in the 1970s and is a widespread medical diagnosis today.

While men and younger women can get *hwa-byung*, and the famous eighteenth-century Korean king Chŏngjo is supposed to have died of it,[13] it is perceived as a disorder that mainly affects middle-aged women. The typical *hwa-byung* patient is a woman over 40 of low socioeconomic status and little education, living in a traditional household. Some researchers have argued that *hwa-byung* patients are more likely than non-sufferers to have egalitarian attitudes—that is, although *hwa-byung* is commonly understood to result from the oppression of women in patriarchal, traditional families, it is the conflict between attitudes and circumstances, and not only patriarchal values themselves, that causes the problem.[14]

Hwa-byung is understood as a profoundly Korean disease, linked to national identity and national history, although many psychiatrists have investigated its relationship to the Western diagnoses of Major Depressive Disorder, Generalized Anxiety Disorder, and Panic Disorder. Classically, it arises from anger. Descriptions often mention the emotion called *haan*, a complex mixture of sadness, hatred, and desire for revenge arising in response to unfair situations that persist over a long period; the word *haan* occurs often in connection with Korea's national history of suffering and violent oppression.[15] When anger is suppressed for a long time for the sake of social harmony, it builds up and can cause a mass to form in the abdomen (*hwa* can mean "mass of heat" as well as "anger" and "fire"). The mass in turn causes many symptoms, in particular gastrointestinal symptoms. Researchers also describe behavioral symptoms such as crying, impulsiveness, and talkative pleading or complaining, and a large number of physical symptoms such as headache, fatigue, palpitations, sensations of intense heat, tightness in the chest, dry mouth, insomnia, indigestion, dizziness, nausea, loss of libido, constipation, blurred vision, and cold sweats. The feelings of heat that are a striking feature of *hwa-byung* are believed to result from the anger or fire that is its root cause. As precipitating factors, Korean patients and physicians point to financial problems; problems with husbands, in-laws, or adult children; traumatic events; and rigid gender roles.[16]

Hwa-byung is normally a chronic disorder, but patients may also have acute panic-like attacks with symptoms like heat sensations, palpitations, tightness in the chest, and a sense of losing control,[17] and these paroxysmal episodes are part of the popular image of the syndrome. In a 2013 episode of the health-based Korean reality show *Vitamin* featuring four middle-aged women with *hwa-byung*, a central scene is a confrontation after a car accident, in which one of the women collapses in a dead faint while arguing with the chauvinistic male driver of the other car.[18]

How is *hwa-byung* related to menopause? In a qualitative study of Korean immigrant women in an anonymous city in the United States published in 2000, the 21 women interviewed did not seem to have adopted a Western view of menopausal syndrome. They answered questions about menopause using the word *gangyunki* (a term that, although likely an adaptation of the Chinese *gengnianqi* or Japanese *konenki*, is given a Korean etymology by the authors of the study), which referred, like *konenki*, not to the end of menstruation but to a period of life experienced by both men and women during which they begin to age. These women called the end of menstruation *pekyungki*, which was thought to occur along with other signs of aging—gray hair, fatigue, wrinkles—as a part of *gangyunki*, but not to cause symptoms itself; the women interviewed did not associate pathological changes with *gangyunki*. Some women who had no sons, or only one son, regretted the loss of fertility that came with *pekyungki*, which they also associated with the end of sexuality, as they believed that after menopause it was healthier to abstain from sex. The study's authors seemed frustrated that their interviewees did not consider menopause very important compared to their other problems, and state more than once their belief that the women "ignored or neglected" their menopausal symptoms, reflecting a stance and set of assumptions typical of Western research on menopause.[19]

When surveyed using the checklist method, majorities of middle-aged Korean immigrants to the United States and of women still living in Korea endorse one or (usually) more of the "symptoms of menopause" included on the checklists; in a comparative study, women living in South Korea reported many more of these symptoms than Korean immigrant women. This latter study's authors attributed their counterintuitive result to the popularization of Western menopausal syndrome in South Korea, from which Korean immigrants in the United States were more insulated.[20] That is, the idea of a menopausal syndrome became part of Korean culture under the influence of Western medicine in the late twentieth century.

Even though the ideas of *hwa-byung* and of Western menopausal syndrome escalated in popularity at about the same time in Korea, the two concepts have not merged. That is, *hwa-byung* is a disorder of middle-aged women, with many of the same symptoms attributed to menopause by modern medicine, including heat sensations, but it is not menopausal syndrome. None of the psychiatric publications I have read has suggested that it has any relationship to menopause, and none of the literature on menopause in Korea published in Western languages mentions *hwa-byung* despite its wide prevalence among middle-aged women. For example, one article about *hwa-byung* in a psychiatric journal describes a 49-year-old woman with self-diagnosed *hwa-byung* without interpreting her sensations of heat as menopausal hot flashes or her feelings of anger as hormonal "irritability," as most of the researchers cited in this part of my book would certainly do:

> The patient was a 49-year-old housewife. She came to us with the chief complaint of pent-up anger, "hwa," which was intermittently accompanied by a hot sensation, that had to be cooled by a fan, and something pushing up in her chest. The other symptoms were "many things accumulated" on the epigastirum [sic] and respiratory stuffiness that used to be relieved by frequent sighing. Sometimes she used to feel so angry and so "ukwool" (a feeling of unfairness) that she felt almost like loosing [sic] control or becoming crazy. Her self-diagnosis was hwa-byung. The reason for her anger was her family situation with her husband and mother-in-law.[21]

The author, Min Sung Kil, one of Korea's leading researchers on *hwa-byung* and the person who has perhaps done the most to define its place in modern psychiatry, continues with a detailed description of the patient's history and family situation going back to childhood and an exegesis of her emotions, without reference to her menopausal status.

The story of *hwa-byung* in traditional Korean medicine is a bit different. Once psychiatrists began describing the syndrome in the late twentieth century, Korean traditional practitioners also entered the competition to treat it, adapting classical Korean medicine to incorporate *hwa-byung*. The explanation of Kim Jong-Woo, a traditional practitioner whose theories have been influential, links *hwa-byung* indirectly to the theory of menopause developed in Traditional Chinese Medicine, in which menopausal syndrome is caused by a depletion of kidney *qi* that also causes other symptoms. In Kim's theory, anger causes liver *qi* to become weakened, and ultimately this substance stagnates and transforms into fire, causing symptoms of *hwa-byung*. The weakened kidney essence typical of menopause contributes by making *qi* more labile and susceptible to agitation.[22] But for the most part, menopause is not part of the discourse about *hwa-byung*; patients and physicians locate its causes in social problems and in emotions. Even for those scholars who interpret *hwa-byung* in the context of Western medicine, ideas of stress and depression are much more important than the notion of menopause, which is almost absent from that conversation. It might seem strange to my readers not to connect *hwa-byung* to hormonal menopause, but based on this example, one could equally argue that it is strange not to invoke culture, social problems, and emotional responses in explanations of menopausal syndrome.

Cultural syndromes are more than the sum of their parts, and this is a point that is simultaneously important to bear in mind and very difficult to articulate clearly. Based only on their symptoms, many Western women would meet criteria for *hwa-byung*, just as many Korean patients with *hwa-byung* meet criteria for *DSM* disorders or Western syndromes like fibromyalgia or chronic fatigue syndrome. It would be wrong, however, to conclude that *hwa-byung* is somehow not a real thing. It has features in common with other anxiety-type disorders, reflecting a basic human physical and mental response to distress. But the more specific features that make *hwa-byung* a compelling force in the lives of so many Korean women, and that partly account for its prevalence, are not easily captured by the checklist method. I think this same logic applies to menopause too. Women in cultures without a word for menopause or a concept of menopausal syndrome, like the Maya or the women of Rabat, may still report symptoms, when asked about them, that Western researchers associate with menopause; some of those symptoms may even correlate with menopause in those populations, as hot flashes do in Rabat. If menopause causes hot flashes in women all over the world— and it is the symptom of menopause with the best claim to this type of universality—that is important to know, but this finding does not invalidate the argument that menopausal syndrome more broadly is an artifact specific to modern Western culture, a perspective that has important implications for how it should be assessed and treated.

Sometimes the idea of a disorder is so pervasive that it becomes a more general "idiom of distress," a metaphor or way of talking or complaining. *Hwa-byung* can have this character in Korea, and some Korean traditional practitioners and popularizing authors equate it with "stress," a common idiom of distress in Western cultures.[23] Westerners also talk casually of "depression" and "being depressed" even as psychiatry recognizes clinical depressive disorders, a usage that can sometimes be confusing when it bleeds into scientific discourse. Menopause, too, seems to function as an "idiom of distress" in the West, invoked in a general sort of way to explain common physical and psychological problems.

I have been describing anxiety-type cultural syndromes because their symptoms are so strikingly similar to those of menopausal syndrome. Is anxiety—or some similar emotion—an important component, then, of menopausal syndrome? In a world influenced by modern medicine—in turn influenced by Western culture's 300-year-old love affair with menopausal syndrome—women experiencing sensations related to menopause might worry that they suffer from a deficiency of a substance critical to their health and gender identity (estrogen); that they are experiencing a loss of youth, femininity, and health; that they are the subjects of social ridicule; that they have an increased risk of health problems ranging from osteoporosis to heart attack; that they will suffer from a loss of emotional control; that they are useless in an existential sense; and they might have a lot of other thoughts that could reasonably cause anxiety. This anxiety might increase if they believe that they now face a difficult, high-stakes decision about medication, balancing risks and benefits about which a great deal of highly technical, contradictory research has been published and about which experts disagree. Even if patients are not fully aware of their anxiety and if its effects are subtle, it could contribute to the escalation of menopausal symptoms, which are mostly of the type easily amplified by anxiety, and to which the culture of menopausal syndrome relentlessly draws their attention.

I do not think that most women have catastrophic cognitions about menopause—fortunately, today's medical and popular literature about menopause stops short of the claim that it will kill you, at least not right away. Despite this, however, I think the role of hot flashes in menopausal syndrome and their relationship to panic attacks deserves further comment. Hot flashes are not the same as panic attacks (fear is more important in panic attacks, and sweating is more important in hot flashes), but they are not that different, either—both are sudden events of short duration, and many of their symptoms, such as heat sensations, sweating, shortness of breath, palpitations, and dizziness, overlap. There are also similarities in what we know about the physiology and biochemistry of the two processes, and they have been found in imaging studies to affect some of the same areas of the brain.[24] Many people experiencing a hot flash as described by Kronenberg or Freedman would technically meet *DSM-5* criteria for Panic Attack. So, while we do not understand the mechanisms of either the hot flash or the panic attack very well, it is possible that some of the same psychological factors that precipitate panic attacks can influence hot flashes too, and in particular can make women feel them more intensely and suffer from them more.[25]

To sum up, menopausal syndrome has many characteristics in common with cultural syndromes. The symptoms are similar—they are nonspecific symptoms of the type made worse by anxious arousal and easily amplified by attention and repeated experience. Menopausal syndrome has deep historical connections to other Western cultural syndromes, like hysteria and hypochondriasis. It affects women differently in different parts of the world, and not all cultures believe in, or experience, a menopausal syndrome. Like many cultural syndromes, it features sudden, brief, recurrent episodes of dramatically elevated symptoms of autonomic nervous system arousal, especially heat, flushing, sweating, and chills. There are many social factors—the availability of a profitable treatment, the widespread notion of estrogen deficiency,

the uncertain status of middle-aged women in a modernizing world—
that work to support the idea of a menopausal syndrome in different
ways. It is a concept familiar to virtually everyone in highly industrialized societies, even where knowledge of modern medicine is limited to
the basics.

I am making a suggestive argument based on menopausal syndrome's
resemblance to other cultural syndromes, and I do not wish to press it
too hard. There is some indirect support for it in recent research, however. Hot flashes and other menopausal symptoms are more serious in
women who also suffer from anxiety or who have high anxiety sensitivity, as well as in those who expect menopause to cause problems, or who
find hot flashes upsetting or embarrassing. Placebo response rates are
high, especially for the therapy most congruent with popular belief about
the cause of menopausal symptoms (hormone therapy). Symptoms tend
to improve with some types of cognitive behavioral therapy, mindful-
ness, and relaxation-based therapy.[26] All of these factors need further
study, and while scientists are investigating cognitive models of menopause, not much research in this area has been published yet.[27] So, my
theory should not be taken too seriously. The most important point
I wish to make is not that menopause *is* a cultural syndrome—which is,
after all, just a made-up name for a certain type of phenomenon—but
that, for some purposes, it is useful to see it as one. With a deep understanding of its historical roots and its relationship to ideas of aging,
midlife, and how the body works (or perhaps even with a shallow one),
we can make better sense of the statistics and checklists produced by
conventional approaches, and perhaps overcome some of the frustra-
tions that have resulted from a focus on symptoms and hormones.

I expect the accusation of being just another dismissive voice saying
that the symptoms of menopause are all in your head. I hope I have
conveyed how far this is from my meaning. The separation between
brain and body that the expression "all in your head" implies is, in the
first place, silly. I think it is likely that hormonal changes cause heat and
other sensations in many women, which are experienced and interpreted differently in different contexts, and I think it is a plausible guess

that the steep decline of fertility in the modern world has made these sensations more noticeable. The symptoms experienced by women going through menopause are real in every sense. But society has had a hand in constructing them for us. This may be hard to believe, and it is only with reservations, and against my better judgment, that I venture an argument that may alienate much of my audience, whose experiences signify an intense and visceral organic condition in which thoughts and beliefs play no role. But I think the evidence and the parallels are compelling. Nothing I say implies that the symptoms of menopause are unserious or the patient's fault—how we interpret bodily sensations is largely an unconscious process shaped by external influences and personal life history. But how we think about menopause is important, and as my research on this book has progressed I have become increasingly impatient with how it is talked about in my society and how this talk unnecessarily demeans a transition and a stage of life that are, by any measure, useful and honorable.

Cultural syndromes are important phenomena in their own right— they themselves, independent of the social factors that create them, can inflict suffering, shape behavior, and influence institutions. They seem to be significant forces in many or most populations. But they also tend to serve certain specific functions that menopausal syndrome might also serve. Syndromes persist not only because they tap into biopsychological feedback loops, because giant corporations can profit from them, or because people who hold power can use them to their advantage, but because they benefit patients in some ways. In particular, that cultural syndromes can be "weapons of the weak"—that is, people who have little power can invoke them to get relief or bring about change in their lives, or simply as a socially acceptable way to complain—is an observation that occurs more than once in the published research. In a groundbreaking paper of 1982, the famous medical anthropologist Arthur Kleinman described the ways in which patients in Hunan Province, China, used a diagnosis of neurasthenia to request job transfers to cities closer to their families, or other accommodations at work or at home. In the case of *hwa-byung*, which is believed to arise from unfair situations, common therapies include listening sympathetically or actually fixing the problem that is bothering the patient. In the Japanese fishing village studied by Rosenberger in the 1980s, some of the women having problems with *chi no michi* used it as leverage, reminding everyone of the sacrifices that led to their condition, to get their children and daughters-in-law to do what they wanted. I think that menopause sometimes plays a similar role in modern Western societies.[28]

Epilogue

GOOD-BYE TO ALL THAT

•

LATER IN LIFE, Simone de Beauvoir changed her mind about growing older. Through middle age, much of her life had been overshadowed by the dread of aging, death, and the loss of sexual attractiveness, and thus of the relationships so important to her—excruciating themes throughout her work. The third volume of her memoir, *Force of Circumstance*, ends in 1962, when she was 54, on a note of paralyzing bitterness. At that age she still believed, as she had all her life, that women past 40 were repulsive, sexless harridans too old to take pleasure in almost anything; thus her famous renunciation:

> Never again! It is not I who am saying good-bye to all those things I once enjoyed, it is they who are leaving me; the mountain paths disdain my feet. Never again shall I collapse, drunk with fatigue into the smell of hay. Never again shall I slide down through solitary morning snows. Never again a man.[1]

It is all the more remarkable that 10 years later, she concluded that old age wasn't so bad after all. Among other things, she had found romantic happiness with a new partner, the young Sylvie de Bon. "I was mistaken . . . in the outline of my future," she wrote in the last volume of her memoir, adding, in what must count as a strikingly optimistic statement from such a depressive writer, "It has been far less sombre than I had foreseen."[2] In *La vieillesse* (*Old Age*), published in 1970, she criticized a modern society that marginalized the aged, a problem to which her own previous works had unfortunately contributed.[3] While her picture of old age in that book is grim, she blames capitalism, materialism, consumerism, and other features of modern culture for the suffering and degradation she describes.

In retrospect, de Beauvoir believed that she had crossed a "frontier" in her early 50s—she uses the word several times—on the other side of which she found peace and satisfaction; indeed, in her 60s, she did not feel that she had changed much since then.[4] If she felt something missing—a sense of progress, or narrative drive in her life[5]—her appreciation of culture, travel, music, food, and life in general all deepened. She experienced a transition to a new stage of life, one characterized mostly by a new way of seeing and defining herself—a transition she does not, in these later writings, call "menopause," perhaps still finding it difficult to apply this word to herself. But in crossing this frontier, and in seeing herself differently afterward, she reflects the historical experience of a large number of women. Midlife was a social frontier long before it acquired the peculiar modern characteristics of "menopausal syndrome."

While doing research for this book, I was struck by a tendency of modern surveys on "attitudes toward menopause" to report that the views of younger women on this stage of life are more negative than those of women who have been through it. The attitudes of the latter tend to be neutral or positive, despite all of the cultural influences described in this last part of my book. Maybe the reason for this is simple: modern society is wrong, and menopause is, after all, a good thing. In our prehistoric past, as lifespans lengthened beyond those of our closest relatives among the great apes, the *usefulness* of the post-reproductive life stage that opened up for women preserved it from an obsolescence that would certainly have occurred had it not been so advantageous. Grandmothers liberated from childbearing gathered food, cared for children, taught skills, shared experiences, and performed all kinds of work for which everyone was grateful. With their help, their daughters and other young women bore more children than they could support on their own and spaced them closer together. With their help, human populations exploited every temporary ecological advantage to boom and spread. Women's transition to post-reproductive life possibly caused them physical discomforts that went mostly unnoticed and unremarked before the modern period, but had it caused serious health problems, it would not have evolved. It is not impossible that in the conditions that prevailed before the Industrial Revolution and the invention of modern medicine, older women without a lot of estrogen circulating in their systems were healthier than those whose reproductive lives were longer.

After humans turned to agriculture and began living in family-centered peasant communities, post-reproductive women dominated the household. Men represented them to the wider community—village, clan, state—and monopolized the public sphere, but at home, older women unencumbered by nursing infants ran the small farms on which their family's survival, and that of their society, depended. Only with the benefit of a life stage, for half the population, of high *productivity* and zero *reproductivity* could the peasant economy work at all. Why menopause acquired its reputation as an ordeal and a health hazard in Europe after about 1700 is harder to explain, and I am not sure that, even now, I know the whole answer to that question or have done it justice in these pages. But it is a view that needs scrutiny and reconsideration.

I have not so far said anything about my own experience of menopause, but here, at the end of my book, is perhaps the place for it. I worry that it will annoy readers to hear that there isn't much to describe. Because I had an IUD that suppressed menstruation, I had no periods to track or to miss. I didn't notice any symptoms not attributable to something else. For a few weeks, I would become hot at night, maybe because of menopause, or maybe because I was overdressed for a not very cold winter. My hair is turning gray, my skin is wrinkling, and some of my joints ache, but I am in my 50s, and it would be strange if that were not happening. I struggle to maintain my weight, as we all do, but in the big picture, is needing only a little food really a bad thing? I have a good sex life, which I intend to hold on to, and learning more about the history of "sexual symptoms" in menopause has made me less apprehensive about whatever changes might be in my future—among other things, I no longer think of sex as the basis of my marriage. My family might say that I'm "irritable," but that is just my personality. I have been "irritable" for decades.

A lot of things have happened in my life over the last 10 years or so. I got divorced. My children became teenagers. I bought a house, sold it, and bought a hobby farm in a rural area. I suffered a serious nerve injury and had a hip replaced. I learned to climb rocks. I tried unsuccessfully to become department chair and lost a humiliating election. I got remarried. My son left for college. I could tell a story about buying a used lawnmower that would make you guffaw in disbelief. All of these everyday changes affected me deeply; some made me sick with anxiety, some caused me pain, and some kept me awake at night. Some still make my hands go clammy and my heart start to pound when I think about them.

Ten years ago, or even five, I might have expected menopause to be another of these stress-inducing challenges. Now—and I am not trying to tell anyone else what to think, but only sharing my own experience—I can't imagine adding it to my list of worries. Midlife is, and has always been, about relationships—about the roles we play in the community and in the family, the sacrifices we make, the experience we bring to bear. We become non-reproductive so that we can do other things. The transition is important, but not because of the symptoms it may or may not cause us to suffer. It is important on a much larger scale, and to reduce it to a medical condition is to trivialize it. The apprehension about menopause, the embarrassment, the tiresome jokes, the judgment and hostility aimed at older women in Western culture and in other cultures, today and in the past, are all unnecessary, but menopause is necessary. Humans have menopause because we need it. The contributions of post-reproductive women have brought us this far and will lead us into whatever future we have.

NOTES

•

PROLOGUE: THE GRANDMOTHER OF US ALL

1. For Hoelun's story, I follow de Rachewiltz's edition and commentary on *The Secret History of the Mongols* (2004–2013). The material is legendary and perhaps not strictly true. Some versions of the tradition, not represented in the *Secret History*, hold that Hoelun remarried after Yisügei's death (de Rachewiltz 2004–2013, 339).

2. Zerjal et al. 2003. In 2018, when this manuscript was in its final stages, a new study, Wei et al. 2018, challenged the claim that the genetic markers identified in 2003 traced back to Genghis Khan. I am retaining the original story because it is likely that geneticists will continue to debate the origins of the markers for some time. The point that Hoelun has a very large number of descendants today is certainly correct.

CHAPTER 1: WHY MENOPAUSE?

1. On chimpanzee longevity, see Levitis and Lackey 2011, fig. 3; Levitis, Burger, and Lackey 2013, 68–69, box 1; Robson and Wood 2008, 398–399; Emery Thompson et al. 2007.

2. Cant and Johnstone 2008, 5332–5333.

3. See Hill et al. 2011; Marlowe 2004. I also discuss this topic further in chapter 2.

4. Wood 1994, 441–443.

5. Gosden 1985, 95.

6. Uematsu et al. 2010.

7. Levitis and Lackey 2011; Levitis, Burger, and Lackey 2013.

8. Gombe, Tanzania: Goodall 1986, ch. 5. Taï Forest, Ivory Coast: Boesch and Boesch-Achermann 2000, chs. 2–4. Mahale Mountains, Tanzania: Nishida et al. 2003. Bossou, Guinea: Sugiyama 2004. Synthetic: Emery Thompson et al. 2007; Levitis and Lackey 2011; Levitis, Burger, and Lackey 2013.

9. E.g. Littleton 2005.

10. Emery Thompson et al. 2007.

11. Levitis, Burger, and Lackey 2013, 68.

12. Wood et al. 2017.

13. Wood et al. have not published their findings on post-reproductive survivorship as of this writing, but according to the life table published, PrR should be about 0.1.

14. Levitis, Burger, and Lackey 2013; Vinicius, Mace, and Migliano 2014; Gurven and Kaplan 2007, fig. 3.

15. Cayo Santiago: Rawlins and Kessler 1986. Florida Keys: Johnson and Kapsalis 1998. Arashiyama: Fedigan and Asquith 1991; Leca, Huffman, and Vasey 2012.

16. Chalmers et al. 2012; see also Pavleka and Fedigan 2012.

17. A study of Rhesus macaques in the Florida Keys (Johnson and Kapsalis 1998) reached similar conclusions, although data were not as exhaustive as for the Arashiyama population, and because several of the oldest monkeys died after a mass relocation in 1995, the researchers could not assess fertility and mortality in females past 25 years of age. See also Cohen 2004, 738; Levitis, Burger, and Lackey 2013, 68.

18. Levitis, Burger, and Lackey 2013, 68.

19. For a more recent review of post-reproductive lifespan across species, see Croft et al. 2015.

20. Lions and baboons: Packer, Tatar, and Collins 1998. Polar bears: Ramsay and Stirling 1988. Ground squirrels: Broussard et al. 2003.

21. Primates: Alberts et al. 2013. On lemurs, see also Wright et al. 2008.

22. On the post-reproductive life stage in whales, see recently Brent et al. 2015; Foster et al. 2012; Johnstone and Cant 2010; Foote 2008; Whitehead and Mann 2000; Croft et al. 2015, 2017. Some dogs, domestic rabbits, and certain strains of laboratory rodents have long post-reproductive lives; see Cohen 2004, table 1.

23. Olesiuk, Bigg, and Ellis 1990.

24. Orcas have never been hunted commercially on a large scale, but many were shot by fishermen and recreational hunters in the decades before the study; since the early 1970s, the species has been protected in Canada and the United States. Dozens of orcas in both communities were apparently captured and removed for aquarium exhibits between 1962 and 1977, but nevertheless both populations grew during the period of the study. This "cropping" by live-capture fisheries affected all the pods of the southern community but only one of the northern pods, which allowed the researchers to check their demographic data by comparing them with data from uncropped pods only.

25. Croft et al. 2015.

26. Baird 2000; Ford 2009.

27. Foster et al. 2012; Croft et al. 2015.

28. Brent et al. 2015.

29. "Orca Calf Shows Signs of Whale Midwifery," *Morning Edition*, Valley Public Radio, NPR, January 3, 2015, http://kvpr.org/post/orca-calf-shows-signs-orca-midwifery.

30. Johnstone and Cant 2010; Croft et al. 2017.

31. Kasuya and Marsh 1984.

32. Croft et al. 2015 calculate PrR at 0.28.

33. The collection of articles in Moss, Croze, and Lee 2011 is highly recommended.

34. Lahdenperä, Mar, and Lumaa 2014.

35. McComb et al. 2001, 2011; Foley, Pettorelli, and Foley 2008; Lahdenperä, Mah, and Lummaa 2016. See also Mutinda, Poole, and Moss 2011 on elephant leadership: older matriarchs are dominant in multifamily groups, and groups follow the oldest matriarch.

36. Mizroch 1981. See also Aguilar and Borrell 1988, 205.

37. Vom Saal, Finch, and Nelson 1994, 1246.

38. On the physiology of reproductive senescence, the standard references are Gosden 1985 and vom Saal, Finch, and Nelson 1994; see also Gosden and Faddy 1998; Wallace and Kelsey 2010. Power curve: Hansen et al. 2008; Knowlton et al. 2014. ADC: Wallace and Kelsey 2010. Average follicle numbers in humans are from Wallace and Kelsey 2010, which synthesizes the results of all histological studies to that date.

39. The theory of postnatal oocyte generation in mammals became more prominent with Johnson et al.'s 2004 publication in *Nature* of the results of experiments on laboratory mice, but the history of this idea is much longer. For a recent review, see Woods and Tilly 2012.

40. Gosden and Telfer 1987.

41. For a recent review on the genetics of age at menopause, see He and Murabito 2014. Estimates of the heritability of age at menopause range from 44 percent to 65 percent (He and Murabito 2014, 768). On variation in age at menopause and at last reproduction, see Hawkes, Smith, and Robson 2009.

42. The main pleiotropic argument besides the Patriarch Hypothesis is Wood et al. 2001.

43. Medawar 1952.

44. Williams 1957.

45. Hamilton 1966.

46. Kirkwood 1977; Kirkwood and Holliday 1979; Kirkwood and Shanley 2010.

47. Peccei 2001.

CHAPTER 2: "THANK YOU, GRANDMA, FOR HUMAN NATURE": THE GRANDMOTHER HYPOTHESIS

1. Hamilton 1966.

2. Hamilton 1966, 37.

3. Hawkes, O'Connell, and Blurton Jones 1989; Hawkes et al. 1998; O'Connell, Hawkes, and Blurton Jones 1999; Hawkes 2003; Hawkes and Paine 2006; Kim, Coxworth, and Hawkes 2012; Hawkes and Coxworth 2013; Kim, McQueen, et al. 2014; Blurton Jones 2016.

4. Charnov 1993.

5. Kaplan et al. 2000, table 1; Robson, van Schaik, and Hawkes 2006, table 2.1.

6. Hrdy 2009 (ch. 8 on grandparenting). On the role of allocare in the evolution of large brains, see also the argument of van Schaik and Burkart 2011; Isler and van Schaik 2012a, 2012b.

7. Kaplan et al. 2010.

8. Wells and Stock 2007; Potts 1996, 2012; Brooke 2014, ch. 2; Antón, Potts, and Aiello 2014. A good introduction to paleoclimatology for nonexperts is Alley 2014. See also Ruddiman 2005, ch. 12, on this point.

9. It is possible that the unusual stability of the Holocene's climate is not a coincidence—that as humans transformed the world's surface after the agricultural revolution, burning forests and irrigating crops, emissions of methane and carbon prolonged a warm interglacial period that would otherwise have ended long ago, and temporarily stalled the wrecking ball at one point in its arc. This is the thesis of Ruddiman 2005, and it is becoming scientific consensus.

10. Wells 2012a.

11. Wells 2010, 2012b.

12. Hill and Hurtado 1996; Blurton Jones 2016, ch. 11.

13. Boone 2002.

14. Boone 2002.

15. Reich et al. 2012; Rasmussen et al. 2014; Llamas et al. 2016; Reich 2018, ch. 7. This simplistic interpretation remains mostly true—genetic evidence reflects the dominant influence of a single founder population for all indigenous peoples south of northern Canada—but evidence that some other populations were displaced by the group from Beringia, or that this group was itself deeply structured, is painting a fascinating (not to say baffling) picture of the settlement of the Americas. See Skoglund et al. 2015; Moreno-Mayar et al. 2018.

16. Gurven et al. 2012.

17. Chu and Lee 2013.

18. Levitis, Burger, and Lackey 2013, 77; O'Connell, Hawkes, and Blurton Jones 1999, 467–468.

19. Hawkes, O'Connell, and Blurton Jones 1997; for the most updated conclusions about grandmothering based on the Hadza, see Blurton Jones 2016, chs. 18–19.

20. Blurton Jones 2016, 367–369.

21. Blurton Jones 2016, fig. 18.1.

22. The Gambia: Sear, Mace, and McGregor 2000; Sear et al. 2002. Finland and Canada: Lahdenperä et al. 2004.

23. See also Winking and Gurven 2011.

24. Fox et al. 2010.

25. Douglass and McGadney-Douglass 2008.

26. Gibson and Mace 2005.

27. Ache: Hill and Hurtado 1996, 162. Hiwi: Hill, Hurtado, and Walker 2007, 449. Hadza: Blurton Jones 2016, 141–142. On rates of maternal mortality in preindustrial populations, see also Mace 2000, 4. Contrary to the assertions of many researchers, Rogers 1993 does not exclude the possibility that menopause is adaptive. He shows that the risk of death in childbirth is unlikely to account for this phenomenon but does not rule out the influence of "opportunity costs," or reduced productivity in the years after the birth of a new child. Pavard, Metcalf, and Heyer 2008 have reevaluated the case for maternal mortality and the Mother Hypothesis, pointing out that rates of death in childbirth, stillbirth, and birth defects are much higher for older women and, without menopause, could continue to increase cumulatively or exponentially; they conclude that this could favor menopause. Their calculations would be undermined if their estimates of increased risk at advanced ages are too high or if, in extending reproductive life, natural selection also extended the viability of oocytes and altered physiological factors that make childbirth more dangerous for older women. Peccei (1995 and 2001) has also supported the Mother Hypothesis.

28. Sear et al. 2002.

29. Sear, Mace, and McGregor 2003.

30. Hill and Hurtado 1991.

31. Lee 2008.

32. Chu, Chien, and Lee 2008.

33. Around the same time that Lee and Chu were developing their theory of transfers, Michael Gurven and Hillard Kaplan began publishing their own similar models of the evolution of post-reproductive life. These are discussed further in the section on "Embodied Capital."

34. Pavard and Branger 2012.

35. Kachel, Premo, and Hublin 2011a.

36. Shanley and Kirkwood 2001.

37. Kim, Coxworth, and Hawkes 2012; Kim, McQueen, et al. 2014.

38. Chan, Hawkes, and Kim 2016.

39. Cf. Coxworth et al. 2015. On human "operational sex ratio," menopause, and factors that may mitigate competition among men for women, see Marlowe and Berbesque 2012.

40. Jones and Bliege Bird 2014.

41. Cant and Johnstone 2008; Croft et al. 2015, 2017.

42. Contrast Lahdenperä et al. 2012, which showed very dramatic declines in child survivorship in preindustrial Finland when two generations of in-laws overlapped in reproduction, with Skjærvø and Røskaft 2013, which showed a modest fitness benefit to overlapping reproduction in preindustrial Norway.

43. Mace and Alvergne 2012.

44. "Grandmother abstinence" (also called "terminal abstinence"): see, for example, Tan 1983; Wood 1994, ch. 1; Menon and Shweder 1998, 69; Bledsoe 2002, 278.

45. Galbarczyk and Jasienska 2013.

46. The classic model is Smith and Fretwell 1974. For a discussion as applied to humans, see Walker et al. 2008.

47. In support of tradeoffs, see recently Hayward, Nenko, and Lummaa 2015 (a study of data from eighteenth- and nineteenth-century Finland). For a critical review of the evidence and discussion of confounding factors, see Gagnon 2015.

48. Gurven et al. 2016.

49. Penn and Smith 2007.

50. Lawson, Alvergne, and Gibson 2012.

51. Borgerhoff Mulder 2007.

52. Borgerhoff Mulder 2007.

53. Sear 2008.

54. See Voland and Beise 2002, 2005.

55. Blurton Jones 2016, 383–400.

56. Gopnik 2014.

CHAPTER 3: PUTTING THE "MEN" IN MENOPAUSE: MALE-CENTERED THEORIES OF HUMAN EVOLUTION

1. Lee and Devore 1969. Paradoxically, many of the papers at this conference challenged the accepted theory, which began to lose credence after the publication of the proceedings; in particular, *Man the Hunter* revealed the importance of plant foods to the diet of most foragers.

2. Washburn 1960; Washburn and Lancaster 1969; Washburn and Moore 1974. Lovejoy 1981 is the seminal article on nuclear family formation.

3. The debate on the role of hunting in the development of foraging economies and in human evolution is intense, as is the closely related debate on pair bonding and parenting. The following is a very selective list of the sources that have been most influential in these debates or that

I have personally found most helpful in composing this section of the book. For what it is worth, my impression is that these debates' highly polemical language obscures the frailty of much of the evidence, which is often statistically weak and open to multiple interpretations. Kelly 1995, ch. 7; Hill and Hurtado 1996, 2009; Kaplan et al. 2000; Bliege Bird, Smith, and Bird 2001; Hawkes, O'Connell, and Blurton Jones 2001; Hawkes and Bliege Bird 2002; O'Connell et al. 2002; Gurven 2004; Smith 2004; Marlowe 2007, 2010; Bliege Bird and Bird 2008; Gurven and Hill 2009; Howell 2010; Codding, Bliege Bird, and Bird 2011; Bliege Bird et al. 2012; Jaeggi and Gurven 2013; Blurton Jones 2016, ch. 14.

4. Bipedality of *A. afarensis*: Haile-Selassie et al. 2010. Female pelvis: Warrener et al. 2015. Pair bonding: Kuhn and Stiner 2006; Plavcan 2012. The bitter debate on chimpanzee brains and development: Vinicius 2005; Robson, van Schaik, and Hawkes 2006; DeSilva and Lesnik 2006; Robson and Wood 2008; Cofran and DeSilva 2015.

5. Aiello and Wheeler 1995.

6. Sayers and Lovejoy 2014; Antón, Potts, and Aiello 2014. The earliest projectile points now date to about 280,000 BCE in Africa (Sahle et al. 2013), although they became common only much later, in the Upper Paleolithic. For an overview of the diet of *Homo erectus*, see Klein 2009, 414–423. For an overview of nutritional hypotheses of brain evolution, see Pontzer 2012.

7. Wrangham 2009.

8. Hardy et al. 2015; see also Sayers and Lovejoy 2014, 334–335.

9. Blurton Jones 2016, 282.

10. Hill and Hurtado 2009; Marlowe 2010, 261.

11. For a recent discussion of research on starvation and famine, see Wells 2010, 153–161. Not only do men require more calories than women, but they also need to maintain a higher BMI in order to survive famine.

12. Howell 2010, ch. 4.

13. Zafon 2006.

14. Goodman and Griffin 1985; Estioko-Griffin and Griffin 1981.

15. Bliege Bird and Bird 2008; Bliege Bird et al. 2012.

16. On this debate, closely related to the debate documented in note 3, and on marriage and mating strategies among humans in general, see Emlen 1995; Hrdy 1999 (ch. 9), 2000, 2009; Blurton Jones et al. 2000; Smith 2004; Marlowe 2007, 2010; Quinlan and Quinlan 2007; Winking et al. 2007; Chapais 2008, ch. 11; Borgerhoff Mulder and Rauch 2009; Gurven et al. 2009; Howell 2010; Codding, Bliege Bird, and Bird 2011; Leonetti and Chabot-Hanowell 2011; Winking and Gurven 2011; Chu and Lee 2012; Marlowe and Berbesque 2012; Coxworth et al. 2015; Blurton Jones 2016.

17. Trivers 1972 is a classic discussion of sexual selection; Barash and Lipton 2001 is also recommended.

18. Chu and Lee 2012.

19. Hrdy 2009, ch. 5; Kramer 2010, 421–422, summarizes results on male parenting effort among foragers.

20. In what follows I outline the basic concepts most often discussed and tested by anthropologists, but a more complex picture of the evolution of cooperation in humans could be presented. See Tomasello 2016 for a highly nuanced model. E. O. Wilson's controversial thesis

that group selection worked on some level to produce humans' unusual patterns of cooperation is most accessible in his 2012 book. This argument relies heavily on the contested thesis that between-group warfare was pervasive in our Paleolithic past. But, like several other theories discussed here, he too identifies cooperation as the most important trait in the human adaptive complex. For an introduction to the history of the study of cooperation, see van Schaik and Kappeller 2013.

21. Zahavi 1975.

22. Marlowe and Berbesque 2012; Coxworth et al. 2015.

23. See references in note 16.

24. Marlowe 2010, 162.

25. Chrisholm and Burbank 1991; Keen 2002, 2006.

26. Marlowe 2000.

27. There is a large scholarly literature on sex differences in mortality. Recommended are Preston 1976, ch. 6; Waldron 1983; Henry 1989; Coale 1991; Kalben 2000; Rogers et al. 2010. Women's advantage is greatest in modern populations, for several reasons: (1) overall mortality is low and social factors that increase male mortality over that of women (increased risk-taking, more dangerous jobs, poorer diet, smoking) have more influence, (2) maternal mortality is very low, and (3) there is less discrimination against women in most modern societies than in agrarian ones. There is a rough consensus that if all social factors could somehow be excluded, mortality would be lower for women than men at all ages; this difference is substantial in infancy, small in young adulthood, and greatest—perhaps 20 or 30 percent—in middle age and later. The model life tables most often used by historians, those published by Coale and Demeny in 1966 and revised in 1983 and based on data from mostly European nations in the later nineteenth and twentieth centuries, reflect this pattern (their Model West life tables are thought to best reflect patterns in nonindustrialized populations with high mortality).

28. See Trivers 1972 on differential mortality and potential adaptive reasons for it; more recently, see Austad 2006 and Austad and Fischer 2016, which emphasize that the picture is complicated.

29. Tuljapurkar, Puleston, and Gurven 2007.

30. Vinicius, Mace, and Migliano 2014.

31. Walter 2006.

32. Vinicius, Mace, and Migliano 2014.

33. Kaplan et al. 2010.

34. Morton, Stone, and Singh 2013.

35. I thank the authors of this model for clarifying its assumptions and methods in personal communication.

36. Marlowe 2010, 160; Howell 2000, 234. Among the northern Ache at contact in 1970, 8 of 21 women over 55 were married: Hill and Hurtado 2009, 3867.

37. Hawkes and Coxworth 2013; Kim, McQueen, et al. 2014.

38. Muller, Thompson, and Wrangham 2006.

39. Marlowe 2010, 184; Hill and Hurtado 1996, 233.

40. Kaplan et al. 2010.

41. Beck and Promislow 2007; Sharma et al. 2015.

42. Kaplan et al. 2010; Winking and Gurven 2011.

43. See also Lee 1992 on the !Kung and Blurton Jones 2016, 303, on the Hadza.

44. Kaplan et al. 2000, 2010; Kaplan and Robson 2002; Robson and Kaplan 2003; Gurven and Walker 2006; Gurven, Kaplan, and Gutierrez 2006; Gurven and Kaplan 2007, 2009; Gurven et al. 2012; Hooper et al. 2015. Recently, Wells 2012b has argued that fat should be considered another kind of capital.

45. Kaplan and Robson 2002; Robson and Kaplan 2003.

46. Kaplan et al. 2010; Gurven et al. 2012; see also Hill, Hurtado, and Walker 2007; Hill and Hurtado 2009.

47. Hill and Hurtado 2009, 3865.

48. Hrdy 2005a, 2005b, 2009.

49. Kaplan et al. 2010.

50. On cooperative breeding among other animals, see the classic discussion of Emlen 1995; see also Clutton-Brock 2006; Hrdy 2009, ch. 6.

51. Cant and Johnstone 2008; Johnstone and Cant 2010; Croft et al. 2015, 2017.

52. Marlowe 2004; Hill et al. 2011.

53. Cf. Kramer 2010, 418–419.

54. Hrdy 2009, 190–197.

CHAPTER 4: FORAGERS TODAY: HUNTING, SHARING, AND SUPER-UNCLES

1. This chapter was written before the publication of Blurton Jones 2016. I have incorporated its contributions where they differ from or add to those of Marlowe.

2. Ruff 2002; Kuzawa and Bragg 2012.

3. Blurton Jones 2006, 242–253.

4. Marlowe 2010, 137–139.

5. Blurton Jones et al. 1992, 171; Marlowe 2010, 152; Blurton Jones 2016, sec. 7.13. Blurton Jones reports that 16 of 39 women over age 40 gave birth after that age.

6. Blurton Jones et al. 1992.

7. Blurton Jones, Hawkes, and O'Connell 2002, 189; see more recently Blurton Jones 2016, sec. 8.1. The Hadza do not practice witchcraft but fear magical attack by neighboring peoples.

8. Marlowe 2010, 65–66.

9. Butovskaya 2013. Because the Hadza population is small, the two homicides that occurred over 30 years, plus a few perpetrated by non-Hadza, translate to a relatively high "homicide rate" by modern standards (measured as deaths per 100,000), but although anthropologists often make this point, I am not sure the comparison is very helpful. See Blurton Jones 2016, 142–143, for a recent discussion of homicide rates among foragers.

10. Konner 2005. For an overview of childcare practices in foraging, horticultural, and agrarian societies, including discussion of "multiage play groups," see Hewlett 1991.

11. Blurton Jones et al. 2000. Blurton Jones 2016, 293, calculates that half of marriages end within about seven years.

12. Marlowe 2005.

13. Blurton Jones 2016, 289, 297.

14. Blurton Jones, Hawkes, and O'Connell 2005; see Marlowe 2004; Hill et al. 2011; Dyble et al. 2015.

15. Crittenden and Marlowe 2008; Marlowe 2005.

16. Blurton Jones 2016, sec. 15.11.

17. Tronick, Morrelli, and Winn 1987; Ivey 2000.

18. See also Hrdy 2005a, 2005b; and see Kramer 2010 on the role of siblings in childcare.

19. Marlowe 2010, 124, 187.

20. Also the finding of Blurton Jones 2016, sec. 15.13.

21. Smith 2004. Blurton Jones 2016, 287–288, finds that good hunters did have a slightly higher number of marriages among the Hadza.

22. Blurton Jones 2016, 302–304.

23. Hawkes, O'Connell, and Blurton Jones 2001.

24. Smith 2004.

25. Blurton Jones 2016, chs. 21–22.

26. Howell 2010.

27. Smith et al. 2010; Borgerhoff Mulder et al. 2009.

28. Fry and Söderberg 2013; and see the collection of essays in Fry 2013.

29. A recent study of the evolution of human morality that stresses the role of cooperation and addresses many of the themes discussed here is Tomasello 2016. A representative of the opposite argument—that great apes in general, not excluding humans, are characterized by brutal aggression and competition for dominance—is Wrangham and Peterson 1996. For a discussion of the debate through the end of the twentieth century, see Boehm 1999, ch. 1.

30. For what follows on the Ache, see Hill and Hurtado 1996.

31. Hill and Hurtado 1996, 55–56, 166–167.

32. It also seems that before contact, the Ache and other South American peoples had a lower disease burden than African foragers, but this is hard to prove, and the greater question of the role of disease in Paleolithic mortality is also fraught. See Gurven and Kaplan 2007; and Hill, Hurtado, and Walker 2007.

33. See Kelly 1995, 233–244. On sex-biased infanticide and male-biased sex ratios, see also Hewlett 1991, 23–28. Schrire and Steiger 1974 show that among some Arctic groups, a classification error is responsible for the impression of widespread femicide—girls, who married earlier, were classified as adults by anthropologists, while boys of the same age were still counted as children—and calculate that Arctic populations in general could not sustain a rate of femicide of more than about 8 percent of female infants born if practiced over a long period.

34. See also Gurven et al. 2012 on the Tsimane.

35. Hill and Hurtado 2009.

36. Kuhn and Stiner 2006, 954.

37. Hrdy 2000.

38. Hill and Hurtado 1996, table 5.1.

39. Simmons 1945, 225–229.

40. Howell 2010, 113–114, 124.

41. Münzel 1973; Arens 1976; Hill and Hurtado 1996, 168–169.

42. Hill and Hurtado 1996, 258, 471; and see chapter 2, note 9.

43. Hill and Hurtado 1996, 276–277.

44. Sommer 2000, ch. 5. Most cases actually prosecuted involved rape and homicide, but the stories clearly illustrate that consensual gay sex was not unusual.

45. Amadiume 1987; Hristova 2013.

46. Young 2000; cf. Dickemann 1997.

47. Vasey, Pocock, and VanderLaan 2007; Vasey and VanderLaan 2007; VanderLaan and Vasey 2011; VanderLaan, Ren, and Vasey 2013.

48. Camperio Ciani, Corna, and Capiluppi 2004.

49. Cf. Kaplan et al. 2010, 237.

50. Dyble et al. 2015.

51. Boehm 1999.

52. Marlowe 2010, fig. 5.11; Howell 2010, tables 5.1, 5.3, and fig. 5.2.

CHAPTER 5: OUR LONG STONE AGE PAST: HOW GRANDMOTHERS (MAYBE) CONQUERED THE WORLD

1. Austad 1994.

2. On short prehistoric lifespans, some of the most influential studies have been those of large burial sites in North America, such as Johnston and Snow 1961 (Indian Knoll, Kentucky, ca. 5000–1500 BCE), Lovejoy et al. 1977 (Libben, Ohio, ca. 800–1100 CE), and Mensforth 1990 (Carlston Annis, Kentucky, ca. 2500–1000 BCE). For synthetic studies, see O'Connor 1995 and Gage 1998.

3. On the problems of paleodemography, see Bocquet-Appel and Masset 1982; Walker, Johnson, and Lambert 1988; O'Connor 1995, chs. 2 and 3; Hoppa and Vaupel 2002; Konigsberg and Herrmann 2006; Pinhasi and Bourbou 2008; Gage and DeWitte 2009.

4. Keckler 1997.

5. Gurven and Kaplan 2007, 331.

6. On forager life history, we are now fortunate to have the comprehensive study of Gurven and Kaplan 2007. For calculations of PrR in several traditional societies, see Levitis, Burger, and Lackey 2013; and Vinicius, Mace, and Migliano 2014. Trinidadian slaves: John 1988; Levitis, Burger, and Lackey 2013. Hiwi mortality: Hill, Hurtado, and Walker 2007.

7. Bamberg Migliano, Vinicius, and Mirazón Lahr 2007. On Agta mortality, cf. Konigsberg and Herrmann 2006, 291; Gurven and Kaplan 2007, 326–330.

8. Ruff 2002.

9. African *Homo erectus* skeletons, dating from about 1.8 to 1.3 million years BCE, look different from Asian *Homo erectus* and are sometimes called *Homo ergaster* instead. Most researchers believe either that *Homo ergaster* is the same species as *Homo erectus* or that it is *Homo erectus'* immediate ancestor.

10. On the controversial topic of speciation early in the history of *Homo*, see recently Lordkipanidze et al. 2013; Antón, Potts, and Aiello 2014; Dembo et al. 2015. For the intriguing "out of Asia" argument, see Dennell 2009, ch. 6.

11. Reich 2018, ch. 3.

12. Brown et al. 2004; Morwood et al. 2005. While initially researchers believed that *Homo floresiensis* lived as late as 12,000 years ago, they have revised that date, also speculating that the arrival of modern humans may account for the extinction of the older species; see Sutikna et al. 2016.

13. For what follows on the natural history of *Homo erectus*, see Potts 1996, 2012; Ruff 2002; Antón 2003; Wells and Stock 2007; Robson and Wood 2008; Klein 2009; Graves et al. 2010; Plavcan 2012; Antón and Snodgrass 2012; Pontzer 2012; Antón, Potts, and Aiello 2014; Brooke 2014.

14. Balzeau, Holloway, and Grimaud-Hervé 2012.

15. See chapter 3.

16. Plavcan 2012.

17. Wells 2010.

18. There is doubt about the date of the latest *Homo erectus* skeletons recovered from Java, which are now thought to be older than previously described; Indriati et al. 2011.

19. Toro-Moyano et al. 2013.

20. For overviews on the evolution of human life history, see Robson and Wood 2008; Zollikofer and Ponce de León 2010; Thompson and Nelson 2011; Schwartz 2012.

21. Caspari, Lee, and Goodenough 2004 used an innovative method to analyze Paleolithic skeletons. Instead of trying to guess the age at death, they examined teeth only and adults only, dividing skulls into two groups: those whose dentition showed that they were adult, and those whose dental wear was twice as great as that of specimens at the age of sexual maturity (that is, if age at maturity was 15, then the two groups represented those aged 15 to 30 and those aged 30 and older, assuming a constant rate of tooth wear). From this they calculated "old-young" (OY) ratios for australopithecines, early *Homo*, Neanderthals, and anatomically modern humans of the later Paleolithic (ca. 30,000 years ago). By these methods they sought to avoid problems caused by the underrepresentation of children in the fossil record and the difficulty of determining older ages at death. They found that OY ratios jumped sharply in the Upper Paleolithic, from 0.25 for early *Homo* and 0.39 for Neanderthals (and just 0.12 for australopithecines) to 2.08 for *Homo sapiens*. These findings are very striking, but they sparked further debate; OY ratios in dead populations (that is, the age structure of the population that dies from year to year rather than that of the population still alive) can vary greatly (in modern populations they can be over 100), and none of the OY ratios reported by Caspari and Lee seem plausible—they are all much too low, even for chimpanzees (Hawkes and O'Connell 2005; Caspari and Lee 2005). We must therefore suspect that the skeletons are not representative, and despite its intriguing conclusion I am not sure this study tells us anything. Using traditional paleodemographic methods, Trinkaus 2011 reached an opposite conclusion: he found no difference in late-age survivorship (past age 40) between Neanderthals and Upper Paleolithic humans and little difference in the samples overall, but he acknowledged that none of the populations would have been viable at the rates of mortality implied in his study.

22. For example, Cofran and DeSilva 2015.

23. Isler and van Schaik 2012b.

24. Thompson and Nelson 2011; Dean et al. 2001; Robson and Wood 2008; Graves et al. 2010; Schwartz 2012; Cofran and DeSilva 2015.

25. See also the Reserve Capacity Hypothesis in Bogin 2009.

26. For what follows on Neanderthals, see Kuhn and Stiner 2006; Klein 2009; Thompson and Nelson 2011; d'Errico and Stringer 2011; Burke 2012; Gunz et al. 2012; Hardy et al. 2012; Bocquet-Appel and Degioanni 2013; Hardy et al. 2013; Paixão-Côrtes et al 2013; Salazar-García et al. 2013; Smith 2013; Neubauer 2014; Villa and Roebroeks 2014; Prüfer et al. 2014; Ermini et al. 2015.

27. Prüfer et al. 2014; Kuhlwilm et al. 2016; Posth et al. 2017.

28. There is currently some debate on the size of the Neanderthal population, with some geneticists arguing for a larger population of tens of thousands, but deeply subdivided into small local populations: see Rogers, Bohlender, and Huff 2017, challenged by Mafessoni and Prüfer 2017.

29. Kuhn and Stiner 2006.

30. Thompson and Nelson 2011, 268–270.

31. Zilhão et al. 2010; Finlayson et al. 2012; Hardy et al. 2012; Hardy et al. 2013; Salazar-García et al. 2013; Beier et al. 2018.

32. Reich et al. 2010; Krause et al. 2010; Meyer et al. 2012, 2014; Ermini et al. 2015.

33. See recently Malaspinas et al. 2016.

34. The cladistic analysis of Dembo et al. 2015 suggests the intriguing possibility that *Homo heidelbergensis* is "really" the Denisovans, a sister taxon to the Neanderthals, though the researchers do not raise this point.

35. Posth et al. 2017.

36. For an accessible introduction to the genetic study of Neanderthals, Denisovans, and early *Homo sapiens*, see Reich 2018, ch. 3.

37. On mitochondrial Eve, see recently Schlebusch et al. 2012; Lippold et al. 2014. For the older date, see Schlebusch et al. 2017. I have continued to use the 200,000-year figure because it seems still to be the consensus, but changing to the older set of dates would not affect the substance of this discussion.

38. Mendez et al. 2013; Harvati et al. 2013.

39. Recently, scientists have identified a skull from Jebel Irhoud, Morocco, as belonging to an "archaic" form of *Homo sapiens* dated to 300,000 years ago, which may lend support to the idea of an earlier speciation event; see Hublin et al. 2017. This very early evidence from Morocco, especially combined with the genetic evidence of Schlebusch et al. 2017 from South Africa, suggests that our species has always been highly mobile, and that efforts to pin our origins geographically to Ethiopia or some other region within Africa may be misconceived.

40. Smith et al. 2007.

41. Mallick et al. 2016.

42. Theories on human dispersal are continually evolving. The recent consensus replaces a two-wave model in which some populations of south Asia, Australia, and Papua New Guinea are remnants of the early wave. See Klein 2009, ch. 7; Petraglia et al. 2007, 2012; Rasmussen et al. 2011, 2014; Reich et al. 2012; Ermini et al. 2015; Skoglund et al. 2015; Pagani et al. 2016; Malaspinas et al. 2016; Mallick et al. 2016. For early *Homo sapiens* in China, see Wu Liu et al. 2015.

43. See Boesch and Boesch 1984; Nelson 1997, ch. 5; Kuhn and Stiner 2006; Wells and Stock 2007.

44. See d'Errico and Stringer 2011; Bocquet-Appel and Degioanni 2013; Villa and Roebroeks 2014; and other references in note 26 in this chapter. Cave paintings: Hoffmann et al. 2018.

45. Gunz et al. 2012; Neubauer and Hublin 2012.

46. Paixão-Côrtes et al. 2013.

47. Mallick et al. 2016.

48. Late Middle Paleolithic "Aterian" technology, including tanged tools, personal ornaments (pierced shells), and possibly projectile points, is older in northern Africa than previously thought; see Richter et al. 2010; Iovitu 2011. South African sites have produced microliths and other technologies characteristic of the Late Stone Age that date to as early as 71,000 BCE; Brown et al. 2012. For a full review of early evidence of technologies and cultural innovations more typical of the Upper Paleolithic or Late Stone Age, see d'Errico and Stringer 2011.

49. For the role of demography in cultural change, see Henrich 2004; Powell, Shennan, and Thomas 2009; Richerson, Boyd, and Bettinger 2009; and Bocquet-Appel and Degioanni 2013.

50. There is, however, debate about the size of the later Neanderthal population. See references cited in note 28.

51. See recently Villa and Roebroeks 2014.

52. For summaries of the research, see Guatelli-Steinberg 2009; Thompson and Nelson 2011, 260–268; Smith 2013; Neubauer and Hublin 2012. The debate on rates of childhood development in Neanderthals continues with no resolution: see Rosas et al. 2017 (based on a newly discovered juvenile skeleton from Spain); DeSilva 2018.

53. Caspari, Lee, and Goodenough 2004; see note 21 in this chapter for further discussion.

54. Henrich 2004. A patchy or sputtering pattern of evidence can also be the result of a weak archaeological signal rather than a discontinuous tradition. While the current trend is to identify short-lived changes that rise and disappear, it is possible that some of these findings will be revised with future discoveries; Brown et al. 2012.

55. Bocquet-Appel and Degioanni 2013; Villa and Roebroeks 2014; d'Errico and Stringer 2011.

56. See chapter 2, note 15 for references and discussion.

57. Lorenzen et al. 2011.

CHAPTER 6: THE AGE OF FARMERS: PATRIARCHY, PROPERTY, AND FERTILITY CONTROL

1. Cohen 2008.

2. Boserup 1965 is the classic description of this process.

3. On the origins of cuneiform writing see Schmandt-Besserat 1996.

4. Of the many theories of social organization and state formation in the traditional world, I recommend Johnson and Earle 1987, though the formulation I present here is different.

5. On egalitarianism and inequality in foraging, horticultural, and early agrarian societies, see Kelly 1995; Boehm 1999, ch. 5; Keen 2006; Borgerhoff Mulder et al. 2009, 2010; Kaplan, Hooper, and Gurven 2009; Shenk et al. 2010. For a brief but comprehensive history of inequality, see Scheidel 2017, part 1.

6. Marciniak and Czerniak 2007; Pilloud and Larsen 2011; Wright 2014.

7. Two English translations of this work, first published in German in 1884, are widely available: Engels 1902 and Engels 1942. For discussions of the intellectual history of this debate, see

Nelson 1997, ch. 6; Knight and Power 2005; Goettner-Abendroth 2012, ch. 1. For updated versions of the Marxist argument, see Meillassoux 1975; and Caldwell 2006, ch. 1.

8. A good introduction to the family in the agrarian period, covering many of the topics in this section, is Dickemann 1979.

9. Ember and Ember 1971; Marlowe 2004.

10. Priest: Le Roy Ladurie 1979, 39. Roman Egypt: Scheidel 1996b; Rowlandson and Takahashi 2009. A good discussion of complex, intergenerational marriage strategies in relation to property in one agrarian society—these strategies included marrying inside the family to consolidate property, marrying outside the family to acquire property, marrying neighbors to consolidate farms or urban property, and adopting kinsmen as heirs—is Cox 1998.

11. Hartung 1976.

12. On polygyny, see Betzig 1986; Scheidel 2009b.

13. Leonetti et al. 2004, 2005; Leonetti, Nath, and Hemam 2007. On matriliny, see Holden, Sear and Mace 2003.

14. On the broad contrast between African and Eurasian systems, see Goody and Tambiah 1973 and Kandiyoti 1988. In-depth studies of marriage and reproduction in West African societies that add nuance to my oversimplified account here include Amadiume 1987 and Bledsoe 2002.

15. Laslett 1972 first undermined the idea of the primeval "extended" family. See also the classic article of Hajnal 1982.

16. For example, Wolf and Huang 1980; Lee and Campbell 1997. But for Chinese populations with small average household size, see Fei Hsiao-Tung 1939; Li Bozhong 1998, 23.

17. Cf. the famous definition of marriage in the sixth and last edition of *Notes and Queries in Anthropology*: "Marriage is a union between a man and a woman such that children born to the woman are recognized as the legitimate offspring of both partners." Royal Anthropological Institute of Great Britain and Ireland 1951.

18. Betzig 1986, 1992; Scheidel 2009a, 2009b.

19. For a good introductory discussion, see Goody and Tambiah 1973.

20. Recommended readings on women's status in classical Athens are Cohen 1991; Hunter 1994; Cox 1998; Patterson 1998. All of these studies emphasize ways in which women exercised power in their own sphere, as is still the trend in scholarship. The standard work on Athenian law is Harrison 1968–1971; for a good discussion of dowry and inheritance, see Hunter 1994, ch. 1. A fascinating introduction to the subject of the physical control of women in ancient Greece is Llewellyn-Jones 2003.

21. See Kandiyoti 1988 for a classic discussion of the "patriarchal bargain."

22. On women's labor in agricultural societies, see Boserup 1970; Goody 1976; Ember 1983; Leacock and Safa 1986; Bradley 1989; Huang 1990; Hudson and Lee 1990; Nelson 1997, ch. 5.

23. A history of textile production for the general reader is Barber 1994.

24. Ember 1983.

25. Furth 1999, ch. 8.

26. Gates 2015; Huang 1990, 55–56; Scheidel 1995, 1996a.

27. Scheidel 1995.

28. *The Secret History* is a difficult source to read; for a recent English translation with commentary, see de Rachewiltz 2004–2013. The most accessible translation is probably Kahn and Cleaves 1984.

29. Le Roy Ladurie 1979, 166. We do not know what words Béatrice used in her native Occitan, but the inquisitor's Latin records "sibi muliebria deffecissent," meaning "her female things [sc. menstrual periods] had stopped." Duvernoy 1965, 1:249.

30. A modern translation of Kempe's autobiography is available in Bale 2015; see also the article of Phillips 2004.

31. Achebe 2011.

32. See the highly recommended work of Dobson 1997.

33. Gurven and Kaplan 2007, 339–340.

34. Bocquet-Appel and Bar-Yosef 2008; Bocquet-Appel 2011.

35. Shennan and Edinborough 2007; Shennan 2009; Downey et al. 2014; Goldberg, Mychajliw, and Hadley 2016. Cohen's response and partial conversion to the Neolithic Demographic Transition theory in Cohen 2008 is interesting but perhaps premature.

36. Good discussions with bibliography are Brown 2015 and Torfing 2015. Summed Calibrated Date Probability Distribution must assume a predictable correlation between the number of datable remains and the size of the population that left them. But systematic sampling error can arise from many sources—permanent settlements are easier to find than clusters of temporary structures, newer sites are better preserved than older ones, coastal sites flooded by rising seas in the Holocene are not detectable at all, and archaeologists are more interested in some sites and locations than others. Many problems also result from the use of radiocarbon methods. Bamforth and Grund 2012 show that summed radiocarbon probability curves produce peaks and troughs as a result of factors inherent to radiocarbon dating and do not necessarily reflect real population events. Contreras and Meadows 2014 show that summed radiocarbon dating methods performed unimpressively when they tried to detect the catastrophic population crashes of fourteenth-century Europe and sixteenth-century Mexico, even when they used a large sample of data, and that different random samples produced quite different results.

37. Hewlett 1991; Bentley, Goldberg, and Jasieńska 1993; Wood 1994, 30–33; Boone 2002; Kramer and Boone 2002. Sellen and Mace 1997 found a linear correlation between dependence on agriculture and fertility.

38. Gurven and Kaplan 2007. On demography in traditional societies more generally, see Wood 1994; Livi-Bacci 2012.

39. Tuberculosis: Brosch et al. 2002. For a fascinating study supporting a very recent origin of smallpox, see Duggan et al. 2016; for an overview, see Harper and Armelagos 2013.

40. See Brooke 2014, especially ch. 8.

41. The catastrophic mortality of the early to mid-twentieth century resulting from two world wars, epidemic influenza, famines in China and the USSR, and genocide, which is horrific in absolute numbers, was less important as a percentage of total population than these previous crises, slowing the rapid growth of the Demographic Transition rather than depleting world population.

42. On mortality crises in early modern England, see Wrigley and Schofield 1981, 332–342 and appendix 10; Hinde 2003, ch. 7.

43. For example, Harvey 1993, ch. 4.

44. Boone 2002.

45. Engels 1975.

46. On the ratchet metaphor, see Wood 1998.

47. For a good introductory discussion, see Caldwell 2006, ch. 4. An influential model that combines Malthusian and Boserupian elements and includes an excellent history of the question to that date is Wood 1998.

48. See Wood 1998.

49. Lee and Tuljapurkar 2008; Puleston and Tuljapurkar 2008; Lee, Puleston, and Tuljapurkar 2009.

50. Unfortunately, this model does not address the role of post-reproductive workers, but the comments on geronticide in Puleston and Tuljapurkar 2008, 153, are interesting. (This model found no long-term advantage to geronticide in space-limited agrarian systems, because older people in this model are not reproducing.)

51. Egypt: Bagnall and Frier 1994. Trinidad: John 1988. Utah: Bean, Mineau, and Anderton 1990. China: Lee and Campbell 1997. Sweden: Low, Clarke, and Lockridge 1991. Japan: Jamison et al. 2002. England: see chapter 7.

52. Henry 1961; see also Wood 1994, 30–47. On the concept of "natural fertility" and debate about fertility control in traditional populations, see Bean, Mineau, and Anderton 1990, ch. 1. For the 24 children figure, see Wood 1994, 31.

53. Bean, Mineau, and Anderton 1990. On North American colonies, see chapter 7.

54. Classic discussions are Bongaarts 1980; Ellison 2001, chs. 5 (puberty) and 6 (fecundity). On nutrition, see also Wood 1994, 522–529. For a recent review of the evidence on age at menarche, see Kuzawa and Bragg 2012.

55. Wood 1994, 441–443.

56. Henry 1961; Caldwell and Caldwell 2003.

57. On *coitus interruptus*, see Musallam 1983; Santow 1995; Han 2007; Davis 1983, 67. On reproduction and child spacing in The Gambia, see Bledsoe 2002.

58. A prominent exponent of the optimistic theory of premodern birth control and abortion is Riddle 1997. For further references and a skeptical view, see Sommer 2010. For an intellectual history of the debate through the 1980s, see Caldwell and Caldwell 2003.

59. Bean, Mineau, and Anderton 1990, 30–32.

60. Bongaarts 1980; Hinde 2003, 132–136.

61. See, for example, Han 2007.

62. Rindfuss and Morgan 1983.

63. On the costs and productivity of children and the life cycle of the peasant farm, see Chayanov 1966, 53–69; Caldwell 1976; Cleland and Wilson 1987; Sanderson and Dubrow 2000; Lee and Kramer 2002; Kramer and Boone 2002; Kramer 2004, 2005; Caldwell 2006, ch. 5; and cf. Lee and Mason 2011. Kaplan 1994 has also been influential, although this argument is based on foraging and horticultural societies; see also Hooper et al. 2015 on the Tsimane, a horticultural society. Specialized studies on family labor in agrarian production include Hanawalt 1986; Huang 1990; Halstead 2014; and Gates 2015.

64. Cf. Mace 2000, 8.

65. Chayanov 1966, 53–69; and see Lee and Kramer 2002; Kramer and Boone 2002.

CHAPTER 7: REPRODUCTION AND NON-REPRODUCTION IN SOME AGRARIAN SOCIETIES

1. Malthus 1798, 22.

2. For what follows on demography, marriage, and family in early modern England, see Laslett 1972; Wrigley and Schofield 1981; Hajnal 1982; Wrigley et al. 1997; Hinde 2003.

3. Hinde 2003, 132–136.

4. There is a great deal of scholarship on the family in early modern England; I have relied most on Laslett 1972, ch. 4; Laslett 1977; Wrigley et al. 1997; Tadmor 2001.

5. Wrigley and Schofield 1981, 257–265.

6. Hindle 1998.

7. Most studies of infanticide in England focus on court cases and do not address the related behavior of abandonment: Hoffer and Hull 1981; Gowing 1997; Roth 2001; Kilday 2013. On infanticide and infant abandonment in Europe, see Langer 1974; Boswell 1988. Dickeman 1975 and Hrdy 1999, chs. 12–14, are recommended as general discussions of infanticide; for more evolutionary context, see the collection of essays in Hausfater and Hrdy 1984. Boswell 1988 argues that most abandoned children in ancient and medieval Europe were rescued, but while his book is an excellent and exhaustive testimony to the frequency of infant abandonment, he underestimates the difficulty of keeping infants alive once separated from their mothers before the advent of modern medicine; see Hrdy 1999, 297–304.

8. On the London foundling hospital, see Levene 2007.

9. Malthus 1826, 2.3.29. On efforts at artificial feeding in Italian foundling hospitals, see Kertzer 1993, 135–137. On mortality in European foundling hospitals, see ibid., 138–144.

10. Levene 2007, 1–6.

11. Malthus 1826, 1.206–229.

12. See Lee and Wang Feng 1999; Pomeranz 2009, ch. 1. These broad studies are both good sources of further references.

13. Lee and Campbell 1997.

14. Rindfuss and Morgan 1983; and cf. Johnson 1975.

15. On infanticide in China, see Lee and Campbell 1997; Lee and Wang Feng 1999; Mungello 2008; King 2014. King argues that in the early imperial period, infanticide may have more closely resembled the European pattern of abandonment (rather than outright killing) of unwanted babies of either sex.

16. Contributing factors: Ember and Ember 1971; Baker and Jacobsen 2007. Das Gupta et al. 2003 and Guilmoto 2009 are good discussions of the modern distribution of femicide and its causes, but they do not attempt to describe why some agrarian societies are more rigidly patrilineal and virilocal than others. See also Caldwell and Caldwell 2005 and Brooks 2012. The question of "missing women" was most famously posed in 1989 by Indian economist Amartya Sen (and more publicly two years later in a *New York Times* article), who estimated that excess mortality due to social discrimination against females had caused a worldwide deficit of about 100 million women, concentrated in East and South Asia; his conclusions remain the subject of much debate and analysis.

17. Panigrahi 1972; Hrdy 1999, 318–340; Das Gupta et al. 2003; Caldwell and Caldwell 2005, 209–212; Dube Bhatnagar, Dube, and Dube 2005. On the Qing lineage, see Lee and Wang Feng 1999, 50.

18. Trivers and Willard 1973. Dickemann 1979 is the classic historical discussion; see more recently Hrdy 1999, 318–340.

19. For recent discussions of sex ratios in China, see Greenhalgh 2012; Tucker and van Hook 2013.

20. King 2014, 15–20. Common motives for infanticide are tabulated in Scrimshaw 1984, table 1.

21. Corsini 1976.

22. For a recent review article on infanticide in modern societies, see Porter and Gavin 2010.

23. Camperio Ciani and Fontanesi 2012; Hrdy 2016.

24. On fertility and the family in traditional China there is a large scholarship and a great deal of controversy. I have relied on Fei Hsiao-Tung 1939; Johnson 1975; Wolf and Huang 1980; Stacey 1983; Lee and Campbell 1997; Lee and Wang Feng 1999; Sommer 2000; and Caldwell and Caldwell 2005. I am aware of the bitter debate over fertility control in premodern China and criticisms of the "revisionist" arguments represented by Lee and Wang Feng 1999 (for example, Wolf 2001; Wolf and Engelen 2008; Sommer 2010). For the purposes of this book, it is enough that scholars agree that fertility within marriage was low to moderate and that infanticide was common.

25. Sommer 2005; Sommer 2015, ch. 1.

26. Lee and Wang Feng 1999.

27. Cf. Linda J. Lee in Haase 2008, 2:641–642.

28. Masvie 2007.

29. Bezner Kerr et al. 2008.

30. Aubel, Touré, and Diagne 2004.

31. Pruitt 1945.

32. On widows in China, see Wolf and Huang 1980, 227–228; Wolf 1981; Bernhardt 1999, ch. 2; Sommer 2000, ch. 5.

33. For this and the following case, see Sommer 2000, ch. 5.

34. Wolf and Huang 1980, 228.

35. Wolf and Huang 1980, 357–364.

36. For example, Hufton 1984; Hill 2001; Froide 2005. On property, see Erikson 1993, ch. 2.

37. Hill 2001, 70–71.

38. On widows in early modern England, see Laslett 1977, ch. 5; Hufton 1984; Erikson 1993, part IV; Whittle 2014.

39. Whittle 2014.

40. Poor relief: Froide 2005, 34–42. Workhouses: Laslett 1977, 199–200. Widows and property: Erikson 1993, part IV.

41. Laslett 1972, 147.

42. United States Department of Labor, Office of Policy Planning and Research 1965. For a brief history of the sociology of the U.S. family in the late twentieth century, see Coontz 1992, ch. 10. On the intellectual history of the "Moynihan Report," see Geary 2015. A recent history of racial wealth inequality is Baradaran 2017.

43. Centers for Disease Control, "Unmarried Childbearing," http://www.cdc.gov/nchs/fastats/unmarried-childbearing.htm; UNICEF Innocenti Research Centre 2012; Haub 2013; Björk Eydal and Gíslason 2014. Iceland's status in the Gender Gap Index can be found at World Economic Forum, "Reports: Economies," http://reports.weforum.org/global-gender-gap-report-2015/economies/#economy=ISL. In 2015 the United States ranked twenty-eighth in this index, just after Mozambique and ahead of Cuba.

44. Stone 1982.

45. Ramakrishna Vedanta Centre 1955; Feldhaus 1982; Bhavalkar 1996; Rosen 1996; Sellergren 1996; Aklujkar 2004; Dube Bhatnagar, Dube, and Dube 2005, chs. 6–7.

46. Tr. Feldhaus 1982, 597.

47. Tr. Aklujkar 2004, 129.

48. Harlan 1992, ch. 7; Mukta 1994; Dube Bhatnagar, Dube, and Dube 2005, chs. 3, 6.

49. Tr. Dube Bhatnagar, Dube, and Dube 2005, 185.

50. Mukta 1994; Harlan 1992, ch. 7.

51. Martin 1996.

52. Lee and Campbell 1997, ch. 7; Low, Clarke, and Lockridge 1991.

53. Greenhalgh 2012. Probably the most influential work in the category of "dire predictions" is Hudson and den Boer 2004. On hostility to bachelors in the mid-twentieth century United States, see Ehrenreich 1983, ch. 2.

54. For violent, lecherous, or sexually predatory monks in the eastern part of the later Roman Empire, see MacMullen 1997, 15–18. In Qing dynasty China, see Sommer 2000, 99–100; Hudson and den Boer 2004, 223–226. On fourteenth-century Dominican friars in Spain, see Vargas 2011. On sixteenth-century Augustinian friars in Hungary, see Erdélyi 2015.

55. Schacht, Rauch, and Borgerhoff Mulder 2014.

56. Mesquida and Wiener 1996; Scheidel 2009b.

57. Duby 1964.

58. Boone 1983; Hudson and den Boer 2004, 212–216; Seaver 2008.

59. Hobsbawm 2000.

60. Hobsbawm 2000; Shaw 1984.

61. Antony 1989; Owenby 2002. On the Republican period, see Billingsley 1988, ch. 4. On the "rootless rascal" stereotype in Qing legal sources, see Sommer 2000, 12–5, 96–101.

62. Le Roy Ladurie 1979, 69–135.

CHAPTER 8: THE MODERN WORLD

1. Accessible introductions to globalization are Derber 2002 and Bourguignon 2015.

2. Ferreira et al. 2015; Bourguignon and Morrisson 2002; Roser 2016.

3. Sea level: Hansen et al. 2016; see also DeConto and Pollard 2016, predicting that the collapse of Antarctic ice sheets is likely to be sudden rather than gradual. Migration: see, for example, Black et al. 2011.

4. A good intellectual history of population theory is Bashford 2014. A summary of the history of population policy is Demeny 2011. On forced sterilization in the United States, see Largent 2007 (p. 7 for the estimate); Stern 2016. There is a large scholarship on Nazi eugenics; see, for example, Proctor 1988.

5. A wide-ranging history of the abuses of population control is Connelly 2008; for a briefer discussion focused on China and India, see Lock and Nguyen 2010, 118–145. For U.S. policy in Puerto Rico, see Briggs 2002.

6. On the recent history of hunger and food insecurity, see Food and Agriculture Organization of the United Nations, International Fund for Agricultural Development, and World Food Programme 2015, which illustrate the highly complex nature of the relationship between hunger and modernization. On agricultural science and geopolitical strategy, see Perkins 1997.

7. These and other modern population statistics I take from the website maintained by the Population Division of the United Nations Department of Economic and Social Affairs, presenting data from their *World Population Prospects, the 2015 Revision*, http://esa.un.org/unpd/wpp/.

8. For a summary of these issues, see Livi-Bacci 2012, ch. 4; Coleman and Rowthorn 2011.

9. Lee and Mason 2011; Rosero-Bixby 2011. Both articles, and several others cited in this chapter, come from a themed supplemental issue of *Population and Development Review* entitled *Demographic Transition and Its Consequences*. Another important collection on age structure in highly industrialized societies is Tuljapurkar, Ogawa, and Gauthier 2010.

10. Scheidel 2017, ch. 10.

11. In defense of population decline, see Coleman and Rowthorn 2011.

12. On English population history, see chapter 7, note 2. Notable advocates of the thesis about New World foods are Ho Ping-ti 1959; Crosby 1972; Langer 1975; and Brooke 2014. That its colonies were the source of the resources that fueled Europe's Industrial Revolution is the argument of Pomeranz 2009.

13. A large body of scholarship addresses the subject of economic development in eighteenth-century China; I have relied on Ho Ping-ti 1959; Li Bozhong 1998; and Lee and Wang Feng 1999.

14. Li Bozhong 1998; Lee and Wang Feng 1999, 37–41, 45–47.

15. For what follows, see the articles collected in Haines and Steckel 2000.

16. Thornton 2000.

17. Gemery 2000; cf. Malthus 1826, 1.5–6.

18. Livi-Bacci 2012, 53–57; see also Emory University's Trans-Atlantic Slave Trade Database, http://www.slavevoyages.org/assessment/estimates.

19. Walsh 2000; Emory University, Trans-Atlantic Slave Trade Database; Kulikoff 1977.

20. For these figures and for what follows, see Livi-Bacci 2012.

21. For what follows on mortality decline and fertility decline I have relied largely on Riley 2001 and Livi-Bacci 2012.

22. Wang Feng 2011.

23. These and the following estimates of average life expectancy are from United Nations, Department of Economic and Social Affairs, Population Division 2015.

24. Pepin 2011.

25. For this and what follows on mortality decline, see Riley 2001, synthesizing a great deal of previous scholarship. The thesis of McKeown (1976) that medicine did not become an important factor in mortality decline until late in the process has been influential in disrupting simplistic interpretations.

26. Cf. Riley 2001, 79. The subject of ethics in public health has grown in the twenty-first century; for a recent review, see Lee 2012. On vaccine resistance, see recently Larson 2016.

27. Of the enormous literature on the fertility transition, I have used here Coale and Watkins 1987; Mace 2000; Tamura 2006; Bryant 2007; Angeles 2010; Bongaarts and Casterline 2012; Qingfeng Wang and Xu Sun 2016.

28. Angeles 2010.

29. Data from World Bank Open Data, https://data.worldbank.org/indicator/SP.DYN .LE00.IN.

30. Some economists are making the argument for more optimism about the human potential for egalitarianism and cooperation. See Bowles 2012.

CHAPTER 9: WOMEN'S HELL: MENOPAUSE AND MODERN MEDICINE

1. Kuriyama 1997.

2. I dislike the term "Western" as an adjective referring to the societies of western Europe and those North American and Australian/Oceanian cultures that most directly descend from them, but since we lack a better word, I will be using "Western" in that sense.

3. The standard references in English on Greek and Roman gynecology are Dean-Jones 1994 and Flemming 2000.

4. Michel et al. 2006.

5. Houck 2006, chs. 1–2.

6. Moschik et al. 2012.

7. World Health Organization 2013, 5.

8. A good introduction to this question, for readers who want a more complete understanding, is Lock and Nguyen 2010.

9. See Amundsen and Diers 1970.

10. Hippocratic Corpus: *Coan Prognostics* 30. Aristotle: *History of Animals* 7.5. Pliny the Elder: *Natural History* 7.14.61. Soranus: *Gynecology* 1.34, 1.20. There is no complete, accessible translation of the *Gnomon* of the *Idios Logos* (*BGU* V 1210). Extracts can be found in Hunt and Edgar 1934, no. 206.

11. Flemming 2000.

12. See Dean-Jones 1994, 105–108; *Diseases of Women* 1.7, 2.137. There is a large literature on hysterical suffocation; see especially King 1993 and Mattern 2015.

13. Mattern 2015.

14. *Gynecology*, 1.26, 1.29.

15. Flemming and Hanson 1998.

16. Dean-Jones 1994, 110–147; Flemming 2000, 159–161, 235–236, 338–339.

17. *On Medical Material* 1.24. The most accessible English translation of Dioscorides is Osbaldeston and Wood 2000. On folk remedies for women's ailments, see von Staden 1992, 2008; King 1998; Flemming 2000, 161–171, 350–358. An interesting argument is made by Riddle 1992, 1997, for the efficacy of traditional contraceptives and abortifacients, but it is controversial; see the rebuttal of King 1998.

18. Amundsen and Diers 1973.

19. *Causes and Cures* 2, *De menstruo* (78 Kaiser), *Quare menstruum* (103–104 Kaiser), *De menstrui defecto* (106 Kaiser), *De menstrui retentione* (107 Kaiser). See also Amundsen and Diers 1973, 607–608.

20. A good English translation of the *Trotula* is Green 2001.

21. Burrow 1986; Engammare 2013.

22. Stolberg 1999. On the origins of the idea of menopause around 1700, see also Wilbush 1979, 1986; and Schäfer 2003. See Crawford 1981 on the absence of references to menopause in seventeenth-century England.

23. Marinello 1563, 87v; Stolberg 1999, 406.

24. Liébault 1598, 332–333.

25. On almanacs, see Weber 2002, 2003.

26. Crawford 1984.

27. Duden 1991, 118. Stolberg 1999 interprets three of Storch's case histories from volume 8 as "menopause-related cases" (411, n. 39). Of the three he cites, the first (case 10, Storch 1753, 52–62) is the most compelling evidence that Storch and his patient attributed her problems to menopause. The woman was 47, had irregular periods, and suffered itching, flushing, toothaches, eye inflammation, and a growth on her leg. Storch treated her with leeches and venesection over a period of several years.

28. Stolberg 1999, 412. Titius's dissertation is accessible at the HathiTrust Digital Library, https://babel.hathitrust.org/cgi/pt?id=ucm.5329206721.

29. Stolberg 1999, 414–415.

30. Stolberg 1999, 421; photographs and a summary of Madame d'Arnoy's letter are available online from the Université de Lausanne at http://tissot.unil.ch/fmi/webd#Tissot. The letter is document IS/3784/II/149.01.02.14.

31. Stolberg 1999, 416–417.

32. On Renaissance and early modern theories of menstruation in Europe, see Crawford 1981 and Stolberg 2000.

33. Wilbush 1988.

34. On symptoms, see Schäfer 2003, 99; Stolberg 1999, 410–414.

35. Ayers Counts 1992, 64.

36. Davis 1983.

37. The thesis of the lost tradition builds on the classic feminist argument of Ehrenreich and English 1973; see Stolberg 1999. Many historians, such as Weber 2003, now question the idea of a strict separation between high and low medical traditions. On the history of gynecology in Renaissance and early modern Europe, see King 2007. On the French Revolution, see Wilbush 1979, 1986, 1988.

38. There is a large literature on both melancholia and hysteria. Recommended on melancholia are Jackson 1986 and Bell 2014; on hysteria, Scull 2009 and Arnaud 2015; on neurasthenia, Taylor 2001.

39. On green sickness, see King 2004.

40. Astruc 1761, 2.309.

41. Tilt 1857, 57, 64–65, 70–71, 132, 156, 184, 199, 242.

42. Tilt 1857, 256.

43. Currier 1897, vii, 187.

44. Currier 1897, 188, 193–194.

45. On sexuality and menopause, see Stolberg 1999, 420, 423–426. "Morte du sexe": de Gardanne 1816, vi.

46. Tilt 1857, 230–231; Anderson et al. 2011, 124–125.

47. For the modern history of menopause, Lock 1993, 301–370 and Houck 2006 are recommended; on the nineteenth century, see also Smith-Rosenberg 1985, 191–196; on the twentieth century, see Banner 1992, ch. 8.

48. Houck 2006, 82.

49. Houck 2006, ch. 5; see also Coontz 1992 on marriage in the twentieth century.

50. Lock 1993, 351–367; see also Houck 2006, ch. 9.

51. Wilson 1966, 43.

52. Lock 1993, 345–346.

53. Deutsch 1944, 2.461.

54. Deutsch 1944, 2.472–474.

55. For what follows on involutional psychosis, see Hirshbein 2009a and 2009b, 81–87.

56. For what follows, see Hirshbein 2009a, ch. 4.

57. For a review of the research in this time period, see Nicol-Smith 1996.

58. For a review of recent research, see Vivian-Taylor and Hickey 2014.

59. Bromberger et al. 2007, 2011.

60. Vivian-Taylor and Hickey 2014.

61. Freeman et al. 2004.

62. Dennerstein et al. 2007

63. Houck 2006, 230.

64. National Institutes of Health 2005.

65. On the Women's Health Initiative, see Prentice and Anderson 2007. On the "timing hypothesis," see Salpeter et al. 2009; Hodis and Mack 2014; Lobo et al. 2016.

66. For an example, see the standard textbook of Lobo et al. 2017, in which the chapter on menopause is subtitled "Endocrinology, Consequences of Estrogen Deficiency, Effects of Hormone Therapy, and Other Treatment Options."

CHAPTER 10: WHAT ARE YOU TALKING ABOUT? MENOPAUSE IN TRADITIONAL SOCIETIES

1. Tr. Scheid 2007, 59.

2. Furth 1999, 286–295; Tan Yunxian 2015.

3. Tan Yunxian 2015, 122, tr. Wilcox.

4. Tr. Furth 2002, 296.

5. For what follows on the history of menopause in TCM, see Scheid 2007.

6. On the development of TCM, see Taylor 2005.

7. Scheid 2007, 59.

8. Marlowe 2010, 152.

9. I am unable to find the reference cited by Marlowe. Several sources cite the abstract of a publication by J. Phillips et al., "New Field Techniques for the Detection of Female Reproductive Status," published in volume 85, no. 2 of the *American Journal of Physical Anthropology*, but it appears not to be there.

10. Hill and Hurtado 1996, 235.

11. Beyene 1989, xiv.

12. Douglas 1966 is the classic introduction to the subject of menstrual taboos.

13. Bledsoe 2002, 174; Beyene 1989, 126; Furth 2002, 297–299; Rice 1995; Chirawatkul and Manderson 1994.

14. Martin et al. 1993; Beyene and Martin 2001.

15. Stewart 2003.

16. Michel et al. 2006.

17. Beyene and Martin 2001, 508.

18. Beyene 1989, 122.

19. Murphy et al. 2013. Medians were much higher than averages: 48.1, 49.2, and 52.8.

20. On age at menopause, see Wood 1994, 420–422; Melby, Lock, and Kaufert 2005, 497–498; Sievert 2006, ch. 4; Murphy et al. 2013.

21. On Puebla, see Sievert and Hautaniemi 2003.

22. Beyene and Martin 2001, 506.

23. Rosenberger 1987.

24. Zeserson 2001a, 2001b.

25. Singh and Sivakami 2014.

26. Bledsoe 2002, quotation on 163.

27. Rice 1995.

28. Bledsoe 2002, 226–227 and ch. 8.

29. Lee 1992.

30. See Brown 1992 for a general discussion of women's midlife (or "mature adulthood") in traditional societies. Collections of examples are Kerns and Brown 1992 and Shweder 1998 (the latter is intended to complicate the picture, but the examples do tend to confirm the pattern described).

CHAPTER 11: SYMPTOMS

1. On the nature of symptoms, see Aronowitz 2001.

2. A good overview of self-report methods and problems is the collection of articles in Stone 2000.

3. Datan, Antonovsky, and Maoz 1981. The question about the importance of menopause might have been more appropriately placed at the beginning of the 23-item list of questions about menopause, rather than the end. Large majorities of women in the most traditional group, Arab Israelis, and also in the most modern group, central European immigrants, answered that menopause was either not important or not very important.

4. Im 2006.

5. Lock 1993, table 1; Melby 2006.

6. National Institutes of Health 2005.

7. Melby, Sievert et al. 2011.

8. Dennerstein et al. 2007.

9. See also Obermeyer et al. 2001.

10. Freeman and Sherif 2007, 207. Reviews of studies of hot flashes include Sievert 2006, ch. 6; Freeman and Sherif 2007; Melby, Anderson, et al. 2011.

11. Spetz, Frederiksson, and Hammar 2003.

12. Freedman 2014.

13. Kronenberg 1990.

14. See Mattern 2015.

15. Melby 2005.

16. See also Zeserson 2001b.

17. Zeserson 2001b also interviewed women about these expressions, with somewhat different conclusions. For the women she talked to, *kaa* expressions were often accompanied by gestures indicating heat rising to the face.

18. Carpenter, Monahan, and Azzouz 2004; Freedman 2010.

19. Sievert et al. 2008.

20. Sievert et al. 2008, 602.

21. Sievert et al. 2016.

22. Boulet et al. 1994.

23. Melby, Sievert, et al. 2011, table 3.

24. Brown et al. 2009.

25. On the physiology of hot flashes, see Freedman 2001, 2014; Sievert 2006, 140–143; Kronenberg 2010. On climate, see Sievert and Flanagan 2005.

26. MacLennan et al. 2004; Freeman et al. 2015.

27. Lock 1993.

28. Melby, Lock, and Kaufert 2005. For a good discussion of the meaning of culture, see Kirmayer and Sartorius 2007.

29. The classic history is Sheehan 1936.

CHAPTER 12: A CULTURAL SYNDROME?

1. Kirmayer and Sartorius 2007.

2. Hinton et al. 2009.

3. Stern et al. 1996.

4. Good general discussions of cultural syndromes are Kirmayer and Sartorius 2007 and Hinton et al. 2009. A discussion with more historical background is Bhugra, Sumathipala, and Siribaddana 2007.

5. Kirmayer and Sartorius 2007.

6. For example, Barsky 2000.

7. On chronic pain syndromes, see Wessely, Nimnuan, and Sharpe 1999; Hennington, Zipfel, and Herzog 2007. On central sensitization, a widely accepted theory of these syndromes' cause, see Bourke, Langford, and White 2015.

8. Hinton et al. 2009; Good and Hinton 2009.

9. On the neurophysiology of anxiety, which is not well understood, a recommended discussion is Barlow 2002, 193–218.

10. Clark 1986.

11. Hinton, Um, and Ba 2001a, 2001b; Hinton and Good 2009.

12. Kim, Hogge, et al. 2014.

13. Suh 2013, 83.

14. Kim, Hogge, et al. 2014.

15. Min 2008; Suh 2013, 87–88.

16. On *hwa-byung* there is now a large body of research. The seminal article in English is Lin 1983. For more on clinical features and diagnostic criteria, see Park et al. 2002; Roberts, Han, and Weed 2006; Min 2008; Min, Suh, and Song 2009; Min and Suh 2010. On the history of the idea, see Suh 2013.

17. Min 2008, 131.

18. KBS World, "비타민 (Vitamin)," August 31, 2013, https://www.youtube.com/watch?v=0T_B_u4-GiY.

19. Im and Meleis 2000.

20. Im 2003.

21. Min 2008, 126.

22. Suh 2013, 91–92.

23. Nichter 1981; Suh 2013, 97–98.

24. Hanisch et al. 2008.

25. Hunter and Mann 2010; Hunter and Chilcot 2013. In the latter study, the authors found that among the factors they studied, beliefs had the strongest influence on how problematic women found their hot flashes to be (though not necessarily on the physiology of hot flashes).

26. Hanisch et al. 2008; Freeman et al. 2015; Muslić and Jokić-Begić 2016; Stefanopoulou and Grunfeld 2017.

27. Hunter and Mann 2010; Hunter and Chilcot 2013.

28. "Weapons of the weak" is the famous phrase of Scott 1985; cited, for example, in Kirmayer and Young 1998; Hinton, Um, and Ba 2001b, 441, 455. On neurasthenia, see Kleinman 1982, 165–169, 175–177. On *chi no michi*, see Rosenberger 1987, 167.

EPILOGUE: GOOD-BYE TO ALL THAT

1. De Beauvoir 1964, 657.

2. De Beauvoir 1974, 116.

3. De Beauvoir 1970; see also Moi 2008, 254–261.

4. De Beauvoir 1974, 30, 116.

5. De Beauvoir 1974, 134.

BIBLIOGRAPHY

•

Achebe, Nwando (2011). *The Female King of Colonial Nigeria: Ahebe Ugbabe.* Bloomington: Indiana University Press.

Adekunle, A. O., A. O. Fawole, and M. A. Okunlola (2000). Perceptions and attitudes of Nigerian women about the menopause. *Journal of Obstetrics and Gynaecology* 20: 525–529.

Aguilar, Alex, and Asunción Borrell (1988). Age- and sex-related changes in organochlorine compound levels in fin whales (*Balaena physalus*) from the eastern North Atlantic. *Marine Environmental Research* 25: 195–211.

Aiello, Leslie C., and Peter Wheeler (1995). The expensive-tissue hypothesis: The brain and the digestive system in human and primate evolution. *Current Anthropology* 36: 199–221.

Aklujkar, Vidyut (2004). Between pestle and mortar: Women in the Marathi Sant tradition. In *Goddesses and Women in the Indic Religious Tradition*, ed. Arvind Sharma, 105–130. Leiden: Brill.

Alberts, Susan C., Jeanne Altmann, Diane K. Brockman, Marina Cords, Linda M. Fedigan, Anne Pusey, Tara S. Stoinski, Karen B. Strier, William F. Morris, and Anne M. Bronikowski (2013). Reproductive aging patterns in primates reveal that humans are distinct. *Proceedings of the National Academy of Sciences* 110: 13,440–13,445.

Alley, Richard B. (2014). *The Two-Mile Time Machine: Ice Cores, Abrupt Climate Change, and Our Future*, 2nd ed. Princeton, NJ: Princeton University Press.

Amadiume, Ifi (1987). *Male Daughters, Female Husbands: Gender and Sex in an African Society.* London: Zed.

Amundsen, Darrel W., and Carol Jean Diers (1970). The age of menopause in classical Greece and Rome. *Human Biology* 42: 79–86.

Amundsen, Darrel W., and Carol Jean Diers (1973). The age of menopause in medieval Europe. *Human Biology* 45: 605–612.

Ande, Adedapo B., Oruerakpo P. Omu, Oluyinka L. Ande, and Nelson B. Olagbuji (2011). Features and perceptions of menopausal women in Benin City, Nigeria. *Annals of African Medicine* 10: 300–304.

Anderson, Debra, Melissa K. Melby, Lynnette Leidy Sievert, and Carla Makhlouf Obermeyer (2011). Methods used in cross-cultural comparisons of psychological symptoms and their determinants. *Maturitas* 70: 120–126.

Angeles, Luis (2010). Demographic transitions: Analyzing the effects of mortality on fertility. *Journal of Population Economics* 23: 99–120.

Antón, Susan C. (2003). Natural history of *Homo erectus*. *Yearbook of Physical Anthropology* 46: 126–170.

Antón, Susan C., Richard Potts, and Leslie Aiello (2014). Evolution of early *Homo*: An integrated biological perspective. *Science* 344: 45 and 1236828. doi: 10.1126/science.1236828.

Antón, Susan C., and J. Josh Snodgrass (2012). Origins and evolution of genus *Homo*: New perspectives. *Current Anthropology* 53, suppl. 6: S479–S496.

Antony, Robert J. (1989). Peasants, heroes, and brigands: The problems of social banditry in early nineteenth-century south China. *Modern China* 15: 123–148.

Arens, Richard, ed. (1976). *Genocide in Paraguay*. Philadelphia: Temple University Press.

Arnaud, Sabine (2015). *On Hysteria: The Invention of a Medical Category Between 1670 and 1820*. Chicago: University of Chicago Press.

Aronowitz, Robert A. (2001). When do symptoms become a disease? *Annals of Internal Medicine* 134: 803–808.

Astruc, Jean (1761). *Traité des maladies des femmes*, 6 vols. Paris: Cavelier.

Aubel, Judi, Ibrahima Touré, and Mamadou Diagne (2004). Senegalese grandmothers promote improved maternal and child nutrition practices: The guardians of tradition are not averse to change. *Social Science & Medicine* 59: 945–959.

Austad, Steven N. (1994). Menopause: An evolutionary perspective. *Experimental Gerontology* 29: 25–263.

Austad, Steven N. (2006). Why women live longer than men: Sex differences in longevity. *Gender Medicine* 3: 79–92.

Austad, Steven N., and Kathleen E. Fischer (2016). Sex differences in lifespan. *Cell Metabolism* 23: 1022–1033.

Ayers Counts, Dorothy (1992). *Tamparonga*: "The big women" of Kalai. In *In Her Prime: New Views of Middle-Aged Women*, 2nd ed., ed. Virginia Kerns and Judith K. Brown, 61–76. Urbana: University of Illinois Press.

Bagnall, Roger S., and Bruce W. Frier (1994). *The Demography of Roman Egypt*. Cambridge: Cambridge University Press.

Baird, Robin W. (2000). The killer whale: Foraging specializations and group hunting. In *Cetacean Societies*, ed. Janet Mann, Richard C. Connor, Peter L. Tyack, and Hal Whitehead, 127–153. Chicago: University of Chicago Press.

Baker, Matthew J., and Joyce P. Jacobsen (2007). A human capital-based theory of postmarital residence rules. *Journal of Law, Economics, & Organization* 23: 208–241.

Bale, Anthony (2015). *The Book of Margery Kempe*. Oxford World's Classics. Oxford: Oxford University Press.

Balzeau, Antoine, Ralph L. Holloway, and Dominique Grimaud-Hervé (2012). Variations and asymmetries in regional brain surface in the genus *Homo*. *Journal of Human Evolution* 62: 696–706.

Bamberg Migliano, Andrea, Lucio Vinicius, and Marta Mirazón Lahr (2007). Life history trade-offs explain the evolution of human pygmies. *Proceedings of the National Academy of Sciences* 104: 20,216–20,219.

Bamforth, Douglas, and Brigid Grund (2012). Radiocarbon calibration curves, summed probability distributions, and early Paleoindian population trends in North America. *Journal of Archaeological Science* 39: 1768–1774.

Banner, Lois W. (1992). *In Full Flower: Aging Women, Power, and Sexuality*. New York: Knopf.

Baradaran, Mehrsa (2017). *The Color of Money: Black Banks and the Racial Wealth Gap*. Cambridge, MA: Belknap Press of Harvard University Press.

Barash, David, and Judith Eve Lipton (2001). *The Myth of Monogamy*. New York: Freeman.

Barber, Elizabeth Wayland (1994). *Women's Work: The First 20,000 Years: Women, Cloth, and Society in Early Times*. New York: Norton.

Barlow, David H. (2002). *Anxiety and Its Disorders: The Nature and Treatment of Anxiety and Panic*, 2nd ed. New York: Guilford Press.

Barsky, Arthur J. (2000). The validity of bodily symptoms in medical outpatients. In *The Science of Self-Report: Implications for Research and Practice*, ed. Arthur A. Stone, 339–362. Mahwah, NJ: Erlbaum.

Bashford, Alison (2014). *Global Population: History, Geopolitics, and Life on Earth*. New York: Columbia University Press.

Bean, Lee L., Geraldine P. Mineau, and Douglas L. Anderton (1990). *Fertility Change on the American Frontier: Adaptation and Innovation*. Berkeley: University of California Press.

Beck, Christopher W., and Daniel E. L. Promislow (2007). Evolution of female preference for younger males. *PLOS ONE* 9: e939.

Beier, Judith, Nils Anthes, Joachim Wahl, and Katerina Harvati (2018). Similar cranial trauma prevalence among Neanderthals and Upper Paleolithic modern humans. *Nature* 563: 686–690.

Bell, Matthew (2014). *Melancholia: The Western Malady*. Cambridge: Cambridge University Press.

Bentley, Gillian R., Tony Goldberg, and Grażyna Jasieńska (1993). The fertility of agricultural and non-agricultural traditional societies. *Population Studies* 47: 269–281.

Bernhardt, Kathryn (1999). *Women and Property in China, 960–1949*. Stanford, CA: Stanford University Press.

Betzig, Laura (1986). *Despotism and Differential Reproduction: A Darwinian View of History*. New York: Aldine.

Betzig, Laura (1992). Roman polygyny. *Ethology and Sociobiology* 13: 309–349.

Beyene, Yewoubdar (1989). *From Menarche to Menopause: Reproductive Lives of Women in Two Cultures*. Albany: State University of New York Press.

Beyene, Yewoubdar, and Mary C. Martin (2001). Menopausal experiences and bone density of Mayan women in Yucatan, Mexico. *American Journal of Human Biology* 13: 505–511.

Bezner Kerr, Rachel, Laifolo Dakishoni, Lizzie Shumba, Rodgers Msachi, and Marko Chirwa (2008). "We grandmothers know plenty": Breastfeeding, complementary feeding and the multifaceted role of grandmothers in Malawi. *Social Science and Medicine* 66: 1095–1105.

Bhavalkar, Tara (1996). Women saint-poets' conception of liberation. In *Images of Women in Maharashtrian Literature and Religion*, ed. Anne Feldhaus, 239–252. Albany: State University of New York Press.

Bhugra, Dinesh, Athula Sumathipala, and Sisira Sirabaddana (2007). Culture-bound syndromes: A re-evaluation. In *Textbook of Cultural Psychiatry*, ed. Dinesh Bhugra and Kamaldeep Bhui, 141–156. Cambridge: Cambridge University Press.

Billingsley, Phil (1988). *Bandits in Republican China*. Stanford, CA: Stanford University Press.

Björk Eydal, Guðný, and Ingólfur Gíslason (2014). Family choices: The case of Iceland. In *Handbook of Family Policies Across the Globe*, ed. Mihaela Robina, 109–124. New York: Springer.

Black, Richard, Stephen R. G. Bennett, Sandy M. Thomas, and John R. Beddington (2011). Climate change: Migration as adaptation. *Nature* 478: 447–449.

Bledsoe, Caroline H. (2002). *Contingent Lives: Fertility, Time, and Aging in West Africa.* Chicago: University of Chicago Press.

Bliege Bird, Rebecca, and Douglas W. Bird (2008). Why women hunt: Risk and contemporary foraging in a western desert Aboriginal community. *Current Anthropology* 49: 655–693.

Bliege Bird, Rebecca, Brooke Scelza, Douglas W. Bird, and Eric Alden Smith (2012). The hierarchy of virtue: Mutualism, altruism and signaling in Martu women's cooperative hunting. *Evolution and Human Behavior* 33: 64–78.

Bliege Bird, Rebecca, Eric Alden Smith, and Douglas W. Bird (2001). The hunting handicap: Costly signaling in human foraging strategies. *Behavioral Ecology and Sociobiology* 50: 9–19.

Blurton Jones, Nicholas G. (2006). Contemporary hunter-gatherers and human life history evolution. In *The Evolution of Human Life History*, ed. Kristen Hawkes and Richard R. Paine, 231–266. Santa Fe, NM: School of American Research Press.

Blurton Jones, Nicholas G. (2016). *Demography and Evolutionary Ecology of Hadza Hunter-Gatherers.* Cambridge: Cambridge University Press.

Blurton Jones, Nicholas G., Kristen Hawkes, and James F. O'Connell (2002). Antiquity of postreproductive life: Are there modern impacts on hunter-gatherer postreproductive life spans? *American Journal of Human Biology* 14: 184–205.

Blurton Jones, Nicholas G., Kristen Hawkes, and James F. O'Connell (2005). Older Hadza men and women as helpers. In *Hunter-Gatherer Childhoods: Evolutionary, Developmental, and Cultural Practices*, ed. Barry S. Hewlett and Michael E. Lamb, 214–236. New Brunswick, NJ: Transaction.

Blurton Jones, Nicholas G., Frank W. Marlowe, Kristen Hawkes, and James F. O'Connell (2000). Paternal investment and hunter-gatherer divorce rates. In *Adaptation and Human Behavior: An Anthropological Perspective*, ed. Lee Cronk, Napoleon Chagnon, and William Irons, 69–90. New York: de Gruyter.

Blurton Jones, Nicholas G., Lars C. Smith, James F. O'Connell, Kristen Hawkes, and C. L. Kamuzora (1992). Demography of the Hadza, an increasing and high density population of savanna foragers. *American Journal of Physical Anthropology* 89: 159–181.

Bocquet-Appel, Jean-Pierre (2011). When the world's population took off: The springboard of the Neolithic Demographic Transition. *Science* 333: 560–561.

Bocquet-Appel, Jean-Pierre, and Ofer Bar-Yosef, eds. (2008). *The Neolithic Demographic Transition and Its Consequences.* London: Springer.

Bocquet-Appel, Jean-Pierre, and Anna Degionanni (2013). Neanderthal demographic estimates. *Current Anthropology* 54, suppl. 8: S202–S213.

Bocquet-Appel, Jean-Pierre, and Claude Masset (1982). Farewell to paleodemography. *Journal of Human Evolution* 11: 321–333.

Boehm, Christopher (1999). *Hierarchy in the Forest: The Evolution of Egalitarian Behavior.* Cambridge, MA: Harvard University Press.

Boesch, Christophe, and Hedwige Boesch (1984). Possible causes of sex differences in the use of natural hammers by wild chimpanzees. *Journal of Human Evolution* 13: 415–440.

Boesch, Christophe, and Hedwige Boesch-Achermann (2000). *The Chimpanzees of the Taï Forest: Behavioural Ecology and Evolution.* Oxford: Oxford University Press.

Bogin, Barry (2009). Childhood, adolescence, and longevity: A multilevel model of the evolution of reserve capacity in human life history. *American Journal of Human Biology* 21: 567–577.

Bongaarts, John (1980). Does malnutrition affect fecundity?: A summary of evidence. *Science* 208: 564–569.

Bongaarts, John, and John Casterline (2012). Fertility transition: Is sub-Saharan Africa different? *Population and Development Review* 38, suppl.: S153–S168.

Boone, James L. (1983). Noble family structure and expansionist warfare in the late Middle Ages: A socioecological approach. In *Rethinking Human Adaptation: Biological and Cultural Models*, ed. Rada Dyson-Hudson and Michael A. Little, 79–98. Boulder, CO: Westview.

Boone, James L. (2002). Subsistence strategies and early human population history: An evolutionary ecological perspective. *World Archaeology* 34: 6–25.

Borgerhoff Mulder, Monique (2007). Hamilton's rule and kin competition: The Kipsigis case. *Evolution and Human Behavior* 28: 299–312.

Borgerhoff Mulder, Monique, Samuel Bowles, Tom Hertz, Adrian Bell, Jan Biese, Greg Clark, Ila Fazzio, et al. (2009). Intergenerational wealth transmission and the dynamics of inequality in small-scale societies. *Science* 326: 682–688.

Borgerhoff Mulder, Monique, Ila Fazzio, William Irons, Richard L. McElreath, Samuel Bowles, Adrian Bell, Tom Hertz, and Leela Hassah (2010). Pastoralism and wealth inequality: Revisiting an old question. *Current Anthropology* 51: 35–48.

Borgerhoff Mulder, Monique, and Kristin Liv Rauch (2009). Sexual conflict in humans: Variations and solutions. *Evolutionary Anthropology* 18: 201–214.

Boserup, Ester (1965). *The Conditions of Agricultural Growth*. Chicago: Aldine.

Boserup, Ester (1970). *Woman's Role in Economic Development*. New York: St. Martin's.

Boswell, John (1988). *The Kindness of Strangers: The Abandonment of Children in Western Europe from Late Antiquity to the Renaissance*. New York: Pantheon.

Boulet, M. J., B. J. Oddens, P. Lehert, H. M. Vemer, and A. Visser (1994). Climacteric and menopause in seven south-east Asian countries. *Maturitas* 19: 157–176.

Bourguignon, François (2015). *The Globalization of Inequality*. Tr. Thomas Scott-Railton. Princeton, NJ: Princeton University Press.

Bourguignon, François, and Christian Morrisson (2002). Inequality among world citizens: 1820–1922. *American Economic Review* 92: 727–744.

Bourke, Julius H., Richard M. Langford, and Peter D. White (2015). The common link between functional somatic syndromes may be central sensitization. *Journal of Psychosomatic Research* 78: 228–236.

Bowles, Samuel (2012). *The New Economics of Inequality and Redistribution*. Cambridge: Cambridge University Press.

Bradley, Harriet (1989). *Men's Work, Women's Work: A Sociological History of the Sexual Division of Labour in Employment*. Minneapolis: University of Minnesota Press.

Brent, Lauren M., Daniel W. Franks, Emma A. Foster, Kenneth C. Balcomb, Michael A. Cant, and Darren P. Croft (2015). Ecological knowledge, leadership, and the evolution of menopause in killer whales. *Current Biology* 25: 1–5.

Briggs, Laura (2002). *Reproducing Empire: Race, Sex, Science, and U.S. Imperialism in Puerto Rico*. Berkeley: University of California Press.

Bromberger, Joyce T., Howard M. Kravitz, Y.-F. Chang, J. M. Cyranowski, C. Brown, and Karen A. Matthews (2011). Major depression during and after the menopausal transition: Study of Women's Health Across the Nation (SWAN). *Psychological Medicine* 41: 1879–1888.

Bromberger, Joyce T., Karen A. Matthews, Laura L. Schott, Sarah Brockwell, Nancy E. Avis, Howard M. Kravitz, Susan A. Everson-Rose, Ellen B. Gold, MaryFran Sowers, and John F. Randolph, Jr. (2007). Depressive symptoms during the menopausal transition: The Study of Women's Health Across the Nation (SWAN). *Journal of Affective Disorders* 103: 267–272.

Brooke, John L. (2014). *Climate Change and the Course of Global History: A Rough Journey.* New York: Cambridge University Press.

Brooks, Robert (2012). "Asia's missing women" as a problem in applied evolutionary psychology? *Evolutionary Psychology* 12: 910–925.

Brosch, R., S. V. Gordon, M. Marmiesse, P. Brodin, C. Buchrieser, K. Eiglmeier, T. Garnier, et al. (2002). A new evolutionary scenario for *Mycobacterium tuberculosis* complex. *Proceedings of the National Academy of Sciences of the United States of America* 99: 3684–3689.

Broussard, D. R., T. S. Risch, F. S. Dobson, and J. O. Murie (2003). Senescence and age-related reproduction of female Columbian ground squirrels. *Journal of Animal Ecology* 72: 212–219.

Brown, D. E., L. L. Sievert, L. A. Morrison, A. M. Reza, and P. S. Mills (2009). Do Japanese American women really have fewer hot flashes than European Americans? The Hilo Women's Health Study. *Menopause* 16: 870–876.

Brown, Judith K. (1992). Lives of middle-aged women. In *In Her Prime: New Views of Middle-Aged Women*, 2nd ed., ed. Virginia Kerns and Judith K. Brown, 17–34. Urbana: University of Illinois Press.

Brown, Kyle S., Curtis W. Marean, Zenobia Jacobs, Benjamin J. Schoville, Simen Oestmo, Erich C. Fisher, Jocelyn Bernatchez, Panagiotis Karkanas, and Thalassa Matthews (2012). An early and enduring advanced technology originating 71,000 years ago in South Africa. *Nature* 491: 590–593.

Brown, P., T. Sutikna, M. J. Morwood, R. P. Soejono, E. Wayhu Saptomo, and Rokus Awe Due (2004). A new small-bodied hominin from the late Pleistocene of Flores, Indonesia. *Nature* 431: 1055–1061.

Brown, William A. (2015). Through a filter, darkly: Population size estimation, systematic error, and random error in radiocarbon-supported demographic temporal frequency analysis. *Journal of Archaeological Science* 53: 133–147.

Bryant, John (2007). Theories of fertility decline and the evidence from development indicators. *Population and Development Review* 33: 101–127.

Burke, Ariane (2012). Spatial abilities, cognition, and the pattern of Neanderthal and modern human dispersals. *Quaternary International* 247: 230–235.

Burrow, J. A. (1986). *The Ages of Man: A Study in Medieval Writing and Thought.* Oxford: Oxford University Press.

Butovskaya, Marina L. (2013). Aggression and conflict resolution among the nomadic Hadza of Tanzania as compared with their pastoralist neighbors. In *War, Peace, and Human Nature: The Convergence of Evolutionary and Cultural Views*, ed. Douglas P. Fry, 278–296. New York: Oxford University Press.

Caldwell, John C. (1976). Toward a restatement of Demographic Transition theory. *Population and Development Review* 2: 321–366.

Caldwell, John C., ed. (2006). *Demographic Transition Theory.* Dordrecht: Springer.

Caldwell, John C., and Bruce K. Caldwell (2003). Pretransitional population control and equilibrium. *Population Studies* 57: 199–215.

Caldwell, John C., and Bruce K. Caldwell (2005). Family size control by infanticide in the great agrarian societies of Asia. *Journal of Comparative Family Studies* 36: 205–226.

Camperio Ciani, Andrea, Francesca Corna, and Claudio Capiluppi (2004). Evidence for maternally inherited factors favouring male homosexuality and promoting female fecundity. *Proceedings of the Royal Society B: Biological Sciences* 271: 2217–2221.

Camperio Ciani, Andrea, and Lilybeth Fontanesi (2012). Mothers who kill their offspring: Testing evolutionary hypothesis in a 110-case Italian sample. *Child Abuse & Neglect* 36: 519–527.

Cann, Rebecca, Mark Stoneking, and Allan C. Wilson (1987). Mitochondrial DNA and human evolution. *Nature* 325: 31–36.

Cant, Michael A., and Rufus A. Johnstone (2008). Reproductive conflict and the separation of reproductive generations. *Proceedings of the National Academy of Sciences* 105: 5332–5336.

Carpenter, Janet S., Patrick O. Monahan, and Faouzi Azzouz (2004). Accuracy of subjective hot flush reports compared with continuous sternal conductance monitoring. *Obstetrics & Gynecology* 104: 1322–1326.

Caspari, Rachel, and Sang-Hee Lee (2005). Are OY ratios invariant? A reply to Hawkes and O'Connell (2005). *Journal of Human Evolution* 49: 654–659.

Caspari, Rachel, Sang-Hee Lee, and Ward H. Goodenough (2004). Older age becomes common late in human evolution. *Proceedings of the National Academy of Sciences* 101: 10895–10990.

Chalmers, Alisa, Michael A. Huffman, Naoki Koyama, and Yukio Takahata (2012). Fifty years of female Japanese macaque demography at Arashiyama, with special reference to long-lived females (>25 years). In *The Monkeys of Stormy Mountain*, ed. Jean-Baptiste Leca, Michael A. Huffman, and Paul L. Vasey, 51–67. Cambridge and New York: Cambridge University Press.

Chan, Matthew H., Kristen Hawkes, and Peter S. Kim (2016). Evolution of longevity, age at last birth, and sexual conflict with grandmothering. *Journal of Theoretical Biology* 383: 145–157.

Chapais, Bernard (2008). *Primeval Kinship: How Pair-Bonding Gave Birth to Human Society*. Cambridge: Harvard University Press.

Charnov, Eric L. (1993). *Life History Invariants: Some Explorations of Symmetry in Evolutionary Ecology*. Oxford and New York: Oxford University Press.

Chayanov, A. V. (1966). *The Theory of Peasant Economy*. Ed. Daniel Thorner, Basile Kerblay, and R.E.F. Smith. Tr. Christel Lane and R.E.F. Smith. Madison: University of Wisconsin Press.

Chirawatkul, Siriporn, and Lenore Manderson (1994). Perceptions of menopause in northeast Thailand: Contested meaning and practice. *Social Science & Medicine* 39: 1545–1554.

Chrisholm, James S., and Victoria K. Burbank (1991). Monogamy and polygyny in southeast Arnhem Land: Male coercion and female choice. *Ethology and Sociology* 12: 291–313.

Chu, C. Y. Cyrus, Hung-Ken Chien, and Ronald D. Lee (2008). Explaining the optimality of U-shaped age-specific mortality. *Theoretical Population Biology* 73: 171–180.

Chu, C. Y. Cyrus, and Ronald D. Lee (2012). Sexual dimorphism and sexual selection: A unified economic analysis. *Theoretical Population Biology* 82: 355–363.

Chu, C. Y. Cyrus, and Ronald D. Lee (2013). On the evolution of intergenerational division of labor, menopause and transfers among adults and offspring. *Journal of Theoretical Biology* 332: 171–180.

Clark, David (1986). A cognitive approach to panic. *Behaviour Research and Therapy* 24: 461–470.

Cleland, John, and Christopher Wilson (1987). Demand theories of the fertility transition: An iconoclastic view. *Population Studies* 51: 321–366.

Clutton-Brock, Tim (2006). Cooperative breeding in mammals. In *Cooperation in Primates and Humans: Mechanisms and Evolution*, ed. Peter M. Kappeler and Carol P. van Schaik, 139–150. Berlin: Springer.

Coale, Ansley J. (1991). Excess female mortality and the balance of the sexes in the population: An estimate of the number of "missing females." *Population and Development Review* 17: 517–523.

Coale, Ansley J., and Paul George Demeny (1983). *Regional Model Life Tables and Stable Populations*, 2nd ed. New York: Academic Press.

Coale, Ansley J., and Susan Cotts Watkins, eds. (1987). *The Decline of Fertility in Europe.* Princeton, NJ: Princeton University Press.

Codding, Brian F., Rebecca Bliege Bird, and Douglas W. Bird (2011). Provisioning offspring and others: Risk-energy trade-offs and gender differences in hunter-gatherer foraging strategies. *Proceedings of the Royal Society B: Biological Sciences* 278: 2502–2509.

Cofran, Zachary, and Jeremy M. DeSilva (2015). A neonatal perspective on *Homo erectus* brain growth. *Journal of Human Evolution* 81: 41–47.

Cohen, Alan A. (2004). Female post-reproductive lifespan: A general mammalian trait. *Biological Reviews of the Cambridge Philosophical Society* 79: 733–750.

Cohen, David J. (1991). *Law, Sexuality, and Society: The Enforcement of Morals in Classical Athens.* Cambridge: Cambridge University Press.

Cohen, Mark Nathan (1977). *The Food Crisis in Prehistory: Overpopulation and the Origins of Agriculture.* New Haven, CT: Yale University Press.

Cohen, Mark Nathan (2008). Implications of the NDT for world wide health and mortality in prehistory. In *The Neolithic Demographic Transition and Its Consequences*, ed. Jean-Pierre Bocquet-Appel and Ofer Bar-Yosef, 481–500. London: Springer.

Coleman, David, and Robert Rowthorn (2011). Who's afraid of population decline? A critical examination of its consequences. *Population and Development Review* 37, suppl.: S217–S248.

Connelly, Matthew (2008). *Fatal Misconception: The Struggle to Control World Population.* Cambridge, MA: Belknap.

Contreras, Daniel A., and John Meadows (2014). Summed radiocarbon calibrations as a population proxy: A critical evaluation using a realistic simulation approach. *Journal of Archaeological Science* 52: 591–608.

Coontz, Stephanie (1992). *The Way We Never Were: American Families and the Nostalgia Trap.* New York: Basic Books.

Corsini, Carlo A. (1976). Materiali per lo studio della famiglia in Toscana. *Quaderni Storici* 11: 998–1052.

Cox, Cheryl Anne (1998). *Household Interests: Property, Marriage Strategies, and Family Dynamics in Ancient Athens.* Princeton, NJ: Princeton University Press.

Coxworth, James E., Peter S. Kim, John S. McQueen, and Kristen Hawkes (2015). Grandmothering life histories and human pair bonding. *Proceedings of the National Academy of Sciences* 112: 11,806–11,811.

Crawford, Patricia (1981). Attitudes to menstruation in seventeenth-century England. *Past & Present* 91: 41–73.

Crawford, Patricia (1984). Printed advertisements for women medical practitioners in London, 1670–1710. *Society for the Social History of Medicine Bulletin* 35: 66–70.

Crittenden, Alyssa N., and Frank W. Marlowe (2008). Allomaternal care among the Hadza of Tanzania. *Human Nature* 19: 249–262.

Croft, Darren P., Lauren J. N. Brent, Daniel W. Franks, and Michael A. Cant (2015). The evolution of prolonged life after reproduction. *Trends in Ecology & Evolution* 30: 407–416.

Croft, Darren P., Rufus A. Johnstone, Samuel Ellis, Stuart Nattrass, Daniel W. Franks, Lauren J. N. Brent, Sonia Mazzi, Kenneth C. Balcomb, John K. B. Ford, and Michael A. Cant (2017). Reproductive conflict and the evolution of menopause in killer whales. *Current Biology* 27: 298–304.

Crosby, Alfred W. (1972). *The Columbian Exchange: Biological and Cultural Consequences of 1492.* Westport, CT: Greenwood.

Currier, Andrew (1897). *The Menopause: A Consideration of the Phenomena Which Occur to Women at the Close of the Child-Bearing Period.* New York: Appleton & Co.

Das Gupta, Monica, Jiang Zhenghua, Li Bohua, Xie Zhenming, Woojin Chung, and Bae Hwa-Ok (2003). Why is son preference so persistent in East and South Asia? A cross-country study of China, India, and the Republic of Korea. *Journal of Development Studies* 40: 153–187.

Datan, Nancy, Aaron Antonovsky, and Benjamin Maoz (1981). *A Time to Reap: The Middle Age of Women in Five Israeli Subcultures.* Baltimore, MD: Johns Hopkins University Press.

Davies, R.P.O., K. Tocque, M. A. Bellis, T. Rimmington, and P.D.O. Davies (1999). Historical declines in tuberculosis in England and Wales: Improving social conditions or natural selection? *International Journal of Tuberculosis and Lung Disease* 3: 1051–1054.

Davis, Dona Lee (1983). *Blood and Nerves: An Ethnographic Focus on Menopause.* St. Johns, Newfoundland: Memorial University of Newfoundland, Institute of Social and Economic Research.

Davis, Kingsley (1945). The world demographic transition. *Annals of the American Academy of Political and Social Science* 237: 1–11.

Dean, Christopher, Meave G. Leakey, Donald Reid, Friedemann Schrenk, Gary T. Schwartz, Christopher Stringer, and Alan Walker (2001). Growth processes in teeth distinguish modern humans from *Homo erectus* and earlier hominins. *Nature* 414: 628–631.

Dean-Jones, Lesley (1994). *Women's Bodies in Classical Greek Science.* Oxford and New York: Oxford University Press.

de Beauvoir, Simone (1964). *Force of Circumstance.* Tr. Richard Howard. New York: G. P. Putnam's Sons.

de Beauvoir, Simone (1970). *La vieillesse.* Paris: Gallimard, 1970.

de Beauvoir, Simone (1974). *All Said and Done.* Tr. Patrick O'Brian. London: André Deutsch and Weidenfeld and Nicholson.

DeConto, Robert M., and David Pollard (2016). The contribution of Antarctica to past and future sea-level rise. *Nature* 531: 591–597.

de Gardanne, Charles Paul Louis (1816). *Avis aux femmes qui entrent dans l'age critique.* Paris: Gabon.

Dembo, Mana, Nicholas J. Matzke, Arne Ø. Mooers, and Mark Colland (2015). Bayesian analysis of a morphological supermatrix sheds light on controversial fossil human relationships. *Proceedings of the Royal Society B: Biological Sciences* 282: 20150943. doi: 10.1098/rspb .2015.0943.

Demeny, Paul (2011). Population policy and the Demographic Transition: Performance, prospects, and options. *Population and Development Review* 37, suppl.: S249–S274.

Dennell, Robin (2009). *The Paleolithic Settlement of Asia.* Cambridge: Cambridge University Press.

Dennerstein, Lorraine, Philippe Lehert, Patricia E. Koochaki, Alessandra Grazziotin, Sandra Leiblum, and Jeanne Leventhal Alexander (2007). A symptomatic approach to understanding women's health experiences: A cross-cultural comparison of women aged 20 to 70 years. *Menopause* 14: 688–696.

de Rachewiltz, Igor (2004–2013). *The Secret History of the Mongols: A Mongolian Chronicle of the Thirteenth Century.* 3 vols. Leiden: Brill.

Derber, Charles (2002). *People Before Profit: The New Globalization in an Age of Terror, Big Money, and Economic Crisis.* New York: St. Martin's.

d'Errico, Francesco, and Chris B. Stringer (2011). Evolution, revolution, or saltation scenario for the emergence of modern cultures? *Philosophical Transactions of the Royal Society B: Biological Sciences* 366: 1060–1069.

DeSilva, Jeremy M. (2018). Comment on "The Growth Pattern of Neandertals, Reconstructed from a Juvenile Skeleton from El Sidrón (Spain)." *Science* 359. doi: 10.1126/science.aar3611.

DeSilva, Jeremy, and Julie Lesnik (2006). Chimpanzee neonatal brain size: Implications for brain growth in *Homo erectus. Journal of Human Evolution* 51: 207–212.

Deutsch, Helene (1944). *The Psychology of Women,* 2 vols. New York: Grune & Stratton.

Dickeman, Mildred (= Mildred Dickemann) (1975). Demographic consequences of infanticide in man. *Annual Review of Ecology and Systematics* 6: 107–137.

Dickemann, Mildred (1979). The ecology of mating systems in hypergynous dowry societies. *Social Science Information* 19: 163–195.

Dickemann, Mildred (1997). The Balkan sworn virgin: A cross-gendered female role. In *Islamic Homosexualities: Culture, History, and Literature,* ed. Stephen O. Murray and Will Roscoe, 197–203. New York: New York University Press.

Dobson, Mary (1997). *Contours of Death and Disease in Early Modern England.* Cambridge: Cambridge University Press.

Douglas, Mary (1966). *Purity and Danger: An Analysis of the Concepts of Pollution and Taboo.* London: Routledge and Kegan Paul.

Douglass, Richard L., and Brenda F. McGadney-Douglass (2008). The role of grandmothers and older women in the survival of children with kwashiorkor in urban Accra, Ghana. *Research in Human Development* 5: 26–43.

Downey, Sean S., Emmy Bocaege, Tim Kerig, Kevan Edinborough, and Stephan Shennan (2014). The Neolithic Demographic Transition in Europe: Correlation with juvenility index supports interpretation of the summed calibrated radiocarbon date probability distribution (SCDPD) as a valid demographic proxy. *PLOS ONE* 9 (8): e105730. doi: 10.1371/journal.pone.0105730.

Dube Bhatnagar, Rashmi, Renu Dube, and Reena Dube (2005). *Female Infanticide in India: A Feminist Cultural History.* Albany: State University of New York Press.

Duby, Georges (1964). Dans la France du nord-ouest au XIIe siècle: Les "jeunes" dans la société aristocratique. *Annales. Economie, société, civilisation* 5: 835–846.

Duden, Barbara (1991). *The Woman Beneath the Skin*. Cambridge, MA: Harvard University Press.

Duggan, Ana T., Maria F. Perdomo, Dario Piombino-Mascali, Stephanie Marciniak, Debi Poinar, Matthew V. Emery, Jan P. Buchmann, et al. (2016). 17th century variola virus reveals the recent history of smallpox. *Current Biology* 26: 3407–3412.

Duvernoy, Jean (1965). *Le registre d'inquisition de Jacques Fournier, évêque de Pamiers (1318–1325)*, 3 vols. Toulouse: Édouard Privat.

Dyble, M., G. D. Salali, N. Chaudhary, A. Page, D. Smith, J. Thompson, L. Vinicius, R. Mace, and A. B. Migliano (2015). Sex equality can explain the unique social structure of hunter-gatherer bands. *Science* 348: 796–798.

Ehrenreich, Barbara (1983). *The Hearts of Men: American Dreams and the Flight from Commitment*. Garden City, NY: Anchor.

Ehrenreich, Barbara, and Deirdre English (1973). *Witches, Midwives, and Nurses: A History of Women Healers*. Old Westbury, NY: Feminist Press.

Ellison, Peter T. (2001). *On Fertile Ground*. Cambridge, MA: Harvard University Press.

Ember, Carol R. (1983). The relative decline in women's contribution to agriculture with intensification. *American Anthropologist* (n.s.) 85: 285–304.

Ember, Melvin, and Carol R. Ember (1971). The conditions favoring matrilocal versus patrilocal residence. *American Anthropologist* (n.s.) 73: 571–594.

Emery Thompson, Melissa, James H. Jones, Anne E. Puysey, Stella Brewer-Marsden, Jane Goodall, David Marsden, Tetsuro Matsuzawa, et al. (2007). Aging and fertility patterns in wild chimpanzees provide insight into the evolution of menopause. *Current Biology* 17: 2150–2156.

Emlen, Stephen T. (1995). An evolutionary theory of the family. *Proceedings of the National Academy of Sciences* 92: 8092–8099.

Engammare, Max (2013). *Soixante-trois: La peur de la grande année climactérique à la Renaissance*. Geneva: Droz.

Engels, Friedrich (1902). *The Origin of the Family, Private Property, and the State*. Tr. Ernest Untermann. Chicago: C. H. Kerr.

Engels, Friedrich (1942). *The Origin of the Family, Private Property, and the State* (*Marxist Library: Works of Marxism-Leninism* 22). Tr. Alick West. New York: International Publishers, 1942.

Engels, Friedrich (1975). Outlines of a critique of political economy. Tr. Martin Milligan. In *Marx/Engels Collected Works*, 3: 418–443. New York: International Publishers.

Erdélyi, Gabriella (2015). *A Cloister on Trial: Religious Culture and Everyday Life in Late Medieval Hungary*. Farnham, UK: Ashgate.

Erikson, Amy Louise (1993). *Women and Property in Early Modern England*. London: Routledge.

Ermini, Luca, Clio der Sarkissian, Eske Willerslev, and Ludovic Orlando (2015). Major transitions in human evolution revisited: A tribute to ancient DNA. *Journal of Human Evolution* 79: 4–20.

Estioko-Griffin, Agnes, and P. Bion Griffin (1981). Woman the hunter: The Agta. In *Woman the Gatherer*, ed. Frances Dahlberg, 121–152. New Haven, CT: Yale University Press.

Fedigan, Linda Marie, and Pamela J. Asquith, eds. (1991). *The Monkeys of Arashiyama: 35 Years of Research in Japan and the West*. Albany: State University of New York Press.

Fei Hsiao-Tung (Fei Xiaotong) (1939). *Peasant Life in China: A Field Study of Country Life in the Yangtze Valley.* London: Routledge and Keegan Paul.

Feldhaus, Anne (1982). Bahiṇā Bāī: Wife and saint. *Journal of the American Academy of Religion* 50: 591–604.

Ferreira, Francisco H. G., Shaohua Chen, Andrew Dabalen, Yuri Dikhanov, Nada Hamadeh, Dean Joliffe, Ambar Narayan, et al. (2015). A global count of the extreme poor 2012: Data issues, methodology, and initial results. Policy Research Working Paper 7432, World Bank Group, Poverty Global Practice Group and Development Data and Research Group.

Finlayson, Clive, Kimberly Brown, Ruth Blasco, Jordi Rosell, Juan José Negro, Gary R. Bortolotti, Geraldine Finlayson, et al. (2012). Birds of a feather: Neanderthal exploitation of raptors and corvids. *PLOS ONE* 7 (9): e45927. doi: 10.1371/journal.pone.0045927.

Flemming, Rebecca (2000). *Medicine and the Making of Roman Women.* Oxford: Oxford University Press.

Flemming, Rebecca, and Ann Ellis Hanson (1998). Hippocrates' *Peri Partheniôn* (*Diseases of Young Girls*): Text and translation. *Early Science and Medicine* 3: 241–252.

Foley, Charles, Nathalie Pettorelli, and Lara Foley (2008). Severe drought and calf survival in elephants. *Biology Letters* 4: 541–544.

Food and Agriculture Organization of the United Nations, International Fund for Agricultural Development, and World Food Programme (2015). *The State of Food Security in the World 2015. Meeting the International Hunger Targets: Taking Stock of Uneven Progress.* Rome: FAO. http://www.fao.org/3/a-i4646e.pdf.

Foote, Andrew D. (2008). Mortality rate acceleration and post-reproductive lifespan in matrilineal whale species. *Biology Letters* 4: 189–191.

Ford, John K. B. (2009). Killer whale (*Orcinus orca*). In *Encyclopedia of Marine Mammals,* 2nd ed., ed. William F. Perrin, Bernd Würsig, and J.G.M. Thewissen, 650–657. London: Academic.

Foster, Emma A., Daniel W. Franks, Sonia Mazzi, Safi K. Darden, Ken C. Balcomb, John K. B. Ford, and Darren P. Croft (2012). Adaptive prolonged postreproductive life span in killer whales. *Science* 337: 1313.

Fox, Molly, Rebecca Sear, Jan Beise, Gillian Ragsdale, Eckart Voland, and Leslie A. Knapp (2010). Grandma plays favourites: X-chromosome relatedness and sex-specific childhood mortality. *Proceedings of the Royal Society B: Biological Sciences* 277: 567–573.

Freedman, Robert R. (2001). Physiology of hot flashes. *American Journal of Human Biology* 13: 453–464.

Freedman, Robert R. (2010). Objective or subjective measurements of hot flashes in clinical trials: Quo vadis. *Maturitas* 67: 99–100.

Freedman, Robert R. (2014). Menopausal hot flashes: Mechanisms, endocrinology, treatment. *Journal of Steroid Biochemistry and Molecular Biology* 142: 115–120.

Freeman, Ellen W., Kristine E. Ensrud, Joseph C. Larson, Katherine A. Guthrie, Janet S. Carpenter, Hadine Joffe, Katherine M. Newton, Barbara Sternfeld, and Andrea Z. LaCroix (2015). Placebo improvement in pharmacological treatment of hot flashes: Time course, duration, and predictors. *Psychosomatic Medicine* 77: 167–175.

Freeman, Ellen W., Mary D. Sammel, Li Liu, Clarisa R. Gracia, Deborah B. Nelson, and Lori Hollander (2004). Hormones and menopausal status as predictors of depression in women in transition to menopause. *Archive of General Psychiatry* 61: 62–70.

Freeman, Ellen W., and K. Sherif (2007). Prevalence of hot flushes and night sweats around the world: A systematic review. *Climacteric* 10: 197–214.

Froide, Amy M. (2005). *Never Married: Singlewomen in Early Modern England.* Oxford: Oxford University Press.

Fry, Douglas P., ed. (2013). *War, Peace, and Human Nature: The Convergence of Evolutionary and Cultural Views.* New York: Oxford University Press.

Fry, Douglas P., and Patrick Söderberg (2013). Lethal aggression in mobile forager bands and implications for the origins of war. *Science* 341: 270–273.

Furth, Charlotte (1999). *A Flourishing Yin: Gender in China's Medical History, 960–1665.* Berkeley: University of California Press.

Furth, Charlotte (2002). Blood, body, and gender: Medical images of the female condition in China, 1600–1850. In *Chinese Femininities/Chinese Masculinities: A Reader*, ed. Susan Brownell and Jeffrey N. Wasserstrom, 291–314. Berkeley: University of California Press.

Gage, Timothy B. (1998). The comparative demography of primates: With some comments on the evolution of life histories. *Annual Review of Anthropology* 27: 197–221.

Gage, Timothy B., and Sharon DeWitte (2009). What do we know about the agricultural Demographic Transition? *Current Anthropology* 50: 649–655.

Gagnon, Alain (2015). Natural fertility and longevity. *Fertility and Sterility* 103: 1009–1016.

Galbarczyk, A., and G. Jasienska (2013). Timing of natural menopause covaries with timing of birth of a first daughter: Evidence for a mother-daughter evolutionary contract? *HOMO: Journal of Comparative Human Biology* 64: 228–232.

Gates, Hill (2015). Footbinding and women's labor in Sichuan. New York: Routledge.

Geary, Daniel (2015). *Beyond Civil Rights: The Moynihan Report and Its Legacy.* Philadelphia: University of Pennsylvania Press.

Gemery, Henry A. (2000). The white population of the colonial United States, 1607–1790. In *A Population History of North America*, ed. Michael R. Haines and Richard H. Steckel, 143–190. Cambridge: Cambridge University Press.

Gibson, Mhairi, and Ruth Mace (2005). Helpful grandmothers in rural Ethiopia: A study of the effect of kin on child survival and growth. *Evolution and Human Behavior* 26: 469–482.

Goettner-Abendroth, Heide (2012). *Matriarchal Societies: Studies on Indigenous Cultures Across the Globe.* New York: Peter Lang.

Goldberg, Amy, Alexis M. Mychajliw, and Elizabeth A. Hadley (2016). Post-invasion demography of prehistoric humans in South America. *Nature* 532: 232–235.

Good, Byron J., and Devon E. Hinton (2009). Introduction: Panic Disorder in cross-cultural perspective. In *Culture and Panic Disorder*, ed. Devon E. Hinton and Byron J. Good, 1–30. Stanford, CA: Stanford University Press.

Goodall, Jane (1986). *The Chimpanzees of Gombe: Patterns of Behavior.* Cambridge, MA: Belknap.

Goodman, Madeleine, and P. Bion Griffin (1985). The compatibility of hunting and mothering among the Agta hunter-gatherers of the Philippines. *Sex Roles* 12: 1119–1209.

Goody, Jack (1976). *Production and Reproduction: A Comparative Study of the Domestic Domain.* Cambridge: Cambridge University Press.

Goody, Jack, and S. J. Tambiah (1973). *Bridewealth and Dowry.* Cambridge: Cambridge University Press.

Gopnik, Alison (2014). Thank you, Grandma, for human nature. American Association for Retired People blog, June 9. https://blog.aarp.org/2014/06/09/thank-you-grandma-for -human-nature/.

Gosden, Roger G. (1985). *Biology of Menopause: The Causes and Consequences of Ovarian Ageing.* London: Academic Press.

Gosden, Roger G, and Malcolm J. Faddy (1998). Biological bases of premature ovarian failure. *Reproduction, Fertility and Development* 10: 73–78.

Gosden, Roger G., and Evelyn Telfer (1987). Numbers of follicles and oocytes in mammalian ovaries and their allometric relationships. *Journal of Zoology: Series A* 211: 169–176.

Gowing, Laura (1997). Secret births and infanticide in seventeenth-century England. *Past & Present* 156: 88–115.

Graves, Ronda R., Amy C. Lupo, Robert C. McCarthy, Daniel J. Wescott, and Deborah I. Cunningham (2010). Just how strapping was KNM-WT 15000? *Journal of Human Evolution* 59: 542–554.

Green, Monica H. (2001). *The Trotula: An English Translation of the Medieval Compendium of Women's Medicine.* Philadelphia: University of Pennsylvania Press.

Greenhalgh, Susan (2012). Patriarchal demographics? China's sex ratio reconsidered. *Population and Development Review* 38, suppl.: S130–S149.

Guatelli-Steinberg, Debbie (2009). Recent studies of dental development in Neandertals: Implications for Neandertal life histories. *Evolutionary Anthropology* 18: 9–20.

Guilmoto, Christophe Z. (2009). The sex ratio transition in Asia. *Population and Development Review* 35: 519–549.

Gunz, Philip, Simon Neubauer, Lubov Golovanova, Vladimir Doronichev, Bruno Maureille, and Jean-Jacques Hublin (2012). A uniquely modern human pattern of endocranial development: Insights from a new cranial reconstruction of the Neandertal newborn from Mezmaiskaya. *Journal of Human Evolution* 62: 300–312.

Gurven, Michael (2004). To give or give not: The behavioral ecology of human food transfers. *Behavioral and Brain Sciences* 27: 543–583.

Gurven, Michael, Megan Costa, Ben Trumble, Jonathan Stieglitz, Bret Beheim, Daniel Eid Rodriguez, Paul L. Hooper, and Hillard Kaplan (2016). Health costs of reproduction are minimal despite high fertility, mortality, and subsistence lifestyle. *Scientific Reports* 6: 30056. doi: 10.1038/srep30056.

Gurven, Michael, and Kim Hill (2009). Why do men hunt?: A reevaluation of "Man the Hunter" and the sexual division of labor. *Current Anthropology* 50: 51–74.

Gurven, Michael, and Hillard Kaplan (2007). Longevity among hunter-gatherers: A cross-cultural examination. *Population and Development Review* 33: 321–365.

Gurven, Michael, and Hillard Kaplan (2009). Beyond the Grandmother Hypothesis: Evolutionary models of human longevity. In *The Cultural Context of Aging: Worldwide Perspectives,* ed. Jay Sokolovsky, 53–66. Westport, CT: Praeger Publishers/Greenwood Publishing.

Gurven, Michael, Hillard Kaplan, and Maguin Gutierrez (2006). How long does it take to become a proficient hunter? Implications for the evolution of extended development and long life span. *Journal of Human Evolution* 51: 454–470.

Gurven, Michael, Jonathan Stieglitz, Paul L. Hooper, Christina Gomes, and Hillard Kaplan (2012). From the womb to the tomb: The role of transfers in shaping the evolved human life history. *Experimental Gerontology* 47: 807–813.

Gurven, Michael, and Robert Walker (2006). Energetic demand of multiple dependents and the evolution of slow human growth. *Proceedings of the Royal Society B: Biological Sciences* 273: 835–841.

Gurven, Michael, Jeffrey Winking, Hillard Kaplan, Christopher von Rueden, and Lisa McAllister (2009). A bioeconomic approach to marriage and the sexual division of labor. *Human Nature* 20: 151–183.

Guttentag, Marcia, and Paul F. Secord (1983). *Too Many Women? The Sex Ratio Question.* Beverly Hills, CA: Sage.

Haase, Donald (2008). *The Greenwood Encyclopedia of Folktales and Fairy Tales*, 3 vols. Westport, CT: Greenwood.

Haile-Selassie, Yohannes, Bruce M. Latimer, Mulugeta Alene, Alan L. Deino, Luis Gilbert, Stephanie M. Melillo, Beverly Z. Saylor, Gary R. Scott, and C. Owen Lovejoy (2010). An early *Australopithecus afarensis* postcranium from Woranso-Mille, Ethiopia. *Proceedings of the National Academy of Sciences* 107: 12,121–12,126.

Haines, Michael R., and Richard H. Steckel, eds. (2000). *A Population History of North America.* Cambridge: Cambridge University Press.

Hajnal, John (1982). Two kinds of preindustrial household formation system. *Population and Development Review* 8: 449–493.

Halstead, Paul (2014). *Two Oxen Ahead: Pre-Mechanized Farming in the Mediterranean.* Malden, MA: Wiley Blackwell.

Hamilton, William D. (1966). The moulding of senescence by natural selection. *Journal of Theoretical Biology* 12: 12–45.

Han, Hua (2007). Under the shadow of the collective good: An ethnographic analysis of fertility control in Xiaoshan, Zhejiang Province, China. *Modern China* 33: 320–348.

Hanawalt, Barbara (1986). *The Ties that Bound: Peasant Families in Medieval England.* Oxford: Oxford University Press.

Hanisch, Laura J., Lisa Hantsoo, Ellen W. Freeman, Gregory M. Sullivan, and James C. Coyne (2008). Hot flashes and panic attacks: A comparison of symptomology, neurobiology, treatment, and a role for cognition. *Psychological Bulletin* 134: 247–269.

Hansen, James, Makiko Sato, Paul Hearty, Reto Ruedy, Maxwell Kelley, Valerie Masson-Delmotte, Gary Russell, et al. (2016). Ice melt, sea level rise and superstorms: Evidence from Paleoclimate data, modeling, and modern observations that 2°C global warming could be dangerous. *Atmospheric Chemistry and Physics* 16: 3761–3812.

Hansen, Karl R., Nicholas S. Knowlton, Angela C. Thyer, Jay S. Charleston, Michael R. Soules, and Nancy A. Klein (2008). A new model of reproductive aging: The decline in ovarian non-growing follicle number from birth to menopause. *Human Reproduction* 23: 699–708.

Hanson, Marta (2013). *Speaking of Epidemics in Chinese Medicine: Disease and the Geographic Imagination in Late Imperial China.* New York: Routledge.

Hardy, Bruce L., Marie-Hélène Moncel, Camille Daujeard, Paul Fernandes, Philippe Béarez, Emmanuel Desclaux, Maria Gema Chacon Navarro, Simon Puaud, and Rosalia Gallotti (2013).

Impossible Neanderthals? Making string, throwing projectiles and catching small game during Marine Isotope Stage 4 (Abri du Maras, France). *Quaternary Science Reviews* 82: 23–40.

Hardy, Karen, Jennie Brand-Miller, Katherine D. Brown, Mark G. Thomas, and Les Copeland (2015). The importance of dietary carbohydrate in human evolution. *Quarterly Review of Biology* 90: 251–268.

Hardy, Karen, Stephen Buckley, Matthew J. Collins, Almudena Estalrrich, Don Brothwell, Les Copeland, Antonio García-Tabernero, et al. (2012). Neanderthal medics?: Evidence for food, cooking, and medicinal plants entrapped in dental calculus. *Naturwissenschaft* 99: 617–626.

Harlan, Lindsey (1992). *Religion and Rajput Women: The Ethic of Protection in Contemporary Narratives.* Berkeley: University of California Press.

Harper, Kristin N., and George Armelagos (2013). Genomics, the origins of agriculture, and our changing microbe-scape: Time to revisit some old tales and tell some new ones. *American Journal of Physical Anthropology* 57: 135–142.

Harrison, A.R.W. (1968–1971). *The Law of Athens,* 2 vols. Oxford: Clarendon.

Hartung, John (1976). On natural selection and the inheritance of wealth. *Current Anthropology* 17: 607–622.

Harvati, Katerina, Chris Stringer, Rainer Grün, Maxime Aubert, Philip Allsworth-Jones, and Caleb Adebayo Folorunso (2013). The later Stone Age calvaria from Iwo Eleru, Nigeria: Morphology and chronology. *PLOS ONE* 6 (9): e24024. doi: 10.1371/journal.pone.0024024.

Harvey, Barbara F. (1993). *Living and Dying in England, 1100–1540: The Monastic Experience.* Oxford: Oxford University Press.

Haub, Carl (2013). Rising trend of births outside marriage. Population Reference Bureau. http://www.prb.org/Publications/Articles/2013/nonmarital-births.aspx.

Hausfater, Glenn, and Sarah Blaffer Hrdy, eds. (1984). *Infanticide: Comparative and Evolutionary Perspectives.* New York: Aldine.

Hawkes, Kristen (2003). Grandmothers and the evolution of human longevity. *American Journal of Human Biology* 15: 380–400.

Hawkes, Kristen (2006). Life history theory and human evolution: A chronicle of ideas and findings. In *The Evolution of Human Life History,* ed. Kristen Hawkes and Richard R. Paine, 45–94. Santa Fe, NM: School of American Research Press.

Hawkes, Kristen, and Rebecca Bliege Bird (2002). Showing off, handicap signaling, and the evolution of men's work. *Evolutionary Anthropology* 11: 58–67.

Hawkes, Kristen, and James E. Coxworth (2013). Grandmothers and the evolution of human longevity: A review of findings and future directions. *Evolutionary Anthropology* 22: 294–302.

Hawkes, Kristen, and James F. O'Connell (2005). How old is human longevity? *Journal of Human Evolution* 49: 650–653.

Hawkes, Kristen, James F. O'Connell, and Nicholas G. Blurton Jones (1989). Hardworking Hadza grandmothers. In *Comparative Socioecology: The Behavioural Ecology of Humans and Other Mammals,* ed. V. Standen and R. A. Foley, 341–366. Oxford: Blackwell Scientific Press.

Hawkes, Kristen, James F. O'Connell, and Nicholas G. Blurton Jones (1997). Hadza women's time allocation, offspring provisioning, and the evolution of long postmenopausal life spans. *Current Anthropology* 38: 551–577.

Hawkes, Kristen, James F. O'Connell, and Nicholas G. Blurton Jones (2001). Hunting and nuclear families: Some lessons from the Hadza about men's work. *Current Anthropology* 42: 681–709.

Hawkes, Kristen, James F. O'Connell, Nicholas G. Blurton Jones, H. Alvarez, and E. L. Charnov (1998). Grandmothering, menopause, and the evolution of human life histories. *Proceedings of the National Academy of Sciences* 95: 1336–1339.

Hawkes, Kristen, and Richard R. Paine, eds. (2006). *The Evolution of Human Life History*. Santa Fe, NM: School of American Research Press.

Hawkes, Kristen, Ken R. Smith, and Shannen L. Robson (2009). Mortality and fertility rates in humans and chimpanzees: How within-species variation complicates cross-species comparison. *American Journal of Human Biology* 21: 578–586.

Hayward, Adam D., Ilona Nenko, and Virpi Lummaa (2015). Early life reproduction is associated with increased mortality risk but enhanced lifetime fitness in pre-industrial humans. *Proceedings of the Royal Society B: Biological Sciences* 282: 20143053. doi: 10.1098/rspb.2014.3053.

He, Chunyan, and Joanne M. Murabito (2014). Genome-wide association studies of age at menarche and age at natural menopause. *Molecular and Cellular Endocrinology* 382: 767–779.

Hennington, Peter, Stephan Zipfel, and Wolfgang Herzog (2007). Management of functional somatic syndromes. *Lancet* 369: 946–955.

Henrich, Joseph (2004). Demography and cultural evolution: How adaptive cultural processes can produce maladaptive losses—the Tasmanian case. *American Antiquity* 69: 197–214.

Henry, Louis (1961). Some data on natural fertility. *Eugenics Quarterly* 8: 81–91.

Henry, Louis (1989). Men's and women's mortality in the past. *Population: An English Selection* 44: 177–201.

Hewlett, Barry S. (1991). Demography and childcare in preindustrial societies. *Journal of Anthropological Research* 47: 1–37.

Hill, Bridget (2001). *Women Alone: Spinsters in England, 1660–1850*. New Haven, CT: Yale University Press.

Hill, Kim, and A. Magdalena Hurtado (1991). The evolution of premature reproductive senescence and menopause in human females: An evaluation of the "Grandmother Hypothesis." *Human Nature* 2: 313–350.

Hill, Kim, and A. Magdalena Hurtado (1996). *Ache Life History: Ecology and Demography of a Foraging People*. New York: de Gruyter.

Hill, Kim, and A. Magdalena Hurtado (2009). Cooperative breeding in South American hunter-gatherers. *Proceedings of the Royal Society B: Biological Sciences* 36: 3863–3870.

Hill, Kim, A. Magdalena Hurtado, and Robert S. Walker (2007). High adult mortality among Hiwi hunter-gatherers: Implications for human evolution. *Journal of Human Evolution* 52: 443–454.

Hill, Kim, Robert S. Walker, Miran Božičević, James Eder, Thomas Headland, Barry Hewlett, A. Magdalena Hurtado, Frank Marlowe, Polly Wiessner, and Brian Wood (2011). Co-residence patterns in hunter-gatherer societies show unique human social structure. *Science* 331: 1286–1289.

Hinde, Andrew (2003). *England's Population: A History Since the Domesday Survey.* London: Hodder Education.

Hindle, Steve (1998). The problem of pauper marriage in seventeenth-century England. *Transactions of the Royal Society* 8: 71–89.

Hinton, Devon E., and Byron J. Good (2009). A medical anthropology of panic sensations: Ten analytic perspectives. In *Culture and Panic Disorder,* ed. Devon E. Hinton and Byron J. Good, 57–81. Stanford, CA: Stanford University Press.

Hinton, Devon E., Lawrence Park, Curtis Hsia, Stefan Hofmann, and Mark H. Pollack (2009). Anxiety disorder presentations in Asian populations: A review. *CNS Neuroscience & Therapeutics* 15: 295–303.

Hinton, Devon E., Khin Um, and Phalnarith Ba (2001a). *Kyol goeu* ("wind overload"), part I: A cultural syndrome of orthostatic panic among Khmer refugees. *Transcultural Psychiatry* 38: 403–432.

Hinton, Devon E., Khin Um, and Phalnarith Ba (2001b). *Kyol goeu* ("wind overload"), part II: Prevalence, characteristics, and mechanisms of *kyol goeu* and near-*kyol goeu* of Khmer patients attending a psychiatric clinic. *Transcultural Psychiatry* 38: 433–460.

Hirshbein, Laura (2009a). Gender, age, and diagnosis: The rise and fall of involutional melancholia in American psychiatry, 1900–1980. *Bulletin of the History of Medicine* 83: 710–745.

Hirshbein, Laura (2009b). *American Melancholy: Constructions of Depression in the Twentieth Century.* New Brunswick, NJ: Rutgers University Press.

Ho Ping-ti (He Bingdi) (1959). *Studies in the Population of China, 1368–1953.* Cambridge, MA: Harvard University Press.

Hobsbawm, E. J. (2000). *Bandits,* 4th ed. New York: New Press.

Hodis, Howard N., and Wendy J. Mack (2014). Hormone replacement therapy and the association with coronary heart disease and overall mortality: Clinical application of the Timing Hypothesis. *Journal of Steroid Biochemistry and Molecular Biology* 142: 68–75.

Hoffer, Peter, and N.E.H. Hull (1981). *Murdering Mothers: Infanticide in England and New England, 1558–1803.* New York: New York University Press.

Hoffmann, D. L., C. D. Standish, M. García-Diez, P. B. Pettitt, J. A. Milton, J. Zilhão, J. J. Alcolea-González, et al. (2018). U-Th dating of carbonate crusts reveals Neandertal origin of Iberian cave art. *Science* 359: 912–915.

Holden, Clare Janaki, Rebecca Sear, and Ruth Mace (2003). Matriliny as daughter-biased investment. *Evolution and Human Behavior* 24: 99–112.

Hooper, Paul, Michael Gurven, Jeffrey Winking, and Hillard S. Kaplan (2015). Inclusive fitness and differential productivity across the life course determine intergenerational transfers in a small-scale society. *Proceedings of the Royal Society B: Biological Sciences* 282: 20142808. doi: 10.1098/rspb.2014.2808.

Hoppa, Robert D., and James W. Vaupel, eds (2002). *Paleodemography: Age Distributions from Skeletal Samples.* Cambridge: Cambridge University Press.

Houck, Judith A. (2006). *Hot and Bothered: Women, Menopause, and Medicine in Modern America.* Cambridge, MA: Harvard University Press.

Howell, Nancy (2000). *Demography of the Dobe !Kung,* 2nd ed. New York: de Gruyter.

Howell, Nancy (2010). *Life Histories of the Dobe !Kung: Food, Fatness, and Well-Being Over the Life Span.* Berkeley: University of California Press.

Hrdy, Sarah Blaffer (1999). *Mother Nature: A History of Mothers, Infants, and Natural Selection.* New York: Pantheon.

Hrdy, Sarah Blaffer (2000). The optimal number of fathers: Evolution, demography, and history in the shaping of female mate preferences. *Annals of the New York Academy of Sciences* 907: 75–96.

Hrdy, Sarah Blaffer (2005a). Comes the child before the man: How cooperative breeding and prolonged postweaning dependence shaped human potential. In *Hunter-Gatherer Childhoods: Evolutionary, Developmental and Cultural Perspectives*, ed. Barry S. Hewlett and Michael E. Lamb, 65–91. New Brunswick, NJ: Aldine.

Hrdy, Sarah Blaffer (2005b). Cooperative breeders with an ace in the hole. In *Grandmotherhood: The Evolutionary Significance of the Second Half of Female Life*, ed. Eckart Voland, Athanasios Chasiotis, and Wulf Schiefenhövel, 295–318. New Brunswick, NJ: Rutgers University Press.

Hrdy, Sarah Blaffer (2009). *Mothers and Others: The Evolutionary Origins of Mutual Understanding.* Cambridge, MA: Belknap Press.

Hrdy, Sarah Blaffer (2016). Variable postpartum responsiveness among humans and other primates with "cooperative breeding": A comparative and evolutionary perspective. *Hormones and Behavior* 77: 272–283.

Hristova, Pepa (2013). *Sworn Virgins.* Heidelberg: Kehrer Verlag.

Huang, Philip C. (1990). *The Peasant Family and Rural Development in the Yangzi Delta, 1350–1988.* Stanford, CA: Stanford University Press.

Hublin, Jean-Jacques, Abdelouahed Ben-Ncer, Shara E. Bailey, Sarah E. Freidline, Simon Neubauer, Matthew M. Skinner, Inga Bergmann, et al. (2017). New fossils from Jebel Irhoud, Morocco and the pan-African origin of *Homo sapiens. Nature* 546: 289–292.

Hudson, Pat, and W. R. Lee, eds. (1990). *Women's Work and the Family Economy in Historical Perspective.* Manchester, UK: Manchester University Press.

Hudson, Valerie M., and Andre M. den Boer (2004). *Bare Branches: Security Implications of Asia's Surplus Male Population.* Cambridge, MA: MIT Press.

Hufton, Olwen (1984). Women without men: Widows and spinsters in Britain and France in the eighteenth century. *Journal of Family History* 9: 355–376.

Hunt, A. S., and C. C. Edgar (1934). *Select Papyri II: Official Documents.* Loeb Classical Library no. 282. Cambridge, MA: Harvard University Press.

Hunter, Myra S., and Joseph Chilcot (2013). Testing a cognitive model of menopausal hot flushes and night sweats. *Journal of Psychosomatic Research* 74: 307–312.

Hunter, Myra S., and Eleanor Mann (2010). A cognitive model of menopausal hot flushes and night sweats. *Journal of Psychosomatic Research* 69: 491–501.

Hunter, Virginia (1994). *Policing Athens: Social Control in the Attic Lawsuits, 420–320 B.C.* Princeton, NJ: Princeton University Press.

Im, Eun-Ok (2003). Symptoms experienced during the menopausal transition: Korean women in South Korea and the United States. *Journal of Transcultural Nursing* 14: 321–328.

Im, Eun-Ok (2006). The Midlife Women's Symptom Index (MSI). *Health Care for Women International* 27: 268–287.

Im, Eun-Ok, and Afaf Ibrahim Meleis (2000). Meanings of menopause to Korean immigrant women. *Western Journal of Nursing Research* 22: 84–102.

Im, Eun-Ok, Young Ko, and Wonshik Chee (2014). Ethnic differences in the clusters of menopausal symptoms. *Health Care for Women International* 35: 549–565.

Indriati, Etty, Carl C. Swisher III, Christopher Lepre, Rhonda L. Quinn, Rusyad A. Suriyanto, Agus T. Hascaryo, Rainer Grün, et al. (2011). The age of the 20 meter Solo River terrace, Java, Indonesia and the survival of *Homo erectus* in Asia. *PLOS ONE* 6 (6): e21562. doi: 0.1371/journal.pone.0021562.

Iovitu, Radu (2011). Shape variation in Aterian tanged tools and the origins of projectile technology: A morphometric perspective on stone tool function. *PLOS ONE* 6 (12): e29029. doi: 10.1371/journal.pone.0029029.

Isler, Karin, and Carol P. van Schaik (2012a). Allomaternal care, life history and brain size evolution in mammals. *Journal of Human Evolution* 63: 52–63.

Isler, Karin, and Carol P. van Schaik (2012b). How our ancestors broke through the gray ceiling: Comparative evidence for cooperative breeding in early *Homo. Current Anthropology* 53, suppl. 6: S453–S465.

Ivey, Paula K. (2000) Cooperative reproduction in Ituri forest hunter-gatherers: Who cares for Efe infants? *Current Anthropology* 41: 856–866.

Ivey Henry, Paula, Gilda A. Morelli, and Edward Z. Tronick (2005). Child caretakers among Efe foragers of the Ituri forest. In *Hunter-Gatherer Childhoods: Evolutionary, Developmental, and Cultural Practices*, ed. Barry S. Hewlett and Michael E. Lamb, 191–213. New Brunswick, NJ: Transaction.

Jackson, Stanley W. (1986). *Melancholia and Depression: From Hippocratic Times to Modern Times*. New Haven, CT: Yale University Press.

Jaeggi, Adrian V., and Michael Gurven (2013). Natural cooperators: Food sharing in humans and other primates. *Evolutionary Anthropology* 22: 186–195.

Jamison, Cheryl Sorenson, Laurel L. Cornell, Paul L. Jamison, and Hideki Nakazato (2002). Are all grandmothers equal?: A review and preliminary test of the "Grandmother Hypothesis" of Tokugawa Japan. *American Journal of Physical Anthropology* 119: 67–76.

John, A. Meredith (1988). *The Plantation Slaves of Trinidad, 1783–1816: A Mathematical and Demographic Inquiry*. Cambridge: Cambridge University Press.

Johnson, Allen W., and Timothy Earle (1987). *The Evolution of Human Societies: From Foraging Group to Agrarian State*. Stanford, CA: Stanford University Press.

Johnson, Elizabeth (1975). Women and childbearing in Kwan Mun Hau Village. In *Women in Chinese Society*, ed. Margery Wolf and Roxane Witke, 215–242. Stanford, CA: Stanford University Press.

Johnson, Joshua, Jacqueline Canning, Tomoko Kaneko, James K. Pru, and Jonathan L. Tilly (2004). Germline stem cells and follicular renewal in the postnatal mammalian ovary. *Nature* 428: 145–150.

Johnson, Rodney L., and Ellen Kapsalis (1998). Menopause in free-ranging Rhesus macaques: Estimated incidence, relation to body condition, and adaptive significance. *International Journal of Primatology* 19: 751–765.

Johnston, F. E., and C. E. Snow (1961). The reassessment of age and sex of the Indian Knoll skeletal population: Demographic and methodological aspects. *American Journal of Physical Anthropology* 19: 237–244.

Johnstone, Rufus A., and Michael A. Cant (2010). The evolution of menopause in cetaceans and humans: The role of demography. *Proceedings of the Royal Society B: Biological Sciences* 277: 3765–3771.

Jones, James Holland, and Rebecca Bliege Bird (2014). The marginal valuation of fertility. *Evolution and Human Behavior* 35: 65–71.

Kachel, A. Friederike, L. S. Premo, and Jean-Jacques Hublin (2011a). Grandmothering and natural selection. *Proceedings of the Royal Society B: Biological Sciences* 278: 384–391.

Kachel, A. Friederike, L. S. Premo, and Jean-Jacques Hublin (2011b). Grandmothering and natural selection reconsidered. *Proceedings of the Royal Society B: Biological Sciences* 278: 1939–1941.

Kahn, Paul, and Francis Woodman Cleaves (1984). *The Secret History of the Mongols: The Origin of Chinghis Khan*. San Francisco: North Point Press.

Kalben, Barbara Blatt (2000). Why men die younger. *North American Actuarial Journal* 4: 83–111.

Kandiyoti, Deniz (1988). Bargaining with patriarchy. *Gender and Society* 2: 264–290.

Kaplan, Hillard (1994). Evolutionary and wealth flows theories of fertility: Empirical tests and new models. *Population and Development Review* 20: 753–791.

Kaplan, Hillard, Michael Gurven, Jeffrey Winking, Paul L. Hooper, and Jonathan Stieglitz (2010). Learning, menopause, and the human adaptive complex. *Annals of the New York Academy of Sciences* 1204: 30–42.

Kaplan, Hillard, Kim Hill, Jane Lancaster, and A. Magdalena Hurtado (2000). A theory of human life history evolution: Diet, intelligence, and longevity. *Evolutionary Anthropology* 9: 156–185.

Kaplan, Hillard, Paul Hooper, and Michael Gurven (2009). The evolutionary and ecological roots of human social organization. *Philosophical Transactions of the Royal Society B*: Biological Sciences 364: 3289–3299.

Kaplan, Hillard, and Arthur J. Robson (2002). The emergence of humans: The coevolution of intelligence and longevity with intergenerational transfers. *Proceedings of the National Academy of Sciences* 15: 10,221–10,226.

Kaplan, Hillard, and Arthur J. Robson (2009). We age because we grow. *Proceedings of the Royal Society B: Biological Sciences* 276: 1837–1844.

Kasuya, T., and H. Marsh (1984). Life history and reproductive biology of the short-finned pilot whale, *Globicephala macrorhynchus*, off the Pacific Coast of Japan. In *Reports of the International Whaling Commission*, special issue 6, ed. William F. Perrin, Robert L. Brownell, and Douglas P. DeMaster, 259–310. Cambridge: International Whaling Commission.

Keckler, Charles N. W. (1997). Catastrophic mortality in simulations of forager age-at-death: Where did all the humans go? In *Integrating Archaeological Demography: Multidisciplinary Approaches to Prehistoric Population*, ed. Richard R. Paine, 205–228. Occasional Paper 24, Southern Illinois University at Carbondale, Center for Archaeological Investigations.

Keen, Ian (2002). Seven Aboriginal marriage systems and their correlates. *Anthropological Forum* 12: 145–157.

Keen, Ian (2006). Constraints in the development of enduring inequalities in late Holocene Australia. *Current Anthropology* 47: 7–37.

Kelly, Robert L. (1995). *The Foraging Spectrum: Diversity in Hunter-Gatherer Lifeways*. Washington, D.C.: Smithsonian Institution Press.

Kerns, Virginia, and Judith K. Brown, eds. (1992). *In Her Prime: New Views of Middle-Aged Women*, 2nd edition. Urbana: University of Illinois Press.

Kertzer, David I. (1993). *Sacrificed for Honor: Italian Infant Abandonment and the Politics of Reproductive Control*. Boston: Beacon.

Keys, Ancel (1950). *The Biology of Human Starvation*, 2 vols. Minneapolis: University of Minnesota Press.

Kilday, Anne-Marie (2013). *A History of Infanticide in Britain c. 1600 to the Present*. London: Palgrave Macmillan.

Kim, Eunha, Ingrid Hogge, Peter Ji, Young R. Shim, and Catherine Lochspeich (2014). Hwabyung among middle-aged Korean women: Family relationships, gender-role attitudes, and self-esteem. *Health Care for Women International* 35: 495–511.

Kim, Peter S., James E. Coxworth, and Kristen Hawkes (2012). Increased longevity evolves from grandmothering. *Proceedings of the Royal Academy of Sciences B: Biological Sciences* 279: 4880–4884.

Kim, Peter S., John S. McQueen, James E. Coxworth, and Kristen Hawkes (2014). Grandmothering drives the evolution of longevity in a probabilistic model. *Journal of Theoretical Biology* 353: 84–94.

King, Helen (1993). Once upon a text: The Hippocratic origins of hysteria. In *Hysteria Beyond Freud*, ed. Sander L. Gilman, Helen King, Roy Porter, G. S. Rousseau, and Elaine Showalter, 3–90. Berkeley: University of California Press.

King, Helen (1998). Reading the past through the present: Drugs and contraception in Hippocratic medicine. In *Hippocrates' Woman: Reading the Female Body in Ancient Greece*, 132–156. London: Routledge.

King, Helen (2004). *The Disease of Virgins: Green Sickness, Chlorosis, and the Problems of Puberty*. New York and London: Routledge.

King, Helen (2007). *Midwifery, Obstetrics, and the Rise of Gynecology*. Aldershot, UK: Ashgate.

King, Michelle (2014). *Between Birth and Death: Female Infanticide in Nineteenth-Century China*. Palo Alto, CA: Stanford University Press.

Kirkwood, Thomas B. L. (1977). The evolution of ageing. *Nature* 270: 301–304.

Kirkwood, Thomas B. L, and R. Holliday (1979). The evolution of ageing and longevity. *Proceedings of the Royal Society B: Biological Sciences* 205: 531–546.

Kirkwood, Thomas B. L., and Daryl P. Shanley (2010). The connections between general and reproductive senescence and the evolutionary basis of menopause. *Annals of the New York Academy of Sciences* 204: 21–29.

Kirmayer, Laurence J., and Norman Sartorius (2007). Cultural models and somatic syndromes. *Psychosomatic Medicine* 69: 832–840.

Kirmayer, Laurence J., and Allan Young (1998). Culture and somatization: Clinical, epidemiological, and ethnographic perspectives. *Psychosomatic Medicine* 60: 420–430.

Klein, Richard G. (2009). *The Human Career: Human Biological and Cultural Origins*, 3rd ed. Chicago: University of Chicago Press.

Kleinman, Arthur (1982). Neurasthenia and depression: A study of somatization and culture in China. *Culture, Medicine, and Psychiatry* 6: 117–190.

Knight, Chris, and Camilla Power (2005). Grandmothers, politics, and getting back to science. In *Grandmotherhood: The Evolutionary Significance of the Second Half of Female Life*, ed. Eckard Voland, Athanasios Chasiotis, and Wulf Schiefenhövel, 81–98. New Brunswick, NJ: Rutgers University Press.

Knowlton, Nicholas S., LaTasha B. Craig, Michael T. Zavy, and Karl R. Hansen (2014). Validation of the power model of nongrowing follicle depletion associated with aging in women. *Fertility and Sterility* 101: 851–856.

Konigsberg, Lyle W., and Nicholas P. Herrmann (2006). The osteological evidence for human longevity in the recent past. In *The Evolution of Human Life History*, ed. Kristen Hawkes and Richard R. Paine, 267–306. Santa Fe, NM: School of American Research Press.

Konner, Melvin (2005). Hunter-gatherer infancy and childhood: The !Kung and others. In *Hunter-Gatherer Childhoods: Evolutionary, Developmental, and Cultural Practices*, ed. Barry S. Hewlett and Michael E. Lamb, 19–64. New Brunswick, NJ: Transaction.

Kramer, Karen L. (2004). Reconsidering the cost of childbearing: The timing of children's helping behavior across the life cycle of Maya families. In *Socioeconomic Aspects of Human Behavioral Ecology*, Research in Economic Anthropology 23, ed. Michael S. Alvard, 335–349. Amsterdam: Elsevier.

Kramer, Karen L. (2005). *Maya Children: Helpers at the Farm*. Cambridge, MA: Harvard University Press.

Kramer, Karen L. (2010). Cooperative breeding and its significance to the demographic success of humans. *Annual Review of Anthropology* 2010: 417–436.

Kramer, Karen L, and James L. Boone (2002). Why intensive agriculturalists have higher fertility: A household energy budget approach. *Current Anthropology* 43: 511–517.

Krause, Johannes, Qiaomei Fu, Jeffrey M. Good, Bence Viola, Michael V. Shunkov, Anatoli P. Derevianko, and Svante Pääbo (2010). The complete mitochondrial DNA genome of an unknown hominin from southern Siberia. *Nature* 464: 894–897.

Kronenberg, Fredi (1990). Hot flashes: Epidemiology and physiology. In *Multidisciplinary Perspectives on Menopause*, ed. Marcha Flint, Fredi Kronenberg, and Wulf Utian, special issue of *Annals of the New York Academy of Sciences* 592: 52–86.

Kronenberg, Fredi (2010). Menopausal hot flashes: A review of physiology and biosociocultural perspective on methods of assessment. *Journal of Nutrition* suppl., *Equol, Soy, and Menopause*: S1380–S1385.

Kuhle, Barry X. (2007). An evolutionary perspective on the origin and ontogeny of menopause. *Maturitas* 57: 329–337.

Kuhlwilm, Martin, Ilan Gronau, Melissa J. Hubisz, Cesare de Filippo, Javier Prado-Martinez, Martin Kircher, Qiaomei Fu, et al. (2016). Ancient gene flow from early modern humans into eastern Neanderthals. *Nature* 530: 429–433.

Kuhn, Steven L., and Mary C. Stiner (2006). What's a mother to do? The division of labor among Neandertals and modern humans in Eurasia. *Current Anthropology* 47: 953–981.

Kulikoff, Allan (1977). A "prolifick" people: Black population growth in the Chesapeake colonies, 1700–1790. *Southern Studies* 16: 391–428.

Kuriyama, Shigehisa (1997). The historical origins of *katakori*. *Japan Review* 9: 127–149.

Kuriyama, Shigehisa (1999). *The Expressiveness of the Body and the Divergence of Greek and Chinese Medicine.* New York: Zone.

Kuzawa, Christopher W., and Jared M. Bragg (2012). Plasticity in human life history strategy: Implications for contemporary human variation and the evolution of the genus *Homo.* *Current Anthropology* 53, suppl. 6: S369–S382.

Lahdenperä, Mirkka, Duncan O. S. Gillespie, Virpi Lummaa, and Andrew F. Russell (2012). Severe intergenerational reproductive conflict and the evolution of menopause. *Ecology Letters* 15: 1283–1290.

Lahdenperä, Mirkka, Virpi Lummaa, Samuli Helle, Marc Tremblay, and Andrew F. Russell (2004). Fitness benefits of prolonged post-reproductive lifespan in women. *Nature* 428: 178–181.

Lahdenperä, Mirkka, Khyne U. Mar, and Virpi Lummaa (2014). Reproductive cessation and post-reproductive lifespan in Asian elephants and pre-industrial humans. *Frontiers in Zoology* 11: 54.

Lahdenperä, Mirkka, Khyne U. Mar, and Virpi Lummaa (2016). Nearby grandmother enhances calf survival and reproduction in Asian elephants. *Scientific Reports* 6: 27213. doi: 10.1038/srep27213.

Langer, William L. (1974). Infanticide: A historical survey. *History of Childhood Quarterly* 1: 353–366.

Langer, William L. (1975). American foods and Europe's population growth, 1750–1850. *Journal of Social History* 8: 51–66.

Largent, Mark (2007). *Breeding Contempt: The History of Coerced Sterilization in the United States.* New Brunswick, NJ: Rutgers University Press.

Larson, Heidi (2016). Vaccine trust and the limits of information. *Science* 353: 1207–1208.

Laslett, Peter, ed. (1972). *Household and Family in Past Time.* Cambridge: Cambridge University Press.

Laslett, Peter (1977). *Family Life and Illicit Love in Earlier Generations.* Cambridge: Cambridge University Press.

Lawson, David W., Alexandra Alvergne, and Mhairi L. Gibson (2012). The life-history trade-off between fertility and child survival. *Proceedings of the Royal Society B: Biological Sciences* 279: 4755–4764.

Leacock, Eleanor, and Helen I. Safa, eds. (1986). *Women's Work: Development and the Division of Labor by Gender.* South Hadley, MA: Bergin & Garvey.

Leake, John (1777). *Medical Instructions Towards the Prevention, and Cure of Chronic or Slow Diseases Peculiar to Women.* London: Baldwin.

Leca, Jean-Baptiste, Michael A. Huffman, and Paul L. Vasey. (2012) *The Monkeys of Stormy Mountain: 60 Years of Primatological Research on the Japanese Macaques of Arashiyama.* Cambridge and New York: Cambridge University Press.

Lee, Charlotte T., Cedric O. Puleston, and Shripad Tuljapurkar (2009). Population and prehistory III: Food-dependent demography in variable environments. *Theoretical Population Biology* 76: 179–188.

Lee, Charlotte T., and Shripad Tuljapurkar (2008). Population and prehistory I: Food-dependent population growth in constant environments. *Theoretical Population Biology* 73: 473–482.

Lee, James Z., and Cameron D. Campbell (1997). *Fate and Fortune in Rural China: Social Organization and Population Behavior in Liaoning, 1774–1873*. Cambridge: Cambridge University Press.

Lee, James Z., and Wang Feng (1999). *One Quarter of Humanity: Malthusian Myths and Chinese Realities*. Cambridge, MA: Harvard University Press.

Lee, Lisa M. (2012). Public health ethics theory: Review and path to convergence. *Journal of Law, Medicine, and Ethics* 40: 85–98.

Lee, Richard B. (1992). Work, sexuality, and aging among !Kung women. In *In Her Prime: New Views of Middle-Aged Women*, 2nd ed., ed. Virginia Kerns and Judith K. Brown, 35–48. Urbana: University of Illinois Press.

Lee, Richard B., and Irven Devore, eds. (1969). *Man the Hunter*. Chicago: Aldine.

Lee, Ronald (2003). Rethinking the evolutionary theory of aging: Transfers, not births, shape senescence in social species. *Proceedings of the National Academy of Sciences* 100: 9637–9642.

Lee, Ronald (2008). Sociality, selection, and survival: Simulated evolution of mortality with intergenerational transfers and food sharing. *Proceedings of the National Academy of Sciences* 105: 7124–7128.

Lee, Ronald, and Karen L. Kramer (2002). Children's economic roles in the Maya family life cycle: Cain, Caldwell, and Chayanov revisited. *Population and Development Review* 28: 475–499.

Lee, Ronald, and Andrew Mason (2011). Generational economics in a changing world. *Population and Development Review* 37, suppl.: S115–S142.

Leonetti, Donna L., and Benjamin Chabot-Hanowell (2011). The foundation of kinship: Households. *Human Nature* 22: 16–40.

Leonetti, Donna L., Dilip C. Nath, and Natabar S. Hemam (2007). In-law conflict: Women's reproductive lives and the roles of their mothers and husbands among the matrilineal Khasi. *Current Anthropology* 6: 861–890.

Leonetti, Donna L., Dilip C. Nath, Natabar S. Hemam, and Dawn B. Neill (2004). Do women really need marital partners for support of their reproductive success? The case of the matrilineal Khasi of N. E. India. In *Socioeconomic Aspects of Human Behavioral Ecology*, Research in Economic Anthropology 23, ed. Michael S. Alvard, 151–174. Amsterdam: Elsevier.

Leonetti, Donna L., Dilip C. Nath, Natabar S. Hemam, and Dawn B. Neill (2005). Kinship organization and the impact of grandmothers on reproductive success among the matrilineal Khasi and patrilineal Bengali of northeast India. In *Grandmotherhood: The Evolutionary Significance of the Second Half of Female Life*, ed. Eckart Voland, Athanasios Chasiotis, and Wolf Schiefenhövel, 194–214. New Brunswick, NJ: Rutgers University Press.

Le Roy Ladurie, Emmanuel (1979). *Montaillou: The Promised Land of Error*. Tr. Barbara Bray. New York: Vintage.

Levene, Alysa (2007). *Childcare, Health, and Mortality at the London Foundling Hospital, 1741–1800*. Manchester, UK: Manchester University Press.

Levitis, Daniel A., and Laurie Bingaman Lackey (2011). A measure for describing and comparing postreproductive life span as a population trait. *Methods in Ecology and Evolution* 2: 446–453.

Levitis, Daniel A., Oskar Burger, and Laurie Bingaman Lackey (2013). The human post-fertile lifespan in comparative evolutionary context. *Evolutionary Anthropology* 22: 66–79.

Li Bozhong (1998). *Agricultural Development in Jiangnan, 1620–1850*. New York: St. Martin's.

Liébault, Jean (1598). *Trois livres appartenans aus aux infirmitez et maladies des femmes*. Lyons, France: Veyrat. http://gallica.bnf.fr/ark:/12148/bpt6k536053/f573.image.

Lin, Keh-Ming (1983). *Hwa-byung*: A Korean culture-bound syndrome? *American Journal of Psychiatry* 140: 105–107.

Lippold, Sebastian, Hongyang Xu, Albert Ko, Mingkun Li, Gabriel Renaud, Anne Butthof, Roland Schröder, and Mark Stoneking (2014). Human paternal and maternal demographic histories: Insights from high-resolution Y chromosome and mtDNA sequences. *Investigative Genetics* 5 (13). doi: 10.1186/2041-2223-5-13.

Littleton, Judith (2005). Fifty years of chimpanzee demography at Taronga Park Zoo. *American Journal of Primatology* 67: 281–298.

Livi-Bacci, Massimo (2012). *A Concise History of World Population*, 5th ed. Chichester, UK: Wiley Blackwell.

Llamas, Bastien, Lars Fehren-Schmitz, Guido Valverde, Julien Soubrier, Swapan Mallick, Nadin Rohland, Susanne Nordenfelt, et al. (2016). Ancient mitochondrial DNA provides high-resolution time scale of the peopling of the Americas. *Science Advances* 2: e1501385. doi: 10.1126/sciadv.1501385.

Llewellyn-Jones, Lloyd (2003). *Aphrodite's Tortoise: The Veiled Woman of Ancient Greece*. Swansea: Classical Press of Wales.

Lobo, Rogerio A., David Gershenson, Gretchen M. Lentz, and Fidel A. Valea (2017). *Comprehensive Gynecology*, 7th ed. Philadelphia: Elsevier.

Lobo, Rogerio A., James H. Pickar, John C. Stevenson, Wendy J. Mack, and Howard N. Hodis (2016). Back to the future: Hormone replacement therapy as part of a prevention strategy for women at the onset of menopause. *Atherosclerosis* 254: 282–290.

Lock, Margaret (1993). *Encounters with Aging: Mythologies of Menopause in Japan and North America*. Berkeley: University of California Press.

Lock, Margaret, and Vinh-Kim Nguyen (2010). *An Anthropology of Biomedicine*. Chichester, UK: Wiley Blackwell.

Longo, Lawrence D. (1979). The rise and fall of Battey's Operation: A fashion in surgery. *Bulletin of the History of Medicine* 53: 244–267.

Lordkipanidze, David, Marcia Ponce de León, Ann Margvelashvili, Yoel Rak, G. Philip Rightmire, Abesalom Vekua, and Christoph P. E. Zollikofer (2013). A complete skull from Dmanisi, Georgia, and the evolutionary biology of early *Homo*. *Science* 342: 326–331.

Lorenzen, Eline D., David Nogués-Bravo, Ludovic Orlando, Jaco Weinstock, Jonas Binladen, Katharine A. Marske, Andrew Ugan, et al. (2011). Species-specific responses of late Quaternary megafauna to climate and humans. *Nature* 479: 359–364.

Lovejoy, C. Owen (1981). The origin of man. *Science* 211: 341–350.

Lovejoy, C. Owen, Richard S. Meindl, Thomas R. Pryzbeck, Thomas S. Barton, Kingsbury G. Heiple, and David Kotting (1977). Paleodemography of the Libben site, Ottawa County, Ohio. *Science* 198: 291–293.

Low, Bobbi S., Alice L. Clarke, and Kenneth A. Lockridge (1991). *Family Patterns in Nineteenth-Century Sweden: Variation in Time and Space*. Reports from the Demographic Data Base, Umeå University 6. Umeå, Sweden: UmU Tryckeri.

Mace, Ruth (1998). The coevolution of human fertility and wealth inheritance strategies. *Philosophical Transactions of the Royal Society B: Biological Sciences* 353: 389–397.

Mace, Ruth (2000). Evolutionary ecology of human life history. *Animal Behaviour* 59: 1–10.

Mace, Ruth, and Alexandra Alvergne (2012). Female reproductive competition within families in rural Gambia. *Proceedings of the Royal Society B: Biological Sciences* 279: 2219–2227.

MacLennan, A. H., J. L. Broadbent, S. Lester, and V. Moore (2004). Oral oestrogen and combined oestrogen/progestogen therapy versus placebo for hot flushes. *Cochrane Database of Systematic Reviews* 4: CD002978. doi: 10.1002/14651858.CD002978.pub2.

MacMullen, Ramsay (1997). *Christianity and Paganism in the Fourth to Eighth Centuries*. New Haven, CT: Yale University Press.

Mafessoni, Fabrizio, and Kay Prüfer (2017). Better support for a small effective population size of Neandertals and a long shared history of Neandertals and Denisovans. *Proceedings of the National Academy of Sciences* 114: E10,256–E10,257.

Malamud, W., S. L. Sands, and I. Malamud (1941). The involutional psychoses: A sociopsychiatric study. *Psychosomatic Medicine* 3: 410–426.

Malaspinas, Anna-Sapfo, Michael C. Westaway, Craig Muller, Vitor C. Sousa, Oscar Lao, Isabel Alves, Anders Bergström, et al. (2016). A genomic history of Aboriginal Australia. *Nature* 538: 207–214.

Mallick, Swapan, Heng Li, Mark Lipson, Iain Mathieson, Melissa Gymrek, Fernando Racimo, Mengyao Zhao, et al. (2016). The Simons Genome Diversity Project: 300 genomes from 142 diverse populations. *Nature* 538: 201–206.

Malthus, Thomas Robert (1798). *An Essay on the Principle of Population*. London: Johnson.

Malthus, Thomas Robert (1826). *An Essay on the Principle of Population*, 2 vols., 6th ed. London: Murray.

Marciniak, Arkadiusz, and Lech Czerniak (2007). Social transformations in the late Neolithic and early Chalcolithic periods in central Anatolia. *Anatolian Studies* 57: 115–130.

Marinello, Giovanni (1563). *Le medicine partenenti alle infirmità delle donne*. Venice: Bonadio.

Marlowe, Frank W. (2000). The Patriarch Hypothesis: An alternative explanation of menopause. *Human Nature* 11: 27–42.

Marlowe, Frank W. (2004). Marital residence among foragers. *Current Anthropology* 5: 277–284.

Marlowe, Frank W. (2005). Who tends Hadza children? In *Hunter-Gatherer Childhoods: Evolutionary, Developmental, and Cultural Practices*, ed. Barry S. Hewlett and Michael E. Lamb, 177–190. New Brunswick, NJ: Transaction.

Marlowe, Frank W. (2007). Hunting and gathering: The human sexual division of foraging labor. *Cross-Cultural Research* 41: 170–195.

Marlowe, Frank W. (2010). *The Hadza: Hunter-Gatherers of Tanzania*. Berkeley: University of California Press.

Marlowe, Frank W., and J. Collette Berbesque (2012). The human operational sex ratio: Effects of marriage, concealed ovulation, and menopause on mate competition. *Journal of Human Evolution* 63: 834–842.

Martin, Mary C., Jon E. Block, Sarah D. Sanchez, Claude D. Arnaud, and Yewoubdar Beyene (1993). Menopause without symptoms: The endocrinology of menopause among rural Mayan Indians. *American Journal of Obstetrics and Gynecology* 168: 1839–1845.

Martin, Nancy (1996). Mīrābāī: Inscribed in text, embodied in life. In *Vaiṣṇavī: Women and the Worship of Krishna*, ed. Steven J. Rosen, 7–46. Delhi: Motilal Banarsidass.

Masvie, Hilde (2007). The role of Tamang grandmothers in perinatal care, Makwanpur District, Nepal. In *Childrearing and Infant Care Issues*, ed. Pranee Liamputtong, 167–184. New York: Nova Science Publishers.

Mattern, Susan P. (2015). Panic and culture: *Hysterike pnix* in the ancient Greek world. *Journal of the History of Medicine and Allied Sciences* 70: 491–515.

McComb, Karen, Cynthia Moss, Sarah M. Durant, Lucy Baker, and Soila Sayialei (2001). Matriarchs as repositories of social knowledge in African elephants. *Science* 292: 491–494.

McComb, Karen, Graeme Shannon, Sarah M. Durant, Katito Sayialei, Rob Slotow, Joyce Poole, and Cynthia Moss (2011). Leadership in elephants: The adaptive value of age. *Proceedings of the Royal Society B: Biological Sciences* 278: 3270–3276.

McKeown, Thomas (1976). *The Modern Rise of Population*. London: Arnold.

Medawar, P. B. (1952). *An Unsolved Problem of Biology*. London: H. K. Lewis.

Meillassoux, Claude (1975). *Maidens, Meal and Money: Capitalism and the Domestic Community*. Cambridge: Cambridge University Press.

Melby, Melissa K. (2005). Vasomotor symptom prevalence and language of menopause in Japan. *Menopause* 12: 250–257.

Melby, Melissa K. (2006). Climacteric symptoms among Japanese women and men: Comparison of four symptom checklists. *Climacteric* 9: 298–304.

Melby, Melissa K., Debra Anderson, Lynette Leidy Sievert, and Carla Makhlouf Obermeyer (2011). Methods used in cross-cultural comparisons of vasomotor symptoms and their determinants. *Maturitas* 70: 110–119.

Melby, Melissa K., Margaret Lock, and Patricia Kaufert (2005). Culture and symptom reporting at menopause. *Human Reproduction Update* 11: 495–512.

Melby, Melissa K., Lynette Leidy Sievert, Debra Anderson, and Carla Makhlouf Obermeyer (2011). Overview of methods used in cross-cultural comparisions of menopausal symptoms and their determinants: Guidelines for Strengthening the Reporting of Menopause and Aging (STROMA) studies. *Maturitas* 70: 99–109.

Mendez, Fernando L., Thomas Krahn, Bonnie Schrack, Astrid-Maria Krahn, Krishna R. Veeramah, August E. Woerner, Forka Leypey Mathew Fomine, et al. (2013). An African American paternal lineage adds an extremely ancient root to the human Y chromosome phylogenetic tree. *American Journal of Human Genetics* 92: 454–459.

Menon, Usha, and Richard A. Shweder (1998). Return of the "white man's burden": The moral discourse of anthropology and the domestic life of Hindu women. In *Welcome to Middle Age! (And Other Cultural Fictions)*, ed. Richard A. Shweder, 139–188. Chicago: University of Chicago Press.

Mensforth, Robert P. (1990). Paleodemography of the Carlston Annis (Bt-5) late archaic skeletal population. *American Journal of Physical Anthropology* 82: 81–99.

Mesquida, Christian G., and Neil I. Wiener (1996). Human collective aggression: A behavioral ecology perspective. *Ethology and Sociobiology* 17: 247–262.

Meyer, Matthias, Qiaomei Fu, Ayinuer Aximu-Petri, Isabelle Glocke, Birgit Nickel, Juan-Luis Arsuaga, Ignacio Martínez, et al. (2014). A mitochondrial genome sequence of a hominin from Sima de los Huesos. *Nature* 505: 403–406.

Meyer, Matthias, Martin Kircher, Marie-Theres Gansauge, Heng Li, Fernando Racimo, Swapan Mallick, Joshua G. Schraiber, et al. (2012). A high-coverage genome sequence from an archaic Denisovan individual. *Science* 338: 222–226.

Michel, Joanna L., Gail B. Mahady, Mari Veliz, Doel D. Soejarto, and Armando Caceres (2006). Symptoms, attitudes, and treatment choices surrounding menopause among the Q'eqchi Maya of Livingston, Guatemala. *Social Science and Medicine* 23: 732–742.

Min, Sung Kil (2008). Clinical correlates of hwa-byung and a proposal for a new anger disorder. *Psychiatry Investigation* 5: 125–141.

Min, Sung Kil, and Shin-young Suh (2010). The anger syndrome hwa-byung and its comorbidity. *Journal of Affective Disorders* 124: 211–214.

Min, Sung Kil, Shin-Young Suh, and Ki-Jun Song (2009). Symptoms to use for diagnostic criteria of hwa-byung, an anger syndrome. *Psychiatry Investigation* 6: 7–12.

Mizroch, S. A. (1981). Analyses of some biological parameters of the Antarctic fin whale (*Balaenoptera physalus*). *Reports of the International Whaling Commission* 31: 425–434.

Moi, Toni (2008). *Simone de Beauvoir: The Making of an Intellectual Woman*, 2nd ed. Oxford: Oxford University Press.

Moreno-Mayar, J. Victor, Lasse Vinner, Peter de Barros Damgaard, Constanza de la Fuente, Jeffrey Chan, Jeffrey P. Spence, Morten E. Allentoft, et al. (2018). Early human dispersals within the Americas. *Science* 362: 1128 and eaav2621. doi: 10.1126/science.aav2621.

Morton, Richard A., Jonathan R. Stone, and Rama S. Singh (2013). Mate choice and the origin of menopause. *PLOS Computational Biology* 9 (6): e1003092. doi: 10.1371/journal. pcbi.1003092.

Morwood, M. J., P. Brown, Jatmiko, T. Sutikna, E. Wahyu Saptomo, K. E. Westaway, Rokus Awe Due, et al. (2005). Further evidence for small-bodied hominins from the late Pleistocene of Flores, Indonesia. *Nature* 437: 1012–1017.

Moschik, E. C., C. Mercado, T. Yoshino, K. Matsuura, and K. Watanabe (2012). Usage and attitudes of physicians in Japan concerning traditional Japanese medicine (kampo medicine): A descriptive evaluation of a representative questionnaire-based study. *Evidence-Based Complementary and Alternative Medicine* 139818. doi: 10.1155/2012/139818.

Moss, Cynthia J. (2001). The demography of an African elephant (*Loxodonta africana*) population in Amboseli, Kenya. *Journal of the Zoological Society of London* 255: 145–156.

Moss, Cynthia J., Harvey Croze, and Phyllis C. Lee, eds. (2011). *The Amboseli Elephants: A Long-Term Perspective on a Long-Lived Mammal*. Chicago: University of Chicago Press.

Mukta, Parita (1994). *Upholding the Common Life: The Community of Mirabai*. Delhi: Oxford University Press.

Mulder-Bakker, Anneke B. (2005). *Lives of the Anchoresses: The Rise of the Urban Recluse in Medieval Europe*. Philadelphia: University of Pennsylvania Press.

Mulder-Bakker, Anneke B. (2011). The age of discretion: Women at forty and beyond. In *Middle-Aged Women in the Middle Ages*, ed. Sue Niebrzydowski, 15–24. Cambridge: D. S. Brewer.

Muller, Martin N., Melissa Emery Thompson, and Richard W. Wrangham (2006). Male chimpanzees prefer mating with old females. *Current Biology* 16: 2234–2236.

Mungello, D. E. (2008). *Drowning Girls in China: Female Infanticide Since 1650*. Lanham, MD: Rowman & Littlefield.

Münzel, Mark (1973). *The Aché Indians: Genocide in Paraguay*. Copenhagen, Denmark: International Work Group for Indigenous Affairs.

Murphy, Lorna, Lynette Leidy Sievert, Khurshida Begum, Taniya Sharmeen, Elaine Puleo, Osul Chowdhury, Shanthi Muttukrishna, and Gillian Bentley (2013). Life course effects on age at menopause among Bangladeshi sedentees and migrants to the UK. *American Journal of Human Biology* 25: 83–93.

Murray, Stephen O. (2000). *Homosexualities*. Chicago: University of Chicago Press.

Musallam, B. F. (1983). *Sex and Society in Islam: Birth Control Before the Nineteenth Century*. Cambridge: Cambridge University Press.

Muslić, Ljiljana, and Nastaša Jokić-Begić (2016). The experience of perimenopausal distress: Examining the role of anxiety and anxiety sensitivity. *Journal of Psychosomatic Obstetrics & Gynecology* 37: 26–33.

Mutinda, Hamisi, Joyce H. Poole, and Cynthia J. Moss (2011). Decision making and leadership in using the ecosystem. In *The Amboseli Elephants: A Long-Term Perspective on a Long-Lived Mammal*, ed. Cynthia J. Moss, Harvey Croze, and Phyllis C. Lee, 246–259. Chicago: University of Chicago Press.

National Institutes of Health (2005). Final panel statement: NIH State-of-the-Science Conference on Management of Menopausal Symptoms. *NIH Consensus State of the Science Statements* 22 (1): 1–38. https://consensus.nih.gov/2005/menopausestatement.htm.

Nelson, Sarah M (1997). *Gender in Archaeology: Analyzing Power and Prestige*. Walnut Creek, CA: AltaMira.

Neubauer, Fernanda (2014). A brief overview of the last 10 years of major late Pleistocene discoveries in the Old World: *Homo floresiensis,* Neanderthal, and Denisovan. Journal of Anthropology 2014: 581689. doi: 10.1155/2014/581689.

Neubauer, Simon, and Jean-Jacques Hublin (2012). The evolution of human brain development. *Evolutionary Biology* 39: 568–586.

Nichter, Mark (1981). Idioms of distress: Alternatives in the expression of psychosocial distress: A case study from South India. *Culture, Medicine, and Psychiatry* 5: 379–408.

Nicol-Smith, Louise (1996). Causality, menopause, and depression: A critical review of the literature. *British Medical Journal* 313: 1229–1232.

Nishida, Toshisada, Nadia Corp, Miya Hamai, Toshikazu Hasegawa, Mariko Hiraiwa-Hasegawa, Kazuhiko Hosaka, Kevin D. Hunt, et al. (2003). Demography, female life history, and reproductive profiles among the chimpanzees of Mahale. *American Journal of Primatology* 59: 99–121.

Obermeyer, Carla Makhlouf, David Reher, and Matilda Saliba (2007). Symptoms, menopause status, and country differences: A comparative analysis from DAMES. *Menopause* 14: 788–797.

Obermeyer, Carla Makhlouf, Michelle Schulein, Najia Hajji, and Mustapha Azelmat (2001). Menopause in Morocco: Symptomology and medical management. *Maturitas* 41: 87–95.

O'Connell, James F., Kristen Hawkes, and Nicholas G. Blurton Jones (1999). Grandmothering and the evolution of *Homo erectus. Journal of Human Evolution* 36: 461–485.

O'Connell, James F., Kristen Hawkes, K. D. Lupo, and Nicholas G. Blurton Jones (2002). Male strategies and Plio-Pleistocene archaeology. *Journal of Human Evolution* 43: 831–872.

O'Connor, Kathleen A. (1995). The age pattern of mortality: A micro-analysis of Tipu and a meta-analysis of twenty-nine paleodemographic samples. Ph.D. diss., SUNY Albany.

Okonofua, F. E., A. Lawal, and J. K. Bamgbose (1990). Features of menopause and menopausal age in Nigerian women. *International Journal of Gynaecology and Obstetrics* 31: 314–345.

Olesiuk, P. K., M. A. Bigg, and G. M. Ellis (1990). Life history and population dynamics of resident killer whales (*Orcinus orca*) in the coastal waters of British Columbia and Washington state. In *Report of the International Whaling Commission*, special issue 12, ed. Philip S. Hammond, Sally A. Mizroch, and Gregory P. Donovan, 209–244. Cambridge: International Whaling Commission.

Omran, Abdel R. (1971). The Epidemiologic Transition: A theory of the epidemiology of population change. *Milbank Memorial Fund Quarterly* 49: 509–538.

Osbaldeston, Tess Anne, and R.P.A. Wood, trs. (2000). *Dioscorides: De materia medica*. Johannesburg: Ibidis.

Otolorin, E. O., I. Adeyefa, B. O. Osotimehin, T. Fatinikun, O. Ojengbede, J. O. Otubu, and O. A. Ladipo (1989). Clinical, hormonal, and biochemical features of menopausal women in Ibadan, Nigeria. *African Journal of Medicine and Medical Sciences* 18: 251–255.

Owenby, David (2002). Approximations of Chinese bandits: Perverse rebels, romantic heroes, or frustrated bachelors? In *Chinese Femininities/Chinese Masculinities: A Reader*, ed. Susan Brownell, Jeffery N. Wasserstorm, and Thomas Laqueur, 226–250. Berkeley: University of California Press.

Packer, Craig, Marc Tatar, and Anthony Collins (1998). Reproductive cessation in female mammals. *Nature* 392: 807–811.

Pagani, Luca, Daniel John Lawson, Evelyn Jagoda, Alexander Mörseburg, Anders Eriksson, Mario Mitt, Florian Clemente, et al. (2016). Genomic analyses inform on migration events during the peopling of Eurasia. *Nature* 538: 238–242.

Paixão-Côrtes, Vanessa R., Lucas Henriques Viscardi, Francisco Mauro Salzano, Maria Cátira Bartolini, and Tábita Hünemeier (2013). The cognitive ability of extinct hominins: Bringing down the hierarchy using genomic evidences. *American Journal of Human Biology* 25: 702–705.

Panigrahi, Lalita (1972). *British Social Policy and Female Infanticide in India*. New Delhi: Munshiram Manoharlal.

Park, Young-Joo, Hesook Suzie Kim, Donna Schwartz-Barcott, and Jong-Woo Kim (2002). The conceptual structure of *hwa-byung* in middle-aged Korean women. *Health Care for Women International* 23: 389–397.

Patterson, Cynthia B. (1998). *The Family in Greek History*. Cambridge, MA: Harvard University Press.

Pavard, Samuel, Jessica E. Metcalf, and Evelyne Heyer (2008). Senescence of reproduction may explain adaptive menopause in humans: A test of the "Mother" Hypothesis. *American Journal of Physical Anthropology* 136: 194–208.

Pavard, Samuel, and Frédéric Branger (2012). Effect of maternal and grandmaternal care on population dynamics and human life-history evolution: A matrix projection model. *Theoretical Population Biology* 82: 364–376.

Pavleka, Mary S., and Linda M. Fedigan. (2012). The costs and benefits of old age reproduction in the Arashiyama west female Japanese macaques. In *The Monkeys of Stormy Mountain*, ed.

Jean-Baptiste Leca, Michael A. Huffman, and Paul L. Vasey, 131–152. Cambridge and New York: Cambridge University Press.

Peccei, Jocelyn Scott (1995). A hypothesis for the origin and evolution of menopause. *Maturitas* 21: 83–89.

Peccei, Jocelyn Scott (2001). Menopause: Adaptation or epiphenomenon? *Evolutionary Anthropology* 10: 43–57.

Penn, Dustin J., and Ken R. Smith (2007). Differential fitness costs of reproduction between the sexes. *Proceedings of the National Academy of Sciences* 104: 553–558.

Pepin, Jacques (2011). *The Origins of AIDS*. Cambridge and New York: Cambridge University Press.

Perkins, John H. (1997). *Geopolitics and the Green Revolution: Wheat, Genes, and the Cold War*. New York: Oxford University Press.

Petraglia, Michael, Peter Ditchfield, Sacha Jones, Ravi Korisettar, and J. N. Pal (2012). The Toba volcanic super-eruption, environmental change, and hominin occupation history in India over the last 140,000 years. *Quaternary Journal* 258: 119–134.

Petraglia, Michael, Ravi Korisettar, Nichole Boivin, Christopher Clarkson, Peter Ditchfield, Sacha Jones, Jinu Koshy, et al. (2007). Middle Paleolithic assemblages from the Indian subcontinent before and after the Toba eruption. *Science* 317: 114–116.

Phillips, Kim M. (2004). Margery Kempe and the ages of woman. In *A Companion to the Book of Margery Kempe*, ed. John Arnold and Katherine J. Lewis, 17–34. Woodbridge, UK: Brewer.

Pilloud, Marin A., and Clark Spencer Larsen (2011). "Official" and "practical" kin: Inferring social and community structure from dental phenotype at Neolithic Çatalhöyük, Turkey. *American Journal of Physical Anthropology* 145: 519–530.

Pinhasi, Ron, and Chryssi Bourbou (2008). How representative are human skeletal assemblages for population analysis? In *Advances in Human Paleopathology*, ed. Ron Pinhasi and Simon Mays, 31–44. Hoboken, NJ: Wiley.

Plavcan, J. Michael (2012). Body size, size variation, and sexual size dimorphism in early *Homo*. *Current Anthropology* 53, suppl. 6: S409–S423.

Pomeranz, Kenneth (2009). *The Great Divergence: China, Europe, and the Making of the Modern World Economy*. Princeton, NJ: Princeton University Press.

Pontzer, Herman (2012). Ecological energetics in early *Homo*. *Current Anthropology* 53, suppl. 6: S346–S358.

Porter, Theresa, and Helen Gavin (2010). Infanticide and neonaticide: A review of 40 years of research literature on incidence and causes. *Trauma, Violence, and Abuse* 11: 99–112.

Posth, Cosimo, Christoph Wißing, Keiko Kitagawa, Luca Pagani, Laura van Holstein, Fernando Racimo, Kurt Wehrberger, et al. (2017). Deeply divergent archaic mitochondrial genome provides lower time boundary for African gene flow into Neanderthals. *Nature Communications* 8: 16,046.

Potts, Richard (1996). *Humanity's Descent: The Consequences of Ecological Instability*. New York: William Morrow and Company.

Potts, Richard (2012). Environmental and behavioral evidence pertaining to the evolution of early *Homo*. *Current Anthropology* 53, suppl. 6: S299–S317.

Powell, Adam, Stephan Shennan, and Mark G. Thomas (2009). Late Pleistocene demography and the appearance of modern human behavior. *Science* 324: 1298–1301.

Prentice, Ross L., and Garnet L. Anderson (2007). The Women's Health Initiative: Lessons learned. *Annual Review of Public Health* 29: 131–150.

Preston, Samuel H. (1976). *Mortality Patterns in National Populations: With Special Reference to Recorded Causes of Death.* New York: Academic Press.

Proctor, Robert (1988). *Racial Hygiene: Medicine Under the Nazis.* Cambridge, MA: Harvard University Press.

Prüfer, Kay, Fernando Racimo, Nick Patterson, Flora Jay, Sriram Sankararaman, Susanna Sawyer, Anja Heinze, et al. (2014). The complete sequence of a Neanderthal genome from the Altai Mountains. *Nature* 505: 43–49.

Pruitt, Ida (1945). *A Daughter of Han: The Autobiography of a Chinese Working Woman.* Stanford, CA: Stanford University Press.

Puleston, Cedric O., and Shripad Tuljapurkar (2008). Population and prehistory II: Space-limited human populations in constant environments. *Theoretical Population Biology* 74: 147–160.

Qingfeng Wang and Xu Sun (2016). The role of socio-economic and political factors in fertility decline: A cross-country analysis. *World Development* 87: 360–370.

Quinlan, Robert J., and Marsha B. Quinlan (2007). Evolutionary ecology of human pair-bonds: Cross-cultural tests of alternative hypotheses. *Cross-Cultural Research* 41: 149–169.

Ramakrishna Vedanta Centre (1955). *Women Saints, East and West.* Hollywood, CA: Vedanta Press.

Ramsay, M. A., and Ian Stirling (1988). Reproductive biology and ecology of female polar bears (*Ursus maritimus*). *Journal of Zoology* 214: 601–634.

Rasmussen, Morten, Sarah L. Anzick, Michael R. Waters, Pontus Skoglund, Michael DeGiorgio, Thomas W. Stafford, Jr., Simon Rasmussen, et al. (2014). The genome of a late Pleistocene human from a Clovis burial site in western Montana. *Nature* 506: 225–229.

Rasmussen, Morten, Xiaosen Guo, Yong Wang, Kirk E. Lohmueller, Simon Rasmussen, Anders Albrechtsen, Line Skotte, et al. (2011). An Aboriginal Australian genome reveals separate human dispersals into Asia. *Science* 334: 94–98.

Rawlins, Richard G., and Matt J. Kessler, eds. (1986). *The Cayo Santiago Macaques: History, Behavior and Biology.* Albany: State University of New York Press.

Reich, David (2018). *Who We Are and How We Got Here: Ancient DNA and the New Science of the Human Past.* New York: Pantheon.

Reich, David, Richard E. Green, Martin Kircher, Johannes Krause, Nick Patterson, Eric Y. Durand, Bence Viola, et al. (2010). Genetic history of an archaic hominin group from Denisova Cave in Siberia. *Nature* 468: 1053–1060.

Reich, David, Nick Patterson, Desmond Campbell, Arti Tandon, Stéphane Mazieres, Nicolas Ray, Maria V. Parra, et al. (2012). Reconstructing Native American population history. *Nature* 488: 370–375.

Rémy, Catherine (2014). "Men seeking monkey-glands": The controversial xenotransplantations of Doctor Voronoff, 1910–1930. *French History* 28: 226–240.

Rice, Pranee Ljamputtong (1995). Pog laus, tsis coj kaub ncaws lawrm: The meaning of menopause in Hmong women. *Journal of Reproductive and Infant Psychology* 13: 72–92.

Richardson, R. C. (2010). *Household Servants in Early Modern England.* Manchester, UK: Manchester University Press.

Richerson, Peter J., Robert Boyd, and Robert L. Bettinger (2009). Cultural innovations and demographic change. *Human Biology* 81: 211–235.

Richter, Daniel, Johannes Moser, Mustapha Nami, Josef Eiwanger, and Abdeslam Mikdad (2010). New chronometric data from Ifri n'Ammar (Morocco) and the chronostratigraphy of the Middle Palaeolithic in the western Maghreb. *Journal of Human Evolution* 59: 672–679.

Riddle, John M. (1992). *Contraception and Abortion from the Ancient World to the Renaissance.* Cambridge, MA: Harvard University Press.

Riddle, John M. (1997). *Eve's Herbs: A History of Contraception and Abortion in the West.* Cambridge, MA: Harvard University Press.

Riley, James C. (2001). *Rising Life Expectancy: A Global History.* Cambridge: Cambridge University Press.

Rindfuss, Ronald R., and S. Philip Morgan (1983). Marriage, sex, and the first birth interval: The quiet revolution in Asia. *Population and Development Review* 9: 259–280.

Roberts, Miguel E., Kyunghee Han, and Nathan C. Weed (2006). Development of a scale to assess *hwa-byung*, a Korean culture-bound syndrome, using the Korean MMPI-2. *Transcultural Psychiatry* 43: 383–400.

Robson, Arthur J., and Hillard S. Kaplan (2003). The evolution of human life expectancy and intelligence in hunter-gatherer economies. *American Economic Review* 93: 150–169.

Robson, Shannen L., Carel P. van Schaik, and Kristen Hawkes (2006). The derived features of human life history. In *The Evolution of Human Life History*, ed. Kristen Hawkes and Richard R. Paine, 17–44. Santa Fe, NM: School of American Research Press.

Robson, Shannen L., and Bernard Wood (2008). Hominin life history: Reconstruction and evolution. *Journal of Anatomy* 212: 394–425.

Rogers, Alan R. (1993). Why menopause? *Evolutionary Ecology* 7: 406–420.

Rogers, Alan R., Ryan J. Bohlender, and Chad D. Huff (2017). Early history of Neanderthals and Denisovans. *Proceedings of the National Academy of Sciences* 114: 9859–9863.

Rogers, Richard G., Bethany G. Everett, Jarron M. Saint Onge, and Patrick M. Krueger (2010). Social, behavioral, and biological factors, and sex differences in mortality. *Demography* 47: 555–578.

Rosas, Antonio, Luis Ríos, Almudena Estalrrich, Helen Liversidge, Antonio García-Tabernero, Rosa Huguet, Hugo Cardoso, et al. (2017). The growth patterns of Neandertals, reconstructed from a juvenile skeleton from El Sidrón (Spain). *Science* 357: 1282–1287.

Rosen, Steven J. (1996). *Vaiṣṇavī: Women and the Worship of Krishna.* Delhi: Motilal Banarsidass.

Rosenberger, Nancy R. (1987). Productivity, sexuality, and ideologies of menopausal problems in Japan. In *Health, Illness, and Medical Care in Japan*, ed. Edward Norbeck and Margaret Lock, 158–188. Honolulu: University of Hawaii Press.

Roser, Max (2016). World poverty. OurWorldInData.org. https://ourworldindata.org/world-poverty/.

Roser, Max, and Esteban Ortiz-Ospina (2018). World population growth. OurWorldInData.org. https://ourworldindata.org/world-population-growth.

Rosero-Bixby, Luis (2011). Generational transfers and population aging in Latin America. *Population and Development Review* 37, suppl.: S143–S157.

Roth, Randolph (2001). Homicide in early modern England, 1549–1800: The need for a quantitative synthesis. *Crime, histoire, et sociétés/Crime, History, and Societies* 5: 33–67.

Rowlandson, Jane, and Ryosuke Takahashi (2009). Brother-sister marriage and inheritance strategies in Greco-Roman Egypt. *Journal of Roman Studies* 99: 104–139.

Royal Anthropological Institute of Great Britain and Ireland (1951). *Notes and Queries on Anthropology*. London: Routledge and Kegan Paul.

Ruddiman, William F. (2005). *Plows, Plagues, and Petroleum: How Humans Took Control of Climate*. Princeton, NJ: Princeton University Press.

Ruff, Christopher B. (2002). Variation in human body size and shape. *Annual Review of Anthropology* 31: 211–232.

Ruff, Christopher B., Erik Trinkaus, and Trenton W. Holliday (1997). Body mass and encephalization in Pleistocene *Homo*. *Nature* 387: 173–176.

Sahle, Yonathan, W. Karl Hutchings, David R. Braun, Judith C. Sealy, Leah E. Morgan, Agazi Negash, and Balemwal Atnafu (2013). Earliest stone-tipped projectiles from the Ethiopian rift date to >279,000 years ago. *PLOS ONE* 8 (11): e78092. doi: 10.1371/journal.pone .0078092.

Salazar-García, Domingo, Robert C. Power, Alfred Sanchis Serra, Valentín Villaverde, Michael J. Walker, and Amanda G. Henry (2013). Neanderthal diets in central and southeastern Mediterranean Iberia. *Quaternary International* 318: 3–18.

Salpeter, Shelley R., Nicholas S. Buckley, Hau Liu, and Edwin E. Salpeter (2009). The cost-effectiveness of hormone therapy in younger and older postmenopausal women. *American Journal of Medicine* 122: 42–52.

Sanderson, Stephen K., and Joshua Dubrow (2000). Fertility decline in the modern world and the original Demographic Transition: Testing theories with cross-national data. *Population and Environment: A Journal of Interdisciplinary Studies* 21: 511–537.

Santow, Gigi (1995). *Coitus interruptus* and the control of natural fertility. *Population Studies* 49: 19–43.

Sayers, Ken, and C. Owen Lovejoy (2014). Blood, bulbs, and bunodonts: On evolutionary ecology and the diets of *Ardepithecus, Australopithecus*, and early *Homo*. *Quarterly Review of Biology* 89: 319–357.

Schacht, Ryan, Kristin Liv Rauch, and Monique Borgerhoff Mulder (2014). Too many men: The violence problem? *Trends in Ecology and Evolution* 29: 214–222.

Schäfer, Daniel (2003). Die alternde Frau in der frühneuzeitlichen Medizin—Ein "vergessene" Gruppe alter Menschen. *Sudhoffs Archiv* 87: 90–108.

Scheid, Volker (2007). Traditional Chinese Medicine—What are we investigating? The case of menopause. *Complementary Therapies in Medicine* 15: 54–68.

Scheidel, Walter (1995). The most silent women of ancient Greece and Rome: Rural labor and women's life in the ancient world (I). *Greece and Rome* 42: 202–217.

Scheidel, Walter (1996a). The most silent women of ancient Greece and Rome: Rural labor and women's life in the ancient world (II). *Greece and Rome* 43: 1–10.

Scheidel, Walter (1996b). The biology of brother-sister marriage in Roman Egypt: An interdisciplinary approach. In *Measuring Sex, Age and Death in the Roman Empire, Journal of Roman Archaeology* Supplementary Series 21, 9–51. Ann Arbor: University of Michigan Press.

Scheidel, Walter (2009a). A peculiar institution? Greco-Roman monogamy in global context. *History of the Family* 14: 280–291.

Scheidel, Walter (2009b). Sex and empire: A Darwinian perspective. In *The Dynamics of Ancient Empires*, ed. Ian Morris and Walter Scheidel, 255–324. New York: Oxford University Press.

Scheidel, Walter (2017). *The Great Leveler: Violence and the History of Inequality from the Stone Age to the Twenty-First Century*. Princeton, NJ: Princeton University Press.

Schlebusch, Carina M., Helena Malmström, Torsten Günther, Per Sjödin, Alexandra Coutinho, Hanna Edlund, Arielle R. Munters, et al. (2017). Southern African ancient genomes estimate modern human divergence to 350,000 to 260,000 years ago. *Science* 358: 652–655.

Schlebusch, Carina, Pontus Skoglund, Per Sjödin, Lucie M. Gattepaille, Dena Hermandez, Flora Jay, Sen Li, et al. (2012). Genomic variation in seven Khoe-San groups reveals adaptation and complex African history. *Science* 338: 374–379.

Schmandt-Besserat, Denise (1996). *How Writing Came About*. Austin: University of Texas Press.

Schrire, Carmel, and William Lee Steiger (1974). A matter of life and death: An investigation into the practice of female infanticide in the Arctic. *Man* (n.s.) 9: 161–184.

Schwartz, Gary T. (2012). Growth, development, and life history throughout the evolution of *Homo*. *Current Anthropology* 53, suppl. 6: S395–S408.

Scott, James C. (1985). *Weapons of the Weak: Everyday Forms of Peasant Resistance*. New Haven, CT: Yale University Press.

Scrimshaw, Susan C. M. (1984). Infanticide in human populations: Societal and individual concerns. In *Infanticide: Comparative and Evolutionary Perspectives*, ed. Glenn Hausfater and Sarah Blaffer Hrdy, 439–462. New York: Aldine.

Scull, Andrew (2009). *Hysteria: The Biography*. Oxford: Oxford University Press.

Sear, Rebecca (2008). Kin and child survival in rural Malawi: Are matrilineal kin always beneficial in a matrilineal society? *Human Nature* 19: 277–293.

Sear, Rebecca, and Ruth Mace (2008). Who keeps children alive? A review of the effects of kin on child survival. *Evolution and Human Behavior* 29: 1–18.

Sear, Rebecca, Ruth Mace, and Ian A. McGregor (2000). Maternal grandmothers improve nutritional status and survival of children in rural Gambia. *Proceedings of the Royal Society B: Biological Sciences* 1453: 1641–1647.

Sear, Rebecca, Ruth Mace, and Ian A. McGregor (2003). The effects of kin on female fertility in rural Gambia. *Evolution and Human Behavior* 24: 25–42.

Sear, Rebecca, Fiona Steele, Ian A. McGregor, and Ruth Mace (2002). The effects of kin on child mortality in rural Gambia. *Demography* 39: 43–63.

Seaver, Paul S. (2008). Apprentice riots in early modern London. In *Violence, Politics, and Gender in Early Modern England*, ed. Joseph P. Ward, 17–40. New York: Palgrave Macmillan.

Sellen, Daniel W., and Ruth Mace (1997). Fertility and mode of subsistence: A phylogenetic analysis. *Current Anthropology* 38: 878–889.

Sellergren, Sarah (1996). Janābāī and Kānhopātrā: A study of two women saints. In *Images of Women in Maharashtrian Literature and Religion*, ed. Anne Feldhaus, 213–239. Albany: State University of New York Press.

Sen, Amartya (1989). Women's survival as a development problem. *Bulletin of the American Academy of Arts and Sciences* 43: 14–29.

Shanley, Daryl P., and Thomas B. L. Kirkwood (2001). Evolution of the human menopause. *BioEssays* 23: 282–287.

Shanley, Daryl P., Rebecca Sear, Ruth Mace, and Thomas B. L. Kirkwood (2007). Testing evolutionary theories of menopause. *Proceedings of the Royal Society B: Biological Sciences* 274: 2943–2949.

Sharma, Rakesh, Ashok Agarwal, Vikram K. Rohra, Mourad Assidi, Muhammad Abu-Elmagd, and Rola F. Turki (2015). Effects of increased paternal age on sperm quality, reproductive outcome and associated epigenetic risks to offspring. *Reproductive Biology and Endocrinology* 13. doi: 10.1186/s12958-015-0028-x.

Shaw, Brent (1984). Bandits in the Roman Empire. *Past & Present* 105: 3–52.

Sheehan, Donal (1936). The discovery of the autonomic nervous system. *Archives of Neurology and Psychiatry* 35: 1081–1115.

Shenk, Mary K., Monique Borgerhoff Mulder, Jan Beise, Gregory Clark, William Irons, Donna Leonetti, Bobbi S. Low, et al. (2010). Intergenerational wealth transmission among agriculturalists: Foundations of agrarian inequality. *Current Anthropology* 51: 65–83.

Shennan, Stephen (2009). Evolutionary demography and the population history of the early European Neolithic. *Human Biology* 81: 339–345.

Shennan, Stephen, and Kevan Edinborough (2007). Prehistoric population history: From the late glacial to the late Neolithic in central and northern Europe. *Journal of Archaeological Science* 34: 1339–1345.

Shumaker, R. W., S. A. Wich, and L. Perkins (2008). Reproductive life history traits of female orangutans (*Pongo* spp.). In *Primate Reproductive Aging: Cross-Taxon Perspectives*, ed. Sylvia Atsalis, Susan W. Margulis, and Patrick R. Hof, 147–161. Basel: Karger.

Shweder, Richard A., ed. (1998). *Welcome to Middle Age! (And Other Cultural Fictions)*. Chicago: University of Chicago Press.

Sievert, Lynnette Leidy (2006). *Menopause: A Biocultural Perspective*. New Brunswick, NJ: Rutgers University Press.

Sievert, Lynnette Leidy, Khurshida Begum, Taniya Sharmeen, Osul Chowdhury, Shanthi Muttukrishna, and Gillian Bentley (2008). Patterns of occurrence and concordance between subjective and objective hot flashes among Muslim and Hindu women in Sylhet, Bangladesh. *American Journal of Human Biology* 20: 598–604.

Sievert, Lynnette Leidy, Khurshida Begum, Taniya Sharmeen, L. Murphy, B. W. Whitcomb, Osul Chowdhury, Shanthi Muttukrishna, and Gillian Bentley (2016). Hot flash report and measurement among Bangladeshi migrants, their London neighbors, and their community of origin. *American Journal of Physical Anthropology* 161: 620–633.

Sievert, Lynnette Leidy, and Erin K. Flanagan (2005). Geographical distribution of hot flash frequencies: Considering climatic influences. *American Journal of Physical Anthropology* 128: 437–443.

Sievert, Lynette Leidy, and Susan I. Hautaniemi (2003). Age at menopause in Puebla, Mexico. *Human Biology* 75: 205–226.

Simmons, Leo W. (1945). *The Role of the Aged in Primitive Society.* New Haven, CT: Yale University Press.

Singh, Vanita, and M. Sivakami (2014). Menopause: Midlife experiences of low socio-economic strata women in Haryana. *Sociological Bulletin* 63: 263–286.

Skjærvø, Gine Roll, and Eivin Røskaft (2013). Menopause: No support for an evolutionary explanation among historical Norwegians. *Experimental Gerontology* 48: 408–413.

Skoglund, Pontus, Swapan Mallick, Maria Cátria Bortolini, Niru Chennagiri, Tábita Hünemeier, Maria Luiza Petzl-Erler, Francisco Mauro Salzano, Nick Patterson, and David Reich (2015). Genetic evidence for two founding populations of the Americas. *Nature* 525: 104–108.

Skutch, Alexander (1935). Helpers at the nest. *Auk* 52: 257–263.

Smith, Christopher C., and Stephen D. Fretwell (1974). The optimal balance between size and number of offspring. *American Naturalist* 108: 499–506.

Smith, Eric Alden (2004). Why do good hunters have higher reproductive success? *Human Nature* 15: 343–364.

Smith, Eric Alden, Kim Hill, Frank W. Marlowe, David Nolin, Polly Wiessner, Michael Gurven, Samuel Bowles, Monique Borgerhoff Mulder, Tom Hertz, and Adrian Bell (2010). Wealth transmission and inequality among hunter-gatherers. *Current Anthropology* 51: 19–34.

Smith, Tanya M. (2013). Teeth and human life-history evolution. *Annual Review of Anthropology* 42: 191–208.

Smith, Tanya M., Paul Tafforeau, Donald J. Reid, Rainer Grün, Stephen Eggins, Mohamed Boutakiout, and Jean-Jacques Hublin (2007). Earliest evidence of modern human life history in North African early *Homo sapiens. Proceedings of the National Academy of Sciences* 104: 6128–6133.

Smith-Rosenberg, Carroll (1985). *Disorderly Conduct: Visions of Gender in Victorian America.* New York: Knopf.

Sommer, Matthew H. (2000). *Sex, Law, and Society in Late Imperial China.* Stanford, CA: Stanford University Press.

Sommer, Matthew H. (2005). Making sex work: Polyandry as a survival strategy in Qing dynasty China. In *Gender in Motion: Divisions of Labor and Cultural Change in Late Imperial and Modern China*, ed. Bryna Goodman and Wendy Larson, 29–54. Lanham, MD: Rowman & Littlefield.

Sommer, Matthew H. (2010). Abortion in late imperial China: Routine birth control or crisis intervention? *Late Imperial China* 31: 97–165.

Sommer, Matthew H. (2015). *Polyandry and Wife-Selling in Qing Dynasty China: Survival Strategies and Judicial Interventions.* Berkeley: University of California Press.

South, Scott J., and Katherine Trent (1988). Sex ratios and women's roles: A cross-national analysis. *American Journal of Sociology* 93: 1096–1115.

Speakman, John R. (2013). Sex- and age-related mortality profiles during famine: Testing the "body fat" hypothesis. *Journal of Biosocial Sciences* 45: 823–840.

Spetz, A.C.E., M. G. Frederiksson, and M. L. Hammar (2003). Hot flushes in a male population aged 55, 65, and 75 years, living in the community of Linkoping, Sweden. *Menopause* 10: 81–87.

Stacey, Judith (1983). *Patriarchy and Socialist Revolution in China.* Berkeley: University of California Press.

Stefanopoulou, Evgenia, and Elizabeth Alice Grunfeld (2017). Mind-body interventions for vasomotor symptoms in healthy menopausal women and breast cancer survivors: A systematic review. *Journal of Psychosomatic Obstetrics & Gynecology* 38: 210–225.

Stern, Alexandra Minna (2016). *Eugenic Nation: Faults and Frontiers of Better Breeding in Modern America*, 2nd ed. Berkeley: University of California Press.

Stern, R. M., S. Hu, S. H. Uijtdehaage, E. R. Muth, L. H. Xu, and K. L. Koch (1996). Asian hypersusceptibility to motion sickness. *Human Heredity* 46: 7–14.

Stewart, Donna E. (2003). Menopause in highland Guatemala women. *Maturitas* 44: 293–297.

Stolberg, Michael (1999). A woman's hell? Medical perceptions of menopause in preindustrial Europe. *Bulletin of the History of Medicine* 73: 404–428.

Stolberg, Michael (2000). The monthly malady: A history of premenstrual suffering. *Medical History* 44: 301–322.

Stone, Arthur A. (2000). *The Science of Self-Report: Implications for Research and Practice.* Mahwah, NJ: Erlbaum.

Stone, Elizabeth (1982). The social role of Nadītu women in Old Babylonian Nippur. *Journal of the Economic and Social History of the Orient* 25: 50–70.

Storch, Johann (1753). *Weiber-kranckheiten*, vol. 8. Gotha: Mevius. https://archive.org/details/bub_gb_o2BWAAAAcAAJ.

Strassmann, Beverly I., and Wendy M. Garrard (2011). Alternatives to the Grandmother Hypothesis: A meta-analysis of the association between grandparental and grandchild survival in patrilineal populations. *Human Nature* 22: 201–222.

Sugiyama, Yukimaru (2004). Demographic parameters and life history of chimpanzees at Bossou, Guinea. *American Journal of Physical Anthropology* 124: 154–165.

Suh, Soyoung (2013). Stories to be told: Korean doctors between *hwa-byung* (fire-illness) and depression, 1970–2011. *Culture, Medicine, and Psychiatry* 37: 81–104.

Sutikna, Thomas, Matthew W. Tocheri, Michael J. Morwood, E. Wahyu Saptomo, Jatmiko, Rokus Due Awe, Sri Wasisto, et al. (2016). Revised stratigraphy and chronology for *Homo floresiensis* at Liang Bua in Indonesia. *Nature* 532: 366–369.

Tadmor, Naomi (2001). *Family and Friends in Eighteenth-Century England: Household, Kinship, and Patronage.* Cambridge: Cambridge University Press.

Tamura, Robert (2006). Human capital and economic development. *Journal of Development Economics* 79: 26–72.

Tan, Jee-Peng (1983). Marital fertility at older ages in Nepal, Bangladesh and Sri Lanka. *Population Studies* 37: 433–444.

Tan Yunxian (2015). *Miscellaneous Records of a Female Doctor.* Tr. Lorraine Wilcox with Yue Lu. Portland, OR: Chinese Medicine Database.

Taylor, Kim (2005). *Chinese Medicine in Early Communist China, 1945–63.* Abingdon, UK: Routledge.

Taylor, Ruth E. (2001). Death of neurasthenia and its psychological reincarnation. *British Journal of Psychiatry* 179: 550–557.

Thompson, Jennifer L., and Andrew J. Nelson (2011). Middle childhood and modern human origins. *Human Nature* 22: 249–280.

Thornton, Russel (2000). Population history of native North Americans. In *A Population History of North America*, ed. Michael R. Haines and Richard H. Steckel, 9–50. Cambridge: Cambridge University Press.

Tilt, Edward John (1857). *The Change of Life in Health and Disease.* London: Churchill.

Tomasello, Michael (2016). *A Natural History of Human Morality.* Cambridge, MA: Harvard University Press.

Torfing, Tobias (2015). Neolithic population and summed probability distribution of 14C-dates. *Journal of Archaeological Science* 63: 193–198.

Toro-Moyano, Isidoro, Bienvenido Martinez-Navarro, Jordi Agustí, Caroline Souday, José María Mermúdez de Castro, María Martinón-Torres, Beatriz Fajardo, et al. (2013). The oldest human fossil in Europe, from Orce (Spain). *Journal of Human Evolution* 65: 1–9.

Trinkaus, Erik (2011). Late Pleistocene mortality patterns and modern human establishment. *Proceedings of the National Academy of Sciences* 108: 1267–1271.

Trivers, Robert L. (1971). The evolution of reciprocal altruism. *Quarterly Review of Biology* 46: 35–57.

Trivers, Robert L. (1972). Parental investment and sexual selection. In *Sexual Selection and the Descent of Man*, ed. Bernard Campbell, 136–179. Chicago: Aldine.

Trivers, Robert L., and Dan E. Willard (1973). Natural selection of parental ability to vary the sex ratio of offspring. *Science* 179: 90–92.

Tronick, Edward Z., Gilda A. Morelli, and Steve Winn (1987). Multiple caretaking of Efe (Pygmy) infants. *American Anthropoligist* 89: 96–106.

Tucker, Catherine, and Jennifer van Hook (2013). Surplus Chinese men: Demographic determinants of the sex ratio at marriageable ages in China. *Population and Development Review* 39: 209–229.

Tuljapurkar, Shripad, Naohiro Ogawa, and Anne H. Gauthier, eds. (2010). *Ageing in Advanced Industrial Societies: Riding the Age Waves*, vol. 3. International Studies in Population 8. Dordrecht: Springer.

Tuljapurkar, Shripad, Cedric O. Puleston, and Michael D. Gurven (2007). Why men matter: Mating patterns drive evolution of human lifespan. *PLOS ONE* 2: e785. doi: 0.1371/journal.pone.0000785.

Turchin, Peter, and Sergey A. Nefedov (2011). *Secular Cycles.* Princeton, NJ: Princeton University Press.

Uematsu, Keigo, Mayako Kutsukake, Takema Fukatsu, Masakazu Shimada, and Harunobu Shibao (2010). Altruistic colony defense by menopausal female insects. *Current Biology* 20: 1182–1186.

UNICEF Innocenti Research Centre (2012). Measuring child poverty: New league tables of child poverty in the world's richest countries. *Innocenti Report Card 10.* Florence: UNICEF Innocenti Research Centre.

United Nations, Department of Economic and Social Affairs, Population Division (2015). *World Population Prospects: The 2015 Revision.* https://esa.un.org/unpd/wpp/.

United States Department of Labor, Office of Policy Planning and Research (1965). *The Negro Family: The Case for National Action.* Washington, D.C.: U.S. Department of Labor.

VanderLaan, Doug P., Zhiyuan Ren, and Paul L. Vasey (2013). Male androphilia in the ancestral environment: An ethnological analysis. *Human Nature* 24: 375–401.

VanderLaan, Doug P., and Paul Vasey (2011). Male sexual orientation in independent Samoa: Evidence for fraternal birth order and maternal fecundity effects. *Archives of Sexual Behavior* 40: 495–503.

VanderLaan, Doug P., and Paul Vasey (2013). Birth order and avuncular tendencies in Samoan men and *fa'afafine*. *Archives of Sexual Behavior* 42: 371–379.

van Schaik, Carol P., and Judith M. Burkart (2011). Social learning and evolution: The cultural intelligence hypothesis. *Philosophical Transactions of the Royal Society B: Biological Sciences* 366: 1008–1016.

van Schaik, Carol P., and Peter M. Kappeller (2013). Cooperation in primates and humans: Closing the gap. In *Cooperation in Primates and Humans: Mechanisms and Evolution*, ed. Peter M. Kappeler and Carol P. van Schaik, 3–24. Berlin: Springer.

Vargas, Michael (2011). Weak obedience, undisciplined friars, and failed reforms in the medieval order of preachers. *Viator* 42: 283–308.

Vasey, Paul L., David S. Pocock, and Doug P. VanderLaan (2007). Kin selection and male androphilia in Samoan *fa'afafine*. *Evolution and Human Behavior* 28: 159–167.

Vasey, Paul L., and Doug P. VanderLaan (2007). Birth order and male androphilia in Samoan *fa'afafine*. *Proceedings of the Royal Society B: Biological Sciences* 274: 1437–1442.

Videan, Elaine N., Jo Fritz, Christopher B. Heward, and James Murphy (2006). The effects of aging on hormone and reproductive cycles in female chimpanzees (*Pan troglodytes*). *Comparative Medicine* 56: 291–299.

Videan, Elaine N., Jo Fritz, Christopher B. Heward, and James Murphy (2008). Reproductive aging in female chimpanzees (*Pan troglodytes*). In *Primate Reproductive Aging: Cross-Taxon Perspectives*, ed. Sylvia Atsalis, Susan W. Margulis, and Patrick R. Hof, 103–118. Basel: Karger.

Villa, Paola, and Wil Roebroeks (2014). Neanderthal demise: An archaeological analysis of the modern human superiority complex. *PLOS ONE* 9 (4): e96424. doi: 10.1371/journal. pone.0096424.

Vinicius, Lucio (2005). Human encephalization and developmental timing. *Journal of Human Evolution* 49: 762–776.

Vinicius, Lucio, Ruth Mace, and Andrea Migliano (2014). Variation in male reproductive longevity across traditional societies. *PLOS ONE* 9 (11): e112236. doi: 10.1371/journal.pone .0112236.

Vivian-Taylor, Josephine, and Martha Hickey (2014). Menopause and depression: Is there a link? *Maturitas* 79: 142–146.

Voland, Eckart, and Jan Beise (2002). Opposite effects of maternal and paternal grandmothers on infant survival in historical Krummhörn. *Behavioral Ecology and Sociobiology* 52: 435–443.

Voland, Eckard, and Jan Beise (2005). "The husband's mother is the devil in the house": Data on the impact of the mother-in-law on stillbirth mortality in historical Krummhörn (1750–1874). In *Grandmotherhood: The Evolutionary Significance of the Second Half of Female Life*, ed. Eckard Voland, Athanasios Chasiotis, and Wulf Schiefenhövel, 239–255. New Brunswick, NJ: Rutgers.

vom Saal, Frederick S., Caleb E. Finch, and James F. Nelson (1994). Natural history and mechanisms of reproductive aging in humans, laboratory rodents, and other selected vertebrates.

In *The Physiology of Reproduction*, 2 vols., 2nd ed., ed. Ernst Knobil and Jimmy D. Neill, 2:1213–1314. New York: Raven Press.

von Staden, Heinrich (1992). Women and dirt. *Helios* 19: 7–30.

von Staden, Heinrich (2008). Animals, women, and *pharmaka* in the Hippocratic Corpus. In *Femmes en médecine en honneur de Danielle Gourevitch*, ed. V. Boudon-Millot, V. Dasen, and B. Maire, 171–204. Paris: De Boccard.

Waldron, Ingrid (1983). Sex differences in human mortality: The role of genetic factors. *Social Science & Medicine* 17: 321–333.

Walker, Phillip L., John R. Johnson, and Patricia M. Lambert (1988). Age and sex biases in the preservation of human skeletal remains. *American Journal of Physical Anthropology* 76: 183–188.

Walker, Robert S., Michael Gurven, Oskar Burger, and Marcus J. Hamilton (2008). The trade-off between number and size of offspring in humans and other primates. *Proceedings of the Royal Society B: Biological Sciences* 275: 827–833.

Wallace, W. Hamish B., and Thomas W. Kelsey (2010). Human ovarian reserve from conception to the menopause. *PLOS ONE* 5 (1): e8772. doi: 10.1371/journal.pone.0008772.

Walsh, Lorena S. (2000). The African American population of the colonial United States. In *A Population History of North America*, ed. Michael R. Haines and Richard H. Steckel, 191–240. Cambridge: Cambridge University Press.

Walter, M. Susan (2006). Polygyny, rank, and resources in Northwest Coast foraging societies. *Ethnology* 45: 41–57.

Wang Feng (2011). The future of a demographic overachiever: Long-term implications of the Demographic Transition in China. *Population and Development Review* 37, suppl.: S173–S190.

Wang, Qingfeng, and Xu Sun (2016). The role of socio-economic and political factors in fertility decline: A cross-country analysis. *World Development* 87: 360–370.

Warrener, Anna G., Kristi L. Lewton, Herman Pontzer, and Daniel E. Lieberman (2015). A wider pelvis does not increase locomotor costs in humans, with implications for the evolution of childbirth. *PLOS ONE* 10 (3): e0118903. doi: 10.1371/journal.pone.0118903.

Washburn, Sherwood L. (1960). Tools and human evolution. *Scientific American* 203 (3): 63–75.

Washburn, Sherwood L., and C. S. Lancaster (1969). The evolution of hunting. In *Man the Hunter*, ed. Richard B. Lee and Irven Devore, 293–303. Chicago: Aldine.

Washburn, Sherwood L., and Ruth Moore (1974). *Ape into Man*. Boston: Little, Brown.

Watkins, Susan Cotts, and Jane Menken (1985). Famines in historical perspective. *Population and Development Review* 11: 647–675.

Webb, James L. A. (2009). *Humanity's Burden: A Global History of Malaria*. New York: Cambridge University Press.

Weber, Alan S. (2002). *Almanacs*. The Early Modern Englishwoman: A Facsimile Library of Essential Works. Aldershot, UK: Ashgate.

Weber, Alan S. (2003). Women's early modern medical almanacs in historical context. *English Literary Renaissance* 33: 358–402.

Wei, Lan-Hai, Shi Yan, Yan Lu, Shao-Qing Wen, Yun-Zhi Huang, Ling-Xiang Wang, Shi-Lin Li, et al. (2018). Whole-sequence analysis indicates that the Y chromosome C2*-Star cluster traces back to ordinary Mongols, rather than Genghis Khan. *European Journal of Human Genetics* 26: 230–237.

Wells, Jonathan C. K. (2010). *The Evolutionary Biology of Human Body Fatness: Thrift and Control.* Cambridge: Cambridge University Press.

Wells, Jonathan C. K. (2012a). Ecological volatility and human evolution: A novel perspective on life history and reproductive strategy. *Evolutionary Anthropology* 21: 277–288.

Wells, Jonathan C. K. (2012b). The capital economy in hominin evolution: How adipose tissue and social relationships confer phenotypic flexibility and resilience in stochastic environments. *Current Anthropology* 53, suppl. 6: S466–S478.

Wells, Jonathan C. K., and Jay T. Stock (2007). The biology of the colonizing ape. *Yearbook of Physical Anthropology* 50: 191–222.

Wessely, S., C. Nimnuan, and M. Sharpe (1999). Functional somatic syndromes: One or many? *Lancet* 354: 936–939.

Whitehead, Hal, and Janet Mann (2000). Female reproductive strategies of cetaceans: Life histories. In *Cetacean Societies*, ed. Janet Mann, Richard C. Connor, Peter L. Tyack, and Hal Whitehead, 219–246. Chicago: University of Chicago Press.

Whittle, Jane (2014). Enterprising widows and active wives: Women's unpaid work in the household economy of early modern England. *History of the Family* 19: 283–300.

Wilbush, Joel (1979). La ménespausie: The birth of a syndrome. *Maturitas* 1: 145–151.

Wilbush, Joel (1986). The climacteric syndrome: Historical perspectives. In *The Climacteric in Perspective*, ed. M. Notelovitz and P. van Keep, 121–130. Lancaster, UK: MTP.

Wilbush, Joel (1988). Menorrhagia and menopause: A historical review. *Maturitas* 10: 5–26.

Williams, George C. (1957). Pleiotropy, natural selection, and the evolution of senescence. *Evolution* 11: 398–411.

Wilson, Edward O. (2012). *The Social Conquest of Earth.* New York: Liveright.

Wilson, Robert A. (1966). *Feminine Forever.* New York: Mayflower.

Wilson, Robert A., and Thelma Wilson (1963). The fate of the nontreated postmenopausal woman: A plea for the maintenance of adequate estrogen from puberty to the grave. *Journal of the American Geriatrics Society* 11: 347–362.

Winking, Jeffrey, and Michael Gurven (2011). The total cost of father desertion. *American Journal of Human Biology* 23: 755–763.

Winking, Jeffrey, Hillard Kaplan, Michael Gurven, and Stacey Lucas (2007). Why do men marry, and why do they stray? *Proceedings of the Royal Society B: Biological Sciences* 274: 1643–1649.

Wittenborn, J. R., and Clark Bailey (1952). The symptoms of involutional psychosis. *Journal of Consulting Psychology* 16: 13–17.

Wolf, Arthur P. (1981). Women, widowhood, and fertility in pre-modern China. In *Marriage and Remarriage in Populations of the Past*, ed. J. Dupâquier, E. Hélin, P. Laslett, M. Livi-Bacci, and S. Songer, 139–150. London: Academic Press.

Wolf, Arthur P. (2001). Is there evidence of birth control in late imperial China? *Population and Development Review* 27: 133–154.

Wolf, Arthur P., and Theo Engelen (2008). Fertility and fertility control in pre-revolutionary China. *Journal of Interdisciplinary History* 38: 345–375.

Wolf, Arthur P., and Chieh-shan Huang (1980). *Marriage and Adoption in China, 1845–1945.* Stanford, CA: Stanford University Press.

Wolfers, Justin, David Leonhardt, and Kevin Quealy (2015). 1.5 million missing black men. *New York Times*, April 20. http://www.nytimes.com/interactive/2015/04/20/upshot/missing-black-men.html.

Wood, Brian M., Kevin E. Langergraber, John C. Mitani, and David P. Watts (2016). Menopause is common among wild female chimpanzees in the Ngogo community: Abstract. *American Journal of Physical Anthropology* 162, suppl. 64: 414–415.

Wood, Brian M., David P. Watts, John C. Mitani, and Kevin E. Langergraber (2017). Favorable ecological circumstances promote life expectancy in chimpanzees similar to that of human hunter-gatherers. *Journal of Human Evolution* 105: 41–56.

Wood, James W. (1994). *Dynamics of Human Reproduction: Biology, Biometry, Demography*. New York: de Gruyter.

Wood, James W. (1998). A theory of preindustrial population dynamics: Demography, economy, and well-being in Malthusian systems. *Current Anthropology* 39: 99–135.

Wood, James W., Kathleen A. O'Connor, Darryl J. Holman, Eleanor Brindle, Susannah H. Barsom, and Michael A. Grimes (2001). The evolution of menopause by antagonistic pleiotropy. Working Paper 01–04, Center for Studies in Demography and Ecology, University of Washington.

Woods, Dori C., and Jonathan L. Tilly (2012). The next (re)generation of ovarian biology and fertility in women: Is current science tomorrow's practice? *Fertility and Sterility* 98: 3–10.

World Health Organization (2013). *WHO Traditional Medicine Strategy 2014–2023*. Geneva: WHO Press. http://who.int/medicines/publications/traditional/trm_strategy14_23/en/.

Wrangham, Richard (2009). *Catching Fire: How Cooking Made Us Human*. New York: Basic Books.

Wrangham, Richard, and Dale Peterson (1996). *Demonic Males: Apes and the Origin of Human Violence*. Boston: Houghton Mifflin.

Wright, Katherine I. (Karen) (2014). Domestication and inequality?: Households, corporate groups and food processing tools at Neolithic Çatalhöyük. *Journal of Anthroplogical Archaeology* 33: 1–33.

Wright, P., S. J. King, A. Baden, and J. Jernvall (2008). Aging in wild female lemurs: Sustained fertility with increased infant mortality. In *Primate Reproductive Aging: Cross-Taxon Perspectives*, ed. Sylvia Atsalis, Susan W. Margulis, and Patrick R. Hof, 17–28. Basel: Karger.

Wrigley, E. A., R. S. Davies, J. E. Oeppen, and R. S. Schofield (1997). *English Population History from Family Reconstitution, 1580–1837*. Cambridge: Cambridge University Press.

Wrigley, E. A., and R. S. Schofield (1981). *The Population History of England, 1541–1871*. Cambridge, MA: Harvard University Press.

Wu Liu, María Martinón-Torres, Yan-jun Cai, Song Xing, Hao-wen Tong, Shu-wen Pei, Mark Jan Sier, et al. (2015). The earliest unequivocally modern humans in southern China. *Nature* 526: 696–700.

Young, Antonia (2000). *Women Who Become Men: Albanian Sworn Virgins*. Oxford: Berg.

Zafon, Carles (2006). Spend less, live longer: The "Thrifty Aged" Hypothesis. *Medical Hypotheses* 67: 15–20.

Zahavi, Amotz (1975). Mate selection—A selection for a handicap. *Journal of Theoretical Biology* 53: 205–214.

Zerjal, Tatiana, Yali Xue, Giorgio Bertorelle, R. Spencer Wells, Weidong Bao, Suling Zhu, Raheel Qamar, et al. (2003). The genetic legacy of the Mongols. *American Journal of Human Genetics* 772: 717–721.

Zeserson, Jan M. (2001a). *Chi no michi* as metaphor: Conversations with Japanese women about menopause. *Anthropology & Medicine* 8: 177–199.

Zeserson, Jan M. (2001b). How Japanese women talk about hot flushes: Implications for menopause research. *Medical Anthropology Quarterly* 15: 189–205.

Zilhão, João, Diego E. Angelucci, Ernestina Badal-García, Francesco d'Errico, Floréal Daniel, Laure Dayet, Katerina Douka, et al. (2010). Symbolic use of marine shells and mineral pigments by Iberian Neanderthals. *Proceedings of the National Academy of Sciences* 107: 1023–1028.

Zollikofer, Christoph P. E., and Marcia S. Ponce de León (2010). The evolution of hominin ontologies. *Seminars in Cell and Developmental Biology* 21: 441–452.

INDEX

•

!Kung foragers, 20, 73, 111, 114, 146; birth intervals, 91; demography, 24; and egalitarianism, 107–8; food consumption, 73; food sharing, 107–8; inequality among, 108; marriage among, 83; Post-Reproductive Representation, 82; and old age, 114; and social status in midlife, 326

abandonment. *See* infanticide/infant abandonment
abortion, 182, 183, 233, 271
abstinence: within marriage, 182, 183, 202; as medical advice, 284; grandmother abstinence/terminal abstinence, 62, 287, 310, 323, 359
Ache people, 94, 108–16, 120, 221; demography of, 55, 113, 115; and disease, 88; fertility of (*see* fertility); and food sharing, 111; and Grandmother Hypothesis, 55; infanticide and child homicide among, 110–11, 121, 123, 179; marriage among, 84, 112–13, 115, 121; maternal mortality, 53; and menopause, 113; and partible paternity, 112; and transgender roles, 115–16, 117, 118
Acheulian technology, 138
adiposity. *See* fat
adolescence, 97, 100, 136, 209, 261; and green sickness, 8, 278; growth during, 97, 151; subfecundity during, 97
adoption/fostering, 4, 6, 45, 102, 164, 168, 199, 202, 204, 208–9, 224, 234
adult-child ratio, 56, 89, 90, 110, 119, 124, 154, 171, 238, 253

adultery, 121, 161
African Americans, 215, 232, 242–43, 296, 297
age at first birth: among Ache, 115; among chimpanzees, 41; among elephants, 30; among Hadza, 97; among humans generally, 41; among Maya, 314
age at last birth: among Ache, 115, 307; among chimpanzees, 41; among Hadza, 97–98, 307; among humans generally, 19, 24, 41, 181; among Hutterites, 19; in Liaoning, 193; in models of human evolution, 59–60; among whales, 28
age at menopause: in ancient Greek and Roman sources, 265–66; in Bangladesh, 318; among Bangladeshi immigrants to London, 318; factors affecting, 62, 181–82; among Hadza, 97; heritability of, 35, 319; among humans generally, 19, 34–35; among Maya, 308, 312, 324, 318; in medieval European sources, 268–69; and modernization, 318–19; among mothers of daughters, 62; in Puebla, Mexico, 319; and smoking, 319; in United States, 319; variation of, 34–35
age-stratified homosexuality: 117
aging. *See* senescence
agrarian era, agrarian societies, 6, 7–8, 9, 12, 129, 154–224, 225; defined, 10; demography of, 131, 172–80; and division of labor, 166–170 (*see also* labor, division by sex); and family, 123, 155, 156–57, 158–66, 208; fertility in (*see* fertility); and gay sex, 116–17; and heritable property (*see* property); household economies in,